ACTIVE TOUCH

The Mechanism of Recognition of Objects by Manipulation:
A multi-disciplinary approach

ACTIVE TOUCH

The Mechanism of Recognition of Objects by Manipulation: A multi-disciplinary approach

PROCEEDINGS OF A SYMPOSIUM HELD AT BEAUNE, FRANCE, JULY 1977 UNDER THE AUSPICES OF THE INTERNATIONAL UNION OF PHYSIOLOGICAL SCIENCES

Editor
GEORGE GORDON
Reader in Sensory Physiology, Oxford University, England

PERGAMON PRESS
OXFORD · NEW YORK · TORONTO · SYDNEY · PARIS · FRANKFURT

U.K.	Pergamon Press Ltd., Headington Hill Hall, Oxford OX3 0BW, England
U.S.A.	Pergamon Press Inc., Maxwell House, Fairview Park, Elmsford, New York 10523, U.S.A.
CANADA	Pergamon of Canada Ltd., 75 The East Mall, Toronto, Ontario, Canada
AUSTRALIA	Pergamon Press (Aust.) Pty. Ltd., 19a Boundary Street, Rushcutters Bay, N.S.W. 2011, Australia
FRANCE	Pergamon Press SARL, 24 rue des Ecoles, 75240 Paris, Cedex 05, France
FEDERAL REPUBLIC OF GERMANY	Pergamon Press GmbH, 6242 Kronberg-Taunus, Pfcrdstrasse 1, Federal Republic of Germany

First edition 1978

British Library Cataloguing in Publication Data

Active touch.
1. Touch - Congresses
I. Gordon, George II. International Union of Physiological Sciences
152.1'82 BF275 77-30729

ISBN 0-08-022647-7 Hardcover
ISBN 0-08-022667-1 Flexicover

In order to make this volume available as economically and as rapidly as possible the authors' typescripts have been reproduced in their original forms. This method unfortunately has its typographical limitations but it is hoped that they in no way distract the reader.

Printed in Great Britain by William Clowes & Sons Limited
London, Beccles and Colchester

CONTENTS

Contents

CONTRIBUTORS AND INVITED PARTICIPANTS

J.D. Armstrong
Blind Mobility Research Unit
Department of Psychology
University of Nottingham
University Park
Nottingham NG7 2RD
U.K.

J.C. Bliss
Telesensory Systems Inc.
3408 Hillview Avenue
Palo Alto
California 94304
U.S.A.

Jacoba Brinkman
Department of Physiology
Monash University
Clayton
Victoria
Australia 3168

A.G. Brown
Department of Veterinary Physiology
University of Edinburgh
Summerhall
Edinburgh EH9 1QH
U.K.

G. Carli
Istituto di Fisiologia umana
dell' Università di Siena
Strada del Laterino 8
53100 Siena
Italy

F. Cervero
Department of Veterinary Physiology
University of Edinburgh
Summerhall
Edinburgh EH9 1QH
U.K.

J.C. Craig
Department of Psychology
Indiana University
Psychology Building
Bloomington
Indiana 47401
U.S.A.

J.M. Coquery
Laboratoire de Psychophysiologie
Université de Lille I
BP 36
59650 Villeneuve d'Ascq
France

J.E. Desmedt
Unité de Recherche sur le cerveau
115 Bd de Waterloo
B 1000 Bruxelles
Belgium

J. Dostrovsky
National Institute for Medical Research
Medical Research Council
Mill Hill
London NW7
U.K.

R. Duclaux
Laboratoire de Physiologie
Université Claude Bernard
Faculté de Médecine Lyon S.O.
Chemin du Petit Revoyet
BP 12, 69600 Oullins
France

P. Dyhre-Poulsen
Institute of Neurophysiology
36 Juliane Mariesvej
Copenhagen
DK 2100
Denmark

O.M. Evans
Department of Biological Sciences
Lincoln Institute
625 Swanston Street
Carlton
Victoria 3053
Australia

C. Fromm
Physiologisches Institut II
Universität Düsseldorf
Moorenstrasse 5
4000 Düsseldorf 1
W. Germany

Y. Galifret
Laboratoire de Psychologie Sensorielle
Université Pierre & Marie Curie
4 Place Jussieu
75230 Paris Cedex 05
France

S.C. Gandevia
School of Physiology & Pharmacology
University of New South Wales
Post Office Box 1
Kensington
New South Wales
2033 Australia

G. Goodwin
University Laboratory of Physiology
Parks Road
Oxford OX1 3PT
U.K.

G. Gordon
University Laboratory of Physiology
Parks Road
Oxford OX1 3PT
U.K.

C.C. Hunt
Department of Physiology
School of Medicine
Washington University
660 S. Euclid Avenue
Saint Louis
Missouri 63110
U.S.A.

A. Iggo
Department of Veterinary Physiology
University of Edinburgh
Summerhall
Edinburgh EH9 1QH
U.K.

Y. Iwamura
Department of Physiology
Toho University
School of Medicine
5-21-16 Omori-Nishi
Ota-Ku
Tokyo 143
Japan

L. Jami
Laboratoire de Neurophysiologie
Collège de France
11, Place Marcelin Berthelot
75231 Paris Cedex 05
France

G. Jansson
Department of Psychology
Uppsala Universitet
Box 227
S-751 04 Uppsala
Sweden

R. Johansson
Department of Physiology
University of Umeå
S-901 87 Umeå
Sweden

R.H. Kay
University Laboratory of Physiology
Parks Road
Oxford OX1 3PT
U.K.

J.M. Kennedy
Scarborough College
University of Toronto
West Hill
Ontario MIC IA4
Canada

D.R. Kenshalo
Department of Psychology
The Florida State University
Tallahassee
Florida 32306
U.S.A.

Y. Laporte
Laboratoire de Neurophysiologie
Collège de France
11 Place Marcelin Berthelot
75231 Paris Cedex 05
France

R.H. LaMotte
Department of Anesthesiology
Yale University
School of Medicine
333 Cedar Street
New Haven
Connecticut 06510
U.S.A.

Susan J. Lederman
Department of Psychology
Queens University
Kingston
Ontario K7L 3N6
Canada

R.N. Lemon
Afdeling Anatomie
Erasmus Universiteit Rotterdam
Faculteit der Geneeskunde
Postbus 1738
Rotterdam
The Netherlands

A. Lequeux
Fondation pour la Réadaptation des déficients visuels
F.R.D.V.
3 rue Lyautey
75016 Paris
France

D.I. McCloskey
School of Physiology & Pharmacology
University of New South Wales
Post Office Box 1
Kensington
New South Wales
2033 Australia

Susanna Millar
Department of Experimental Psychology
University of Oxford
South Parks Road
Oxford OX1 3UD
U.K.

U. Norrsell
Department of Physiology
University of Göteborg
Fack
S-400 33 Göteborg
Sweden

J. Paillard
Département de Psychophysiologie générale
C.N.R.S. - INP4
31 Chemin Joseph-Aiguier
13274 Marseille
France

A.S. Paintal
Vallabhbhai Patel Chest Institute
University of Delhi
Delhi 110007
Post Office Box 2101
India

J. Petit
Laboratoire de Neurophysiologie
Collège de France
11 Place Marcelin Berthelot
75231 Paris Cedex 05
France

C.G. Phillips
Department of Human Anatomy
University of Oxford
South Parks Road
Oxford OX1 3QX
U.K.

B.H. Pubols
College of Medicine
Pennsylvania State University
Hershey
Pennsylvania 17033
U.S.A.

Lillian M. Pubols
Department of Anatomy
Medical College of Pennsylvania
Philadelphia
Pennsylvania
U.S.A.

T.A. Quilliam
Department of Anatomy and Embryology
University College London
Gower Street
London WC1E 6BT
U.K.

F. de Ribeaupierre
Institut de Physiologie
Faculté de Médecine
Université de Lausanne
7 rue du Bugnon
CH-1011 Lausanne
Switzerland

H. Sakata
Tokyo Metropolitan Institute for Neurosciences
2-6 Musashidai
Fuchu-City
Tokyo
Japan

R.F. Schmidt
Physiologisches Institut der Universität Kiel
23 Kiel
Olshausenstrasse 40/60
W. Germany

E.G. Walsh
Department of Physiology
University Medical School
Teviot Place
Edinburgh EH8 9AG
U.K.

J.F. Stein
University Laboratory of Physiology
Parks Road
Oxford OX1 3PT
U.K.

H.E. Torebjörk
Department of Clinical Neurophysiology
Academic Hospital
750 14 Uppsala
Sweden

Å.B. Vallbo
Department of Physiology
University of Umeå
S-901 87 Umeå
Sweden

C.J. Vierck
Department of Neurosciences
University of Florida
College of Medicine
Box 723 JHM Health Center
Gainesville, Florida 32610
U.S.A.

EDITOR'S PREFACE

This volume is based on a meeting sponsored by the International Union of Physiological Sciences' Commission on Somatosensory Physiology, as a satellite symposium of the XXVII International Congress of Physiological Sciences. As the proposer of this particular topic, I found myself accepting the heavy responsibility of arranging the programme and in this way determining its scope. It was a stimulating and rewarding experience; and I should like here to acknowledge the valuable help and advice of many colleagues at home and abroad.

I have tried in an introductory chapter to give an outline of the issues which are prominent in this subject at its present stage of development, and to relate them to matters discussed in the remaining chapters. This outline necessarily reflects a personal view; but I have tried also to highlight some questions of special interest which arose during discussion, and here I must particularly thank my colleagues in Oxford who were present, for supplementing my own records and impressions with theirs.

It was specially appropriate that this symposium was held in Beaune, the birthplace of Etienne-Jules Marey (1830-1904), a pioneer in the analysis of human and animal movement, many of whose remarkable instruments and records are displayed there in the Musée Marey. The chairman was Yves Laporte, Professor at the Collège de France and honorary Director of the Institut Marey, and everyone who took part would wish me to thank both him and the Municipality of Beaune for making it possible to work in such congenial and beautiful surroundings.

INTRODUCTION

George Gordon

University Laboratory of Physiology, Oxford, England

'If I could only move my hand about I should know what the things were'. This was said to Victor Horsley (1909) by a patient who was asked to identify various familiar objects placed in his hand three weeks after the arm area of his precentral gyrus (the 'so-called motor area') had been removed surgically. He could feel himself being touched; he knew approximately where he was touched, and the position of his fingers; but he had an enduring disability in moving them. This showed 'what the real basis of the stereognostic sense is - namely, merely a complex of tactile, muscular and arthric memories of movements, which are, in fact, the compound experiences of grasping and feeling objects'. This is a view few would disagree with today; saving only the word 'merely', because the detailed mechanisms underlying manual recognition, some of which we now have the methods to study, are by no means simple. They form the subject of this book.

Thanks to modern electrophysiological studies on single peripheral nerve-fibres in animals, and more recently to analogous studies on man with microneurography (Vallbo and Johansson in this book), a quite precise account can now be given of the responses of four anatomically defined types of cutaneous or subcutaneous sense organ to deformation of the skin. What is still not clear is their relative roles in different facets of perception. The articles on skin in this book deal almost entirely with the human fingers. Quilliam considers how fingertip skin is structurally adapted to its role in exploration, laying special emphasis on the role of the sweat-lubricated ridges as pliant and non-slipping structures which may create a vibratory effect as they move across the explored surface. Petit and Galifret studied the mechanical properties of finger skin directly with stationary displacements normal to the surface, measuring applied force and displacement simultaneously, and use their data to show how skin properties interact with receptor properties in determining the overall coupling function that relates mechanical input to nervous output. Susan Lederman considers another aspect of skin properties by distinguishing between normal and shearing forces as the skin slides over a rough surface. She describes experimental techniques which either lessened or virtually eliminated shearing forces and shows how this heightened the perception of roughness. This leads her, using structural arguments, to give special emphasis to the Meissner corpuscles in roughness perception. LaMotte and Mountcastle also emphasise the importance of Meissner corpuscles in mediating vibratory 'flutter' sensation in the 5 to 40 Hz range, and report parallel experiments in monkey and man implicating these sense organs in roughness discrimination through the clues of amplitude and frequency of vibration. The question of the vibrations created in the ridged skin provoked discussion several times during the meeting; and amongst others the question was raised whether pacinian corpuscles had not been too readily dismissed as agents in roughness discrimination simply because their receptive fields are too big and the frequency band in which they work optimally (40 to 400 Hz) too high. It was pointed out that even if the frequencies expected to be generated in an exploring finger were lower than this, the skin might transform the vibration into something with higher frequency components. There does in fact seem to be a need to measure vibration frequencies in skin directly under appropriate conditions before this can be answered. Psychologists will recall the preoccupation of David Katz (see Krueger, 1970) with the sense of vibration and its links with movement.

While pacinian corpuscles responding to vibration might contribute to discrimination of textural features of a surface during movement, it is clear, as Vallbo and Johansson show, that the large receptive fields of 'PC' units, and also those of the 'SA II' units with sense organs responding directionally to longitudinal stretch of the skin, exclude both these from consideration in accounting for two-point acuity. It was suggested in discussion that the properties and distribution of SA II units are better suited to a proprioceptive role, in association with sense organs in muscles and joint capsules. Vallbo and Johansson provide strong evidence that both RA (presumably Meissner) and SA I (presumably Merkel) units are concerned in two-point acuity. The receptive fields of the former are smallest where acuity is high and the density of distribution of both is well correlated with acuity, RA units in particular having a strikingly high density in the fingertips.

It is accepted that the discrimination of, for example, differently textured surfaces is improved when the finger moves relative to the surface. This will use more nervous channels and exploit the dynamic characteristics of the sense organs. But it seems there is no inherent perceptual advantage in the movement being active - that is, carried out by the subject. Katz (1925) found active movement no advantage in texture discrimination; and during discussion at the meeting it appeared that roughness perception was not improved if the sensing finger moved actively, nor was the recognition of printed letters with the Optacon (q.v.) when the subject moved the sensing camera himself. It is well known that simple shapes and letters can be recognised when presented passively especially if they are traced out on the skin.

Recognising three-dimensional shapes must present more difficulty; and it is interesting that Sakata and Iwamura describe cells in the monkey's parietal cortex which only responded when a spherical or nearly spherical object, for example, was actively grasped by the hand, passive presentation being ineffective. The discrimination of a tennis ball from a metal boule certainly requires muscular activity. The temperature sense may admittedly help to detect sensory attributes that depend on the two having different thermal capacities, but not the fact that they differ in hardness and weight. Hardness will be estimated by grasping and probing; and estimates of heaviness, as McCloskey and Gandevia show here, depend on the 'sense of effort' (efference copy, corollary discharge) associated with supporting each ball in the hand.

Beyond some degree of irregularity in the shape or texture, active exploration must become necessary - for instance the most successful of Armstrong's subjects reading embossed tactual maps were those who made extensive fine exploratory movements. If a complex object was passively moved about in the hand of a blindfold subject it would produce unintelligible signals because the programme of movement would be known only to the experimenter, not to the subject. The subject needs a record of his own programme of movement against which to interpret what he feels. As Sakata and Iwamura show, many cells, especially in the posterior parietal cortex, have properties suitable for extracting information about objects moving across the skin in different directions while the parts of the limb are in specified relationships to each other. The shifting pattern of activity among such cells must determine each succeeding step in the movement sequence that constitutes active exploration. But how does the brain acquire this record of the movements that are made? It is clear that it depends in part on monitoring the instructions it gives to the muscles - even if this central monitoring or 'corollary discharge' may itself give rise to no sensation of movement - but this corollary discharge mechanism is not independent of information from sense organs in skin, joints and muscles in the arm and hand, since these interact with it, as McCloskey and Gandevia show. The article by Paillard, Brouchon-Viton and Jordan refers to this sort of interaction. It might be expected that an evaluation of the peripheral information against the corollary discharge would provide the most accurate available record of the movements, and that the sensory information generated by the movement would in turn be evaluated against this record.

The papers in the second section of this book are very much concerned with a number of areas in the cerebral cortex that differ in cell-architecture and are nowadays increasingly associated with different facets of function. In interdisciplinary meetings a body of knowledge that is taken for granted in one discipline may be a source of difficulty for others. It was probably in the matter of central neuroanatomy that the use of specialised scientific shorthand was most obvious during our meeting, and it may help the reader not specialised in this field if some of the principal points are clarified by reference to diagrams (Fig. 1). Fig. 1a is a side view of the brain of a monkey, showing the position and extent of the parietal lobe. It is bounded in front by the central sulcus, separating it from the motor cortex which can thus be called precentral. The existence of these sulcal folds makes any presentation inadequate which only views the surface; so the finer detail is shown both on a surface view (Fig. 1c), and on a section (Fig. 1b) cut at right-angles to the surface along the line A-B in Fig. 1a. This section emphasises how large a part of the cortex is buried from view. The postcentral gyrus contains, in its anterior part, the first somatosensory area (S I), referred to by Sakata and Iwamura and labelled as such in Fig. 1c; and the cytoarchitectural areas it comprises are numbered in Brodmann's terminology 3, 1 and 2 from anterior to posterior. S I receives a massive ascending projection of tactile and proprioceptive inputs from the opposite side of the body.

The posterior part of the postcentral gyrus contains area 5 in its medial (upper) part and the second somatosensory area (S II) more laterally (below). Their position can be seen in Fig. 1c, though S II is in fact almost entirely concealed in a sulcus and is therefore shown stippled; the buried part of area 5 can be seen in Fig. 1b. The lesion (1) of LaMotte and Mountcastle removed the whole postcentral gyrus (areas 3, 1, 2, 5 and S II). Behind and below area 5 lies area 7 (see the related diagram in Stein's Fig. 1): their lesion (2) removed this also and was a complete parietal lobectomy. Areas 5 and 7 are often called together the 'posterior parietal association cortex'. S I projects to area 5, and both area 5 and the visual system project to area 7;

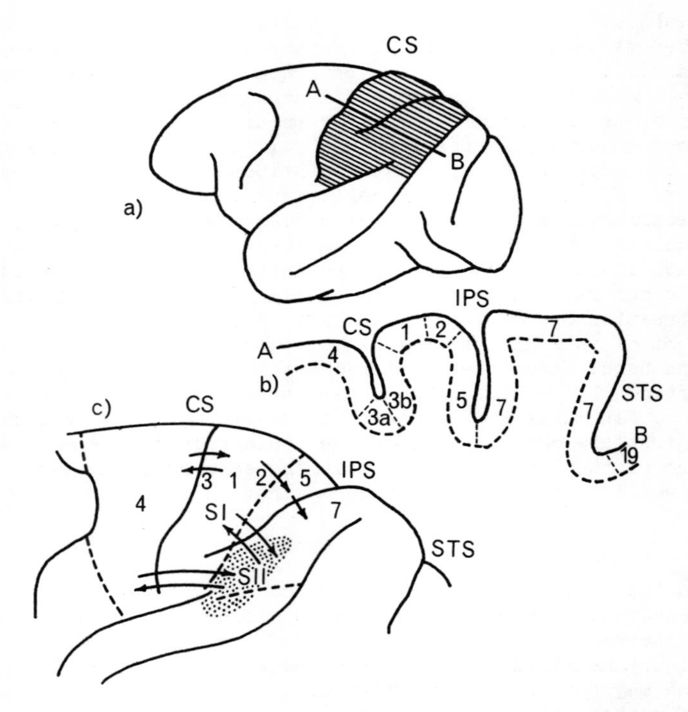

Fig. 1. Semi-diagrammatic views of the areas of sensory and motor cortex referred to in this book, showing known connections between them. In each the frontal direction is to the left.

(a) Left side view of a monkey's brain, showing the central sulcus, and the parietal lobe (shaded).

(b) Section at right-angles to the surface along the line A-B in (a), showing the different areas of cortex numbered according to Brodmann. Area 3a, which receives both tactile and proprioceptive inputs, is not referred to explicitly in the text.

(c) Enlarged side view of the precentral and parietal areas ('motor' and 'sensory') showing the visible parts of area 4, S I (comprising areas 3, 1 and 2), areas 5 and 7. S II is buried from view and is represented by shading. Arrows indicate connections between these regions.

Abbreviations :- CS, central sulcus; IPS, intraparietal sulcus; STS, superior temporal sulcus; S I, first somatosensory area; S II, second somatosensory area.

(Adapted from various sources, including Jones & Powell (1969) and Mountcastle et al. (1975)).

and it is to these connections that we would attribute the development of complexity which Sakata and Iwamura describe in the receptive properties of the cells in these posterior areas. This question is also discussed by Stein, in relation to his experiments on reversible cooling of the cortex in either area 5 or area 7: he shows that the control of tactile manipulation in the absence of visual clues requires the integrity of area 5, while visually guided manipulation requires area 7.

The precentral gyrus contains Brodmann's area 4, corresponding to the classical 'motor cortex'. It is reciprocally connected with S I in the postcentral gyrus. Many of its cells of output project in the pyramidal tract directly to the spinal cord. Three articles in this book are concerned with the connections and properties of cells in this region (Fromm and Evarts; Brinkman and Porter; and Lemon and Porter). Many of these cells were pyramidal tract neurones (PTN) - many presumably influencing motoneurones. Some fibres descending as far as the medullary pyramidal tract influence sensory relay nuclei (e.g. the cuneate nucleus) in the medulla: some of these descend no further, but others, by branching, project both to these nuclei and to the spinal cord. All the 'sensory' areas of the parietal lobe also contribute to this descending path and may thus influence movement directly, and they also may influence the signals ascending in sensory tracts. There was discussion of the likelihood of a projection from area 5 to area 4.

One would expect exploratory movements made by the hand to be mechanically efficient in exerting a force appropriate to overcome resistances they encounter. The gearing of many precentral neurones to the control of force has been emphasised in the past work of Evarts. It has become clear that there is a fast, and so presumably fairly direct path from peripheral sense organs - some tactile, but most proprioceptive - to the precentral cortex, as Lemon and Porter show here. The anatomy of this path is still not clear; but it is reasonable to suppose that it could be the afferent arm of the force-controlling (or load-compensating) loop which is referred to here by Fromm and Evarts, McCloskey and Gandevia, and by Torebjörk, Hagbarth and Eklund. The two latter articles refer to cutaneous contributions to motor control in man which may use this loop. The tactile finger-flexion reflex shown by Torebjörk et al. calls to mind the property of some precentral cortical neurones with tactile receptive fields on the hand of increasing their discharge markedly when active movement brings the field into contact with an external object. The peripheral fields of such cells are spatially related to the 'field of influence' of each neurone in motor performance (Lemon and Porter). It is interesting that perceived roughness of a textured surface also increases when the subjects press harder (see the article by Susan Lederman).

Although there was acceptance of such a load-compensating mechanism, involving an interplay between tactile and proprioceptive inputs, there was apparent conflict about the behavioural contexts in which it would be active. Fromm and Evarts show that load-compensating responses of precentral neurones are evident during small precise movements but not during ballistic movements, and suggest that they are therefore highly relevant to the topic of fine manipulative activity. Brinkman and Porter show that when the peripheral input to precentral neurones is removed by section of the cuneate fascicle of the dorsal column, the responses of the cells during the performance of skilled forelimb movements nevertheless appear normal and impairment of motor performance is minimal or non-existent. The impairment seen in operated animals in extracting food rewards from gutters in a test board was attributed to diminished ability to recognise the texture of the food. While these contrasting interpretations of different kinds of experiments cannot be resolved here, an appreciation of the extreme difficulty in analysing the deficits resulting from section of a tract such as the dorsal column can be gathered from the article by Vierk, who points out that the contributions made by the dorsal column to sensation or motor control are undoubted, but are highly specialised. It is therefore essential to design tests specific enough to identify them.

The subject of load-compensation is clearly interlocked with the broader one referred to earlier, of the programming of exploratory movements that lead to recognising objects. A serious problem in studying stereognosis is that it is axiomatic that true exploratory movements are naive, and therefore not reproduced in the movements of an animal trained to a particular task even if these are fine and skilful, nor even in free actions that need good sensorimotor coordination, if these have become habitual. Stereognosis can properly be investigated only in man. Just the same, studies in man preclude detailed investigation of possible mechanisms; and a combination of quantitative studies on mechanism in animals with parallel psychophysical human studies provides another approach.

This approach is represented here in three articles (LaMotte and Mountcastle; Coquery; and Dyhre-Poulsen). The two latter deal with changes in sensory transmission towards the cortex, and in psychophysical threshold, during active movement, which bring them close to the central theme of the book. It has been known for several years that ascending activity (recorded as the mass response in the cat's medial lemniscus to stimulating a peripheral nerve) is attenuated, presumably by the descending projections from the cortex to sensory relay nuclei already mentioned, when the animal makes various sorts of active movement. Dyhre-Poulsen uses both electrical and tactile stimuli to elicit ascending activity in the monkey's midbrain, and shows here that this mass response is depressed both before and during a lever-pressing movement made by the limb receiving the stimulus. Coquery reports similar results in cats trained to press a lever, and adds evidence of the actions occurring at different points along the ascending path. Coquery also shows that suprathreshold electrical stimuli to the skin of human fingers becomes imperceptible just before and during active flexion of the fingers, and discusses the relative contributions of peripherally and centrally induced inhibition in producing these effects. Dyhre-Poulsen describes elevation of the human tactile threshold in both ballistic and tracking movements.

One of the problems in relating these striking phenomena to tactile recognition is that the movements themselves were not of an exploratory character, controlled by tactile and proprioceptive feedback. One's naive expectation would be that tactile acuity would if anything be improved during exploratory movement; and much discussion centred round this question, which is not resolved by observations of overall reduction of ascending activity as made so far in animals, since this can conceal the existence of more than one pattern of selective depression. It is easy to see different contexts in which selective depression might be advantageous which are not mutually exclusive. In exploratory movement the spatial acuity of perception might be improved by increasing lateral inhibition in the relay nuclei of the ascending path through cortical control. Improved acuity is not inconsistent with a raised detection threshold - in fact, in a simple model system incorporating lateral inhibition

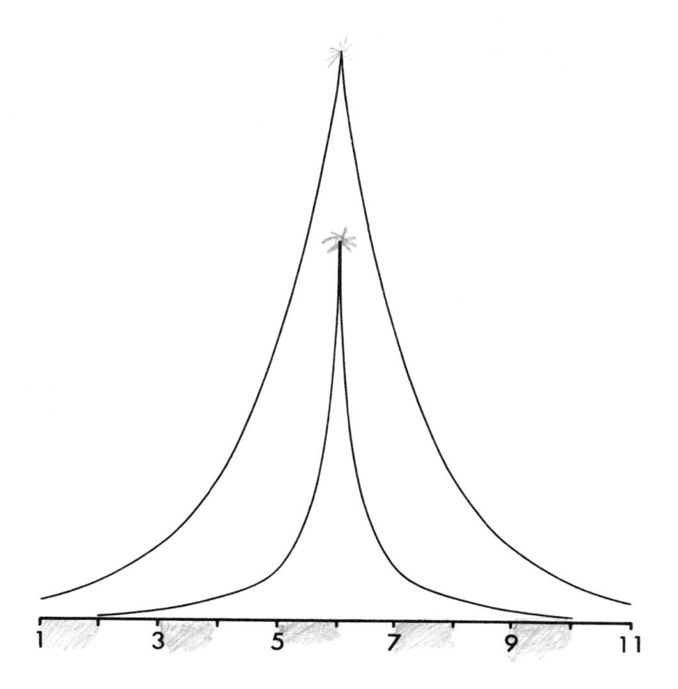

Fig. 2. Representations of calculated frequencies of discharge among a group of nerve-cells (numbered 1 to 11 along abscissa) whose receptive fields lie in regular serial order in a line across the skin. Ordinate: frequency of discharge. The cells are to be thought of as relay cells in e.g. the cuneate nucleus. The skin is momentarily distorted by a 'punctate' stimulus, declining exponentially with distance. In the upper curve, frequency of discharge in each cell is in linear proportion to deformation of its field. In the lower curve, each cell is in addition receiving inhibition from each of its two neighbouring peripheral fields, inhibitory action being taken arbitrarily to be one-third as effective as excitatory action (cf. Kay, 1964).

The upper curve reproduces the actual form of the stimulus. The lower one shows it 'sharpened' by inhibition; at the same time peak impulse frequency has fallen, so under experimental conditions perceptual threshold may have risen. Above threshold, however, acuity should have increased.

which is described in the legend to Fig. 2 (q.v.) such an increase would actually be predicted so long as lateral inhibition operates under threshold conditions. This prediction could be tested experimentally. Alternatively, selective depression could be valuable in removing ascending information which was redundant; or was inappropriate to the type of movement programmed, as in the ballistic movements of Fromm and Evarts. Lemon and Porter also describe considerable modification in the peripheral influences on precentral neurones during various voluntary movements.

The last six articles are valuable not only as perceptual and cognitive studies of touch but also in providing data which might be used in the design or improvement of tactile substitution devices to help the blind. Susan Lederman, whose article on a method for heightening impressions of roughness has been mentioned earlier in other contexts, speculates on its usefulness to the blind - for example in reading the textured areas of tactile maps. Texture is also emphasised by Susanna Millar, who argues that it may be in textural rather than in spatial terms that tactual shapes are initially coded in memory. Craig uses perceptual masking as a useful weapon for finding how the skin processes tactile stimuli. He examines both detection of vibration, and ability to recognise symbols, using both forward and backward masking. Since more backward than forward masking of letter recognition was found, and more forward than backward masking of detection, he pursues the possibility that detection and recognition involve different perceptual processes. The target letters used by Craig were generated by the vibrotactile array (Optacon) developed by Bliss, whose own article reports on its use as a reading machine for the blind. Craig's work demonstrates two facts which should be important in the design and use of cutaneous communication systems - that a single vertical line was nearly as effective as a letter-sized rectangle in backward masking of letter recognition, showing the power of the effect; and that backward masking is effective over at least 150 milliseconds after the offset of the target letter. Such quantification of the period in which a cutaneous pattern of activity remains fragile in the nervous system would need to be taken into account in, for example, neurophysiological models of cortical sensory processing. Bliss deals specifically with reading machines. The special merit of the Optacon is that, unlike braille, it enables the blind to read a wide variety of printed material; its limitation, that the average reading rate is slow. Bliss describes various attempts to improve reading rates, including variation in the width of the 'window' on to the skin, and presentation across more than one finger. These were unsuccessful; spatial integration across fingers does not occur as it does with 'real' objects contacted normally by the hand. But the main problem being not with letter recognition but with integration into words, training which emphasised language skills has been found to improve reading rates. The augmentation of a tactile by an auditory display in synthetic speech also seems promising.

Jansson and Armstrong both deal with devices to help mobility in the blind. Armstrong describes the stages involved in the design, production and evaluation of tactile maps, and shows the remarkable three-dimensional understanding of spatial layout that blind people can achieve as a result of using several maps simultaneously. With these maps the tactile sense is used normally. In Jansson's experiments, a visual-to-tactile conversion is used, as with the Optacon, but the skin stimuli are distributed over a spatial mosaic of electrically stimulated points on the abdomen, and the camera is worn on the head, whose active movements can, with profit, 'scan' the environment. He describes how some form of externalisation is possible with this equipment, and shows that with single targets it can be used with considerable accuracy with the subject free to walk. Jansson takes the view that it is only through a clearer knowledge of what kind of information the tactile system can handle that we can advance our techniques for sensory substitution. And it is clear that information acquired from those concerned in sensory substitution is itself contributing to this body of fundamental knowledge. The development of the Optacon to make new material available to the blind has in itself provided the most sophisticated instrumental device so far known for delivering patterned stimuli in psychophysical experiments.

The expression 'Active Touch' and the concepts underlying it have a long and honourable history, particularly among psychologists. Katz (1925; see also Krueger, 1970) regarded movement as essential to recognition and the hand as an exploring sense-organ which searches for information; and Gibson (1962) entitled his well-known review 'Observations on active touch'. Physiologists have not been unaware of the importance of the active element in the sense of touch: the first use of the expression I have seen, in explicit contrast with 'passive touch', was made by Sherrington (1900) in his exhaustive review of cutaneous sensations, though the underlying sensorimotor problem had been recognised before that. Through many of the intervening years physiology has been preoccupied with evolving biophysical concepts of transduction, impulse conduction and synaptic action and in so doing has generated powerful techniques for recording impulses from single nerve cells and fibres. Until recently these techniques were limited to use with inactive anaesthetised animals; but now, as a number of articles here will show, they can be used in conscious moving animals. The sensorimotor question is therefore accessible to experiment on several fronts; and that is why it was possible to bring together so much related material for the first time, in this book.

REFERENCES

J. J. Gibson, Observations on active touch, Psychol. Rev. 69, 477-491 (1962).

V. A. H. Horsley, The function of the so-called motor area of the brain, Brit. Med. J. Vol. II, pp. 125-132 (1909).

E. G. Jones & T. P. S. Powell, Connections of the somatic sensory cortex of the rhesus monkey I: ipsilateral cortical connections, Brain 92, 477-502 (1969).

D. Katz, Der Aufbau der Tastwelt. Zeitschrift für Psychologie, Ergänzungsband 11, 270 pp. (1925).

R. H. Kay, Experimental Biology, Measurement and analysis. Reinhold: New York (1964).

L. E. Krueger, David Katz's Der Aufbau der Tastwelt (The world of touch): A synopsis. Perception & Psychophysics 7, 337-341 (1970).

V. B. Mountcastle, J. C. Lynch, A. Georgopoulos, H. Sakata & C. Acuna, Posterior parietal association cortex of the monkey: command functions for operations within extrapersonal space, J. Neurophysiol. 38, 871-908 (1975).

C. S. Sherrington, Cutaneous sensations. In E. A. Schäfer (Ed), Text Book of Physiology, Vol. II, 920-1001. Pentland: Edinburgh & London (1900).

THE STRUCTURE OF FINGER PRINT SKIN

T. Andrew Quilliam

Department of Anatomy, University College London

INTRODUCTION

This opening contribution deals with the structure of the human skin in general, the modifications to its various components that occur in the digital region and the role that these features may play in the various tactile manoeuvres habitually employed by man in order to achieve recognition of nearby objects by manipulation alone - a process referred to as stereognosis by clinical neurologists.

BASIC STRUCTURE

General Considerations

The skin or integument is the most extensive and, functionally speaking, perhaps the most versatile of all the body's organs. Having increased sevenfold since birth, its surface area in the adult approaches two square meters. (This is just about enough to "carpet" a very small bathroom if you do not mind a great many seams!) Despite a somewhat deceptively inert appearance, the skin is far from being a passive barrier type organ only. However, while no attempt will be made here to enumerate or enlarge upon its many activities which are not involved in tactile sensibility, it is of some interest to note that it possesses more than two million sweat glands and five million hairs which may be of either the fine (down) or the coarser (terminal) types. Indeed, the scalp alone carries 100,000 hairs.

An area of skin devoid of hair is often referred to as being glabrous in character. Since glabrous means bare and smooth, in man and primates this is a convenient - rather than an entirely accurate - descriptive term. For example, the large areas of skin on the palm and sole - although bare of hair - do not possess an absolutely smooth surface.

Skin is usually described as consisting of an outer, thin, waterproof stratified squamous epithelium derived from ectoderm and called the epidermis, cuticle or scarf skin plus an inner, relatively thick, flexible, supporting layer of dense connective tissue, derived from mesoderm and called dermis, corium or cutis vera. In spite of their considerable extent, these two layers together only weigh 3 to 5 Kgm because they are relatively thin. In fact, they can vary in total thickness from 0.5 mm over the eardrum and the eyelid to 5 mm or more over the palm of the hand the sole of the foot.

However, deep to and co-extensive with the epidermis and dermis lies a closely related layer called the hypodermis or superficial fascia which merits consideration in all discussions about skin function. It is loosely attached both to the under surface of the dermis above and to the outer aspect of the underlying muscles and bones. Together, these three layers can form as much as 15 to 20% of the total body weight - a proportion which exceeds that fraction contributed by the whole of the skeleton or which amounts to from one third to one half as much as does the entire muscular system.

The Epidermis

Although the epidermis is very tough, it is unexpectedly thin being only about 0.1 mm thick in many parts of the body. It is non-vascular and, microscopically, it can be divided up into a series of cellular layers or strata. Whereas the inner layers consist of living cells, by way of complete contrast, the outer layers are composed entirely of dead cells. Intervening between the two extremes lie several layers of cells which, visually, exhibit a whole spectrum of intermediate appearances. Indeed, the fact that the skin of man is so distinctly multilayered may be responsible for the ease with which blistering can be effected in contrast to the situation in many other species.

Collectively, the living cells of the epidermis are referred to as the Malpigian layer, stratum malpigii or rete mucosum but there is an innermost, single lamella of darkly staining and actively dividing columnar cells known as the stratum basale or stratum germinativum. The more superficial cells of the stratum malpigii are several lamellae thick and they are somewhat polyhedral in outline. Their boundary membranes - though separated from those of their neighbours by small gaps - seem to be closely linked with one another in a punctate manner by minute tonofibrils related to desmosomes. This appearance has given rise to two further descriptive names for this region - the stratum spinosum and the prickle cell layer.

The next layer from within outwards is the stratum granulosum or layer of Langerhans which is usually 2 or 3 cells thick. Here, the cells are often diamond shaped and are darkly staining as intracellular granules start to appear in their cytoplasm. Presumably these granules contain the precursor of keratin (keratohyalin) which must not be confused with the skin colouration pigment called melanin that is produced by special cells called melanoblasts in the basal layers of the epidermis. In the stratum granulosum not only is cellular tight packing becoming apparent but cell death seems imminent as is indicated by the beginnings of loss of nuclear integrity. In the skin of the palms and soles, a narrow, clear, homogeneous layer - the stratum lucidum - can be easily discerned next. Little remains of the nuclei in this layer. There are no gaps between neighbouring cells although there may be (? artefactual) clefts between successive horizontal lamellae. Desmosomes are less easy to distinguish. Hereabouts, the skin becomes relatively impermeable to water and eleidin replaces the keratohyalin granules.

Superficial to the stratum lucidum is the stratum corneum which consists of many lamellae of flat, elongated cells whose cytoplasm has been largely replaced by keratin. It is highly variable in thickness depending on local environmental conditions and use. Finally, at the skin surface, yet another layer is sometimes distinguishable. This is the epitrichium or stratum disjunctum which consists of dried horny plates. Here sheets of cell "corpses" maintain a precarious attachment to the skin surface for a while before, ultimately, being completely desquamated.

This description of epidermal cytomorphosis provides no more than a series of vignettes depicting successive stages in the various continuing, complex and concurrent processes involved in the transformation of "tender" newly formed cells emanating from the stratum germinativum into a final, tough surface barrier (stratum corneum) intervening between an individual and his environment. The whole cycle of events occupies from about 20 to 30 days so the cells must undergo their obligatory centrifugal migration at an average rate of between 25 and 50 μm per day. It is remarkable that, in a non vascular tissue, the various processes that are concerned can be harmonized so precisely that a dynamic equilibrium is achieved in health which results in overall skin thickness remaining constant. Disturbance of any one or more of the processes involved or faulty sequencing can give rise to a multitude of changed skin appearances which Clinical Dermatologists recognize and treat as distinct disease entities.

The cellular migration itself is induced and maintained by physical means. Thus the repeated emergence of new generations of cells from the stratum germinativum continually thrusts existing cells further and further peripherally and causes the latter to become progressively more and more flattened. As this change in shape is occurring, the cells also become more translucent to transmitted light - an alteration in appearance described by microscopists as hyalinisation. This is due to the deposition of more and more keratin (or its precursors) intracellularly. This is so fundamental a change that, eventually, cell death occurs but so gradual is the build up of keratin that the cell walls are retained intact and the keratin thus remains securely packaged. Functionally, these changes result in the cells becoming tough and trauma resistant (i.e. horny) in character, a change described as cornification. At the same time, the water content of the cells is reduced from 70-80% in the region of the stratum germinativum to about 10% in the region of the stratum corneum. Indeed, in a damp atmosphere the more superficial layers of the skin may absorb a considerable amount of water which will be lost just as quickly again if the individual moves into a dry environment. Only at this stage do the dead cells of the superficial layers of the epidermis achieve full functional maturity - surely a remarkable situation very seldom found elsewhere than in this tissue which also bears hairs and nails whose outer parts are, likewise, composed of dead cells (cf. mature red blood corpuscles).
Eventually, the physical bonds between neighbouring cells become so weakened that, imperceptably, flakes of dried cell corpses packed with keratin are shed from the skin surface to become significant only in so far as they contribute to house dust!

Immediately below and closely following the contours of the stratum germanitivum lies a continuous, rather amorphous looking, non-cellular layer called the basement membrane, or the basal or adepidermal lamina. It is thought to be supportive in function and to consist of an extracellular network of very fine fibrils plus glyco-proteins. Since, in most histological preparations, a narrow gap appears between the plasma membranes of the stratum germinativum above and the more clearly defined region of the basement membrane below, it is presumed that the latter has a dual composition and even, perhaps, a dual origin. No nerve fibre can be considered to be truly intra-epithelial unless it can be demonstrated to have pierced the basement membrane at some point along its course. Elsewhere (e.g. in the walls of capillaries), it has been shown that the texture of basement membranes is sufficiently open to allow the fairly free diffusion of a variety of small and moderately sized ions and molecules.

The epidermis is traversed by sweat ducts and hair shafts. Details of its innervation will be considered later.

The Dermis

The dermis consists of dense connective tissue varying in thickness from about 0.4 mm in the eyelid and prepuce to 3.0 mm or more in the palms and soles with an all over average of 1.0-2.0 mm. Generally speaking, it is thicker over the back of the body than the front and thinner in women than in men. It is highly vascular, contains lymph vessels, nerve fibres and specialised nerve endings together with the greater part of certain skin appendages (e.g. sweat glands and hair follicles). Besides an abundance of collagen and elastic fibres, some smooth muscle fibres (e.g. for hair erection and involuntary skin wrinkling) and striated muscle fibres (e.g. for facial expression) may be found in it but only in certain parts of the body. The striated muscle fibres represent the panniculus carnosus of animals which can "flick" a fly off from inaccessible areas of the skin with a maximum of local movement and the minimum of over all effort!

The interface between the epidermis and dermis - where the two tissues are firmly adherent to one another - is easily identifiable. It is of considerable importance especially in regions of special tactile sensitivity where the interface is usually highly folded. Those parts of the dermis that lie in close proximity to the epidermis - especially where "fingers" of dermis project upwards into hollowed out "pockets" on the underside of the epidermis - are known as the papillary layers. Elsewhere, the remaining parts of the dermis are referred to as the reticular layers. Except that the latter contain a denser network of collagen and, possibly, some fat cells, it is very difficult to delineate a precise plane separating these two regions of the same tissue.

The Superficial Fascia

This tissue intervenes between the epidermis and dermis superficially and the muscles and skeleton below and it allows a certain amount of gliding of the one over the other with the minimum degree of friction. This fascia also consists of connective tissue and it is in direct continuity with the deepest layers of the reticular regions of the dermis. Indeed, its elastic and collagenous fibres mingle with both those of the dermis and those of deeper structures. Obviously, the actual extent of the mobility of the skin over its underlying structures varies from region to region and is dictated by the numbers and arrangement of these fibrous attachments locally. In many animals, these attachments are much looser than in man. Consequently, not only is the skinning of many animals relatively easy but any skin scarring subsequent to trauma is much less obvious to the eye in animals than it is in man.

Where the fascia contains much fat, it is often called the panniculus adiposus and its distribution endows an individual's figure with its characteristically rounded outline that aids recognition even when the face is hidden. Over the abdomen, the skin is both thick and mobile and here the fascia may be as much as 3 cms deep, a dimension to which fatty tissue contributes signific-antly. Over the eyelid and prepuce this fascia never contains fat cells.

Large blood vessels, nerve trunks and nerve endings abound.

DIGITAL SKIN

When compared with the epidermis covering most of the other parts of the body, that of the finger is very thick (i.e. circa 1.00 mm). This increase in thickness is both real and apparent because not only is there an increased depth to several of the various cellular layers but there is also a characteristic folding of the whole epidermis. This latter effect results in the formation on the outer surface of the epidermis of permanent, low, linear eminences just visible to the naked eye. Variously called sweat, primary or papillary ridges, these are composed of externally projecting folds of stratum corneum arranged in parallel whorled groupings. They are disposed in particularly well ordered arrays in the region of the finger pulp where they are unobscured by skin creases and form curvilinear patterns with a distinct inter-unit periodicity. Popularly, these patterns are known as finger prints although, strictly speaking, this term ought to be applied only to facimile reproductions of such ridge patterns obtained by "inking" the finger tip and using it in a standardized manner like a printer's block in order to obtain on paper a flat (two dimensional) likeness of a natural pattern that actually possesses an additional element which is orientated in a third dimension. The very practical medico-legal importance of finger prints should not be allowed to overshadow the role that papillary ridges may play in the tactile processes associated with stereognosis.

The Dermo-Epidermal Interface

There is a vertical correspondence between the tracks of these papillary ridges and two series of rather more bulky downwardly projecting ridges (whose profiles when seen in vertical sections under the light microscope are known as "rete pegs") situated on the under surface of the epidermis. Thus, larger interpapillary or intermediate ridges directly underlie the papillary ridges and it is through them that sweat gland ducts pass to the skin surface to open on the summits of the papillary ridges. Somewhat smaller 'limiting' ridges are located below the grooves intervening between adjacent papillary ridges.

Since, at the dermo-epidermal interface, the two opposed tissues are intimately adherent to one another, the upper surface of the dermis faithfully replicates the undulations on the under surface of the epidermis in a reciprocal fashion. As a result of this arrangement, the upper surface of the dermis is moulded into a series of upwardly projecting folds whose profiles when seen in vertical section are called primary dermal papillae. These interdigitate with the downwardly projecting intermediate and limiting ridges on the under surface of the epidermis. Furthermore, this relatively simple degree of coadaptation at the dermo-epidermal interface is complicated by the presence of smaller upwardly directed extensions of the comparatively large primary dermal papillae to form very much smaller secondary papillae that occupy dome shaped niche-like excavations hollowed out from the under surface of the epidermis usually in the concave inter-ridge regions especially in or near the broad bases of the intermediate ridges. Although some secondary dermal papillae appear to be empty of large complex nerve endings, a considerable proportion house a Meissner type corpuscle (see later). These receptors are found in large numbers only in the skin of the fingers and toes of man and of simians although, singly, they can occur elsewhere.

Mechanisms Resisting Skin Displacement

Shearing forces applied to the finger tip which elsewhere in the body would result in shifting and/or wrinkling of the skin are resisted in this region partly by the buttressing effect of the corrugations described above on the inner and outer aspects of the epidermis and partly by the very firm attachment by means of a basement membrane of the epidermis to the dermis at their mutual interface. The strength of this latter attachment is augmented by the physical interlocking of the two tissues which results from the reciprocal folds situated on their opposed surfaces. This arrangement also prevents any tendency of the epidermis to slide over the dermis when the former is subjected to tangentially applied forces. It is also possible that the great number of sweat gland ducts passing from the dermis to the epidermis contributes to the binding of these two tissues firmly together.

Furthermore, numerous fibrous trabeculae passing through the subcutaneous tissue anchor the dermis to the periostium of the terminal phalanx. In the subcutaneous tissue, these same trabeculae help to enmesh groups of fat cells which endow the finger pad with both a firm rounded outline and a cushion-like ability to adapt its shape snuggly - by means of temporary deformation rather than by displacement - to the contours of any object currently being held firmly between the fingers or being pressed upon.

Receptor Identification

Despite the widely held belief that structure is closely related to function, attempts to classify the receptors present in any particular skin area using histological and neuro-physiological criteria do not always result in a direct correspondence between the two disciplines being achieved at once. However, the degree of unanimity in what might otherwise be regarded as a somewhat sterile "demarcation dispute" can usually be increased by a twin attack which gradually tries to reduce the areas of current uncertainty inherent in the two approaches and which carefully evaluates the interplay of relevant data, interpretations and opinions which emerge from interdisciplinary discussion.

In human digital skin, the sensory nerve endings visible to the microscopist fall into five distinct categories. These consist of "free" (bare) intra-epidermal nerve endings, Merkel type "discs", papillary nerve endings, Meissner type corpuscles, and Pacinian type corpuscles.

Intra-Epidermal Receptors

The first two varieties of receptor mentioned above belong to this group. The free nerve endings, in particular, are very difficult to demonstrate convincingly with any regularity in human or in most animal material. On the other hand, Merkel's discs are somewhat easier to find in a few carefully selected animal tissues. Whether this situation is a reflection of some unsuspected limitation of current histological techniques as applied to human tissue or whether it truly represents a comparatively low incidence of both these particular kinds of nerve endings even in human finger tip skin still remains an unresolved problem. In this respect, post mortem material derived from subjects who have died from accidental electrocution is said to be particularly suitable for demonstration purposes but the very infrequency with which such material comes to hand suggests that even if the association were not fortuitous, other and more controllable methods of revealing intra-epidermal receptors will have to be found before definitive studies can be pursued with any expectation of universal acceptance!

Prior to becoming intra-epidermal, the neurites that supply these two types of nerve ending can be traced as myelinated fibres located in primary dermal papillae having arrived there from the superficial dermal nerve plexus. These fibres approach the basement membrane and travel along it for some way in the dermis before actually entering the epidermal domain. In the case of a free nerve ending, this part of its course may take it as far as the dome of a papilla. On penetrating through the basement membrane, both the Schwann cell sheath and the myelin coat are discarded and a thin tapering bare axon - usually unbranched - threads its way for a variable distance between the cells of the stratum malpigii to be lost to view as the stratum granulosum is approached. Consequently, the ultra-terminal parts of these endings may be located only a few hundred µm from the actual skin surface.

Merkel's discs are seen by the light microscopist in or near the stratum germinativum as occasional, large cells which are less densely stained than are conventional epidermal cells (keratinocytes). Special stains reveal that not only are several neighbouring Merkel's discs innervated by fine branches of a single papillary nerve fibre but that each disc appears to possess an intracellular terminal nerve network. It has been postulated that Merkel's discs - just like keratinocytes - undergo an obligatory centrifugal migration, become keratinized (and thus lose their separate morphological identity) die and become disconnected from their erstwhile axons of supply. When these axons survive this disconnection for a variable period they are believed to constitute the scant population of bare intra-epidermal nerve endings that have hitherto been accorded independent status in the roll call of cutaneous receptors (Quilliam et al., 1973a).

When viewed under the electronmicroscope, other aspects of the morphology of Merkel's discs tend to be emphasized. Thus, it is quickly apparent that, in reality, the so called disc consists of a specialized non-keratinocytic type of cell partially embraced by one or more axon terminals. Perhaps the most interesting feature of the cytoplasm of a typical cell is the presence of many rather large, electron dense, membrane bounded lysosome-like granules which may possibly contain A.T.P. or its precursor. Occasional desmosomes join the walls (plasma membranes) of these cells to those of neighbouring keratinocytes but hemi-desmosomes have not been identified joining the Merkel's cell to its adjacent basement membrane as is the case where conventional keratinocytes are concerned (Quilliam, 1966; Quilliam et al., 1973b).

Papillary Nerve Endings

These axons which are freely ending are to be found exclusively in the papillary layers of the dermis and particularly within primary dermal papillae. They are derived from myelinated fibres in the superficial dermal plexus and they may be seen to taper into a very thin ultraterminals to branch or to form either loops, simple spirals or lattice like networks. The term Hederiform expansion is a name given to a terminal ball of neurofilaments sometimes seen in this region which is unrelated to any particular component of the connective tissue. In normal skin, they must be differentiated, in the dermis, from casual irregularities in diameter occurring along the course of papillary nerve endings and, in the epidermis, from Merkel's discs. Where the section involved is a relatively thick one, this is not always an easy task.

The Meissner Corpuscle

Apart from epidermal folding, the presence of many Meissner corpuscles in a section of skin viewed under a low power objective immediately identifies the specimen concerned as having been obtained from the digital region of the hand or foot. Also, all species save man and some higher apes can be confidently excluded as the likely source of the material as these corpuscles only occur sparingly in other tissues and in other species (Griffin, 1977).

The corpuscles which are ovoid in shape are situated very near the dermo-epidermal interface in either a primary or a secondary dermal papilla. Their overall average length is usually of the order of 80 µm and their transverse diameter about 30 µm. They are almost invariably oriented so that their long axes are disposed at right angles to the skin surface with their upper pole about 500 µm from the individual's actual interface with the environment and as little as 100 µm from the outermost boundary of the most superficial living cells. Usually a single primary or secondary dermal papilla contains only one corpuscle but sometimes two corpuscles may occupy the same primary papilla.

The corpuscle itself consists of a richly innervated cellular inner core which is totally enclosed in a well defined multi laminated capsule probably of connective tissue origin. The upper pole of the corpuscle often abuts snugly onto the under surface of the dermo-epidermal junction at the apex of the roof of a dermal papilla and may even be attached to it by elastic fibrils but the junction does not involve desmosomes so, presumably, cannot be particularly strong or taut. At its lower pole, the capsule is not so well defined and some of its laminae blend imperceptibly with the connective tissue sleeve accompanying and enwrapping the fascicle containing the several nerve fibres supplying the corpuscle.

Most corpuscles are innervated by 2 or 3 myelinated fibres but it is not very uncommon for as many as 4 or 5 or even 6 or 7 fibres to pass to a single corpuscle. All the fibres to a particular corpuscle come from the superficial dermal plexus but since each leaves the plexus at a separate point before assembly into a single fasciculus of supply, it is assumed that each of the fibres involved is a branch of a different stem fibre in the plexus. Furthermore, it would seem that other branches of these same stem fibres help to supply other nearby Meissner corpuscles as well as some of the other nerve endings of various types which happen to be located in the near neighbourhood. Usually, a nerve fibre in any one fasciculus is much thinner than the rest but, on entering the corpuscle, it does not appear to be distributed any differently to its companions. The fasciculus joins the corpuscle at its lower pole and the nerve fibres it contains all pierce the capsule of the corpuscle still retaining their myelinated sheaths for a variable distance within.

Inside the capsule, the core cells are grouped into 3 or 4 lobules arranged vertically one above the other. The cells of the basal lobule are irregularly shaped and possess a central nucleus. The cells of the apical lobule are distinctly disc shaped and they possess a peripherally situated nucleus. The cells themselves are arranged one on top of another in a reasonably orderly pile rather like a stack of coins. Elsewhere (e.g. Cauna & Ross, 1960) these cells have been referred to as "laminar" cells but, in light microscopy at any rate, 'discoid' appears to be a more accurate description. In fact, their characteristic shape has not yet been determined precisely. The cells of the remaining lobule or lobules (i.e. intermediate or upper and lower intermediate lobules) have a somewhat similar overall appearance and arrangement to those of apical lobule but there is a tendency amongst the lowermost representatives to revert to the irregular outlines characteristic of the cells of the basal lobule (Quilliam, 1975a).

Inside the capsule, the nerve fibres retain their myelinated sheaths only whilst in passage to a lobule. Thus, one or more fibres pass to a basal lobule, lose their myelin sheaths, enter the lobule and weave a sinuous course between the processes of the irregularly shaped cells of this lobule. These neurites do not branch or terminate whilst within this lobule but eventually leave it to pass to and then enter an intermediate lobule. This latter lobule is usually also supplied by a separate myelinated fibre that takes a direct route from the fasciculus. This fibre also loses its myelin sheath on entering the lobule. Once inside, the unmyelinated neurites from both sources behave similarly in that they divide up into a series of branches that pass between the opposed surfaces of adjacent pairs of the discoid cells that collectively make up the lobule concerned. Such neurites may branch and then terminate finally within an intermediate lobule or they may pass further on to help innervate a lobule situated at a higher level.

As far as the apical lobule is concerned, it receives some innervation from neurites that have previously supplied branches to intermediate lobules but, almost invariably, the apical lobule also receives a myelinated fibre direct from the fasciculus as well. As it enters the lobule, this fibre loses its myelin sheath and, in a similar manner to other neurites supplying the same lobule, it divides into a number of branches which insinuate themselves between the opposed surfaces of neighbouring cells.

Attempts to correlate the appearances of the intralobular nerve endings as seen in light microscopy and in electron microscopy have not yet proved successful. Indeed, this is a highly complex field that it is more appropriate to explore in detail elsewhere. However, it has not yet been possible to exclude with any degree of certainty the possibility that arborizations derived from one neurite supplying a particular lobule may be directly connected to (or may intimately interdigitate with) arborizations derived from another neurite (Quilliam, 1975a).

With advancing years, the corpuscles become increasingly loculated and distorted in overall shape and the regular arrangement of the discoid cells they contain is gradually lost. This change spreads upwards from the basal lobule and often becomes so marked that, in persons over the age of 60 years, many corpuscles have completely lost their typical ovoid shape and appear to consist only of a tangle of axons ramifying among very irregularly shaped inner core cells. Eventually, especially when the axons are silver stained they come to resemble a tangled skein of wool and the corpuscles thus lose all those classical structural characteristics by which a Meissner's type of end organ can be identified (Ridley, 1968; Ridley, 1969). Only their presence and their position within digital skin enables their true identity to be suspected. Indeed, at this stage, the histological appearances seen can sometimes resemble those of classical Krause type corpuscles or even of Golgi-Mazzoni type corpuscles which, typically, are not present to any conspicuous degree in digital skin.

Blood capillaries have not been observed within the Meissner corpuscle, but numerous capillary loops lie in close relationship to the outer aspect of its capsule within the dermal papilla. Deeper within the dermis, blood sinuses are abundant.

The Pacinian Corpuscle

Not only is this ovoid corpuscle much larger (i.e. circa 1.5 mm x 0.5 mm) than any of the previously mentioned receptors but it is also situated more deeply. Typically, it lies in the subcutaneous tissue at least 2-3 mm from the skin surface and, especially in the digits, often seemingly in relation to tendons and joints. It is believed that of a total of about 2000 such corpuscles in the human skin, approximately one third are located in the digits. Elsewhere, these corpuscles can be found comparatively sparingly (Wood Jones, 1941) in a number of non cutaneous situations which include the mesentery and the pancreas. They may also occur near blood vessels, nerves, periosteum joints and inter-osseous membranes. Neurophysiological experiments on corpuscles obtained from non-human sources and indirect evidence from behavioural studies involving the much smaller but structurally somewhat similar Herbst corpuscles clearly indicate that these receptors are sensitive to vibration stimuli over a circumscribed frequency range.

The corpuscle consists of a thin capsule, a wide outer core made up of many, very thin, cytoplasmic "onion skin" coverings alternating with separate tissue fluid filled lakes together with a much more compact inner core. This latter is composed of two bilaterally symmetrically arranged stacks of thin, cytoplasmic, closely applied, hemi-lamellae separated from each other by two radial clefts (Pease & Quilliam, 1957). The nerve terminal, which will be described in detail later, occupies a central compartment embraced by the inner core. Occasionally, specialized membrane thickenings are present (Quilliam, 1966).

Each corpuscle is supplied by a single medium sized myelinated nerve fibre (diameter approximately 4-7 μm) which is unbranched either centrally or peripherally so that there is a direct line of communication with the spinal cord and possibly with the brain itself. This is in contrast to all the other types of receptors considered so far which receive either a single branch of a stem fibre derived from the superficial dermal plexus or, in the case of the Meissner corpuscles, several branches each from different stem fibres deriving from the same plexus. In all these cases, the other branches of the stem fibres concerned supply neighbouring receptors as well which may be either of a similar or of various different varieties.

On entering the corpuscle at its proximal pole, the myelinated nerve fibre of supply is somewhat convoluted as it transverses the outer core surrounded by connective tissue and Schwann cell sheaths. Here it usually exhibits one or more Nodes of Ranvier. On entering the inner core, the axon loses its connective tissue, myelin and Schwann cell sheaths. It becomes temporarily constricted and then proceeds in a straight line in the longitudinal axis of the corpuscle in the centre of the inner core for about 5-600 μm (Quilliam & Sato , 1955). In this part of its course, the nerve terminal is somewhat oval in cross section with its longer radius being in the same plane as the clefts of the inner core. Typically, its peripheral axoplasm exhibits a considerable concentration of mitochondria. At the far end of the inner core, the terminal may show some minor elaboration such as bifurcation or vesiculation. This region is conveniently referred to as the ultraterminal. However, in the region of the corpuscle's distal pole, the ultraterminal never penetrates back again into the outer core.

Satisfactory interpretation of occasional appearances which are apparently at variance with the above description - especially the suspicion of multiple innervation - rests on the basis of a knowledge of minor localized morphological variation occurring among individual corpuscles of a normal population (Quilliam, 1975b) and an appreciation of the effects of re-innervation.

The Role of Skin Structure in the Peripheral Mechanisms of Sensation

At present, it is not possible to associate the conscious perception of any given specific modality of sensation with the activation of a particular histological type of cutaneous receptor. Indeed, with one notable exception (i.e. the Pacinian corpuscle), it is not yet known for certain to what kinds of physical stimuli cutaneous receptors will respond *in vitro*. Consequently, it is not proposed to attempt to address this particular problem in the present contribution. However, there are certain structural aspects of digital skin which suggest that the skin itself may be significantly involved in the processes that accompany and follow closely upon the application of an effective external stimulus and which preceed the final transduction processes occurring within the receptors themselves immediately prior to impulse formation. For instance, the various physical components of an "omnibus" type of externally applied stimuli may be "filtered" differentially (i.e. speedily damped or gradually attenuated to extinction or otherwise modified) as they are propagated through the skin away from their point of application. In addition, the act of digital exploration may of itself generate physical parameters and thus add certain enriching dimensions to the primary sensory input. Another possibility is that the receptors themselves may be "tuned" locally so as to become more (or less) sensitive as their tissue microenvironment is altered slightly in response to changing circumstances. The significance of receptor numbers, size, orientation and spatial relationships also demands attention.

Indirect support for the idea that the sensitivity of a particular region of skin depends in some small part on the properties and dimensions of the various non-nervous cutaneous components represented locally is provided by the work of Jänig (1971a & b). Thus, not only was this worker able to show that, using graded mechanical stimuli on the intact skin of the cats' paw, the receptive fields of individual receptors were very much larger than the cross sectional areas of the receptors themselves, but also that the fields usually possessed two or more points of minimum threshold which themselves extended over several outer skin ridges. He also demonstrated that when most of the stratum corneum was removed, the thresholds of both rapidly adapting and slowly adapting receptors were reduced by from 10 to 30% of their initial values. At the same time, he found that the areas of the corresponding receptive fields had become smaller. These findings can be partially explained by consideration of the enhanced resistance to deformation exhibited by the relatively stiff and horny stratum corneum. In practical terms this means that during any but the very lightest of stimuli, besides a primary skin indentation of maximum depth occurring precisely at the point of contact with an applied probe, a much wider but shallower, saucer shaped, secondary depression surrounds the primary indentation. Consequently the number of cutaneous receptors that might be stimulated by even a slender probe will be considerably greater than a strict mathematical comparison between the face size of the probe and the density of occurrence of histologically recognisable receptors in the skin locally might at first suggest. In the human finger, this inevitable disproportion between probe size and area of receptor stimulation tends to be accentuated by the extra thickness of the stratum corneum. However, this effect may be mitigated to a certain degree by the constant moistening of the skin surface. Nowhere is there a higher concentration of sweat glands than on the fingers and the channelling effect of the papillary ridges distributes the sweat evenly over the whole skin surface and so renders it more compliant than might otherwise be the case. Evidently, then, in the richly innervated human finger or in the calloused skin of the monkey hand, it is most unlikely that during conventional probe type experiments a single receptor only will be stimulated. However, if a sufficiently delicate discriminatory "touch" test could be devised in the future, it would prove most instructive to determine precisely how skin thickness and induced skin dryness might affect sensory performance.

In theory, there may well be some justification for the suggestion that, for analytical purposes, the tactile activities undertaken by the fingers should be considered to occur in two distinct phases even though, in practice, the one phase often follows the other without pause in such rapid succession that the movements associated with each appear to constitute a single, smoothly coordinated and uninterrupted, purposeful act. For the sake of argument, the first or preliminary phase could be thought of as a momentary gentle contact brought about largely by antecedent arm movement (Quilliam, 1971). This might be described as a variety of passive touch. Immediate detection of the presence of nearby objects in the immediate environment with their speedy preliminary classification (presumably on a pain or thermal basis) into those that are potentially harmful and those that seem relatively innocuous would appear to be the objective of this simple probing type of activity. The second more delicate and deliberate phase involves actual digital manipulation of the object under examination. Grasping and picking up the object enables an immediate judgement as to its relative weight to be made. Typically, subsequent surface exploration is accomplished by the finger tips lightly stroking and then, if necessary, more firmly rubbing the surface of the object held. During this activity, the palm usually still faces downwards if the object is small and/or light but the palm is turned upwards if the object is large and/or heavy in order to release for exploration purposes two or more of the five digits that would otherwise be involved in tight gripping. This stroking may be accompanied by the backwards and forwards rolling movement of the object between the fingers and/or palm. The extent of this latter movement is usually controlled by the thumb. The whole of this aspect of the tactile process can be thought of as active touch. Not only can the object's overall size, shape and special contours be rapidly assessed by these processes but such information as the position of the object's centre of gravity and its surface texture can be obtained. Comparisons between two small and light objects can be made simultaneously in one hand alone. Comparisons of larger and heavier objects require the separate use of both hands at the same time or sequentially. During such fine movements of the fingers, the very stiffness of the external papillary ridges (cf. a washboard) may well create vibratory effects which - although originating in the outer layers of the skin - may be propagated more widely through its other layers to be detected, perhaps, not only by relatively superficial receptors but, in all probability, by the more deeply-placed Pacinian Corpuscles as well. This could give rise to nervous signals that could supplement those arising primarily from skin indentation and so hasten recognition when comparisons with previously stored tactile patterns in the brain's memory bank are made.

Because they are so large, so complex, so well innervated and are present in such large numbers in the skin of the fingers it is difficult to escape the conclusion that Meissner's corpuscles play a dominant role in tactile discrimination, a task in which the fingers are constantly involved and which they perform so accurately and quickly. Whether these corpuscles are unimodality or multimodality receptors is not known. Also whether they (or any other histologically distinct type of receptor) participate exclusively in the processes associated with passive touch or in those of active touch is still a matter for conjecture. Perhaps the separation between the two phases is, in behavioural terms, an artificial one. However, the incidence of Meissner corpuscles per unit area of skin varies from place to place on the palmar aspect of the hand being maximal at the finger tip and minimal towards the wrist with a somewhat comparable distribution on the sole of the foot. Such quantitative estimates as are available tend to suggest a frequency of occurrence in the region of 25 to 50 corpuscles per mm^2 of finger tip skin. These suggest that - supposing the array of corpuscles to be arranged in a fairly regular semi geometrical pattern - the distance between any one Meissner type receptor and any of its immediate neighbours of the same variety would seldom exceed more than about 200 μm. Thus, in view of what has already been said about the physical properties of the stratum corneum, even when a slender bristle like probe with a diameter of about 100 μm were applied to the surface of the finger tip, probably no less than 4 Meissner corpuscles would be activated. Since the corpuscles are believed to be rapidly adapting, the maximum amount of tactile information about an object under examination could only be obtained through their agency if a sequence of fleeting sensory nerve impulse "vignettes" were made available by them to the brain. The latter could then extract, store and base future activity upon those features which constantly recurred and extract and reject items that appeared inconstantly and which could therefore be classified as being artefactual (i.e. having been randomly or spontaneously generated and thus of doubtful significance for recognition purposes).

The idea that the sensitivity of a receptor can to some extent be "tuned" even in primates whose deep body temperatures are held virtually constant throughout the 24 hours is an intriguing one for which several mechanisms have been suggested (Quilliam, 1966, Santini, 1976). As far as the effects of temperature are concerned, there is no evidence to suggest that, in man, during illness induced hyperthermia, there is an increased cutaneous sensitivity. Experiments have not yet been performed on healthy volunteers in whom controlled hyperthermia has been artifically induced. However, even without delicate instrumentation, it is a common experience that in cold weather, when the hands only are exposed, sensory discrimination in the digits rapidly becomes less efficient and that this occurs long before general body hypothermia occurs. Consequently, it would seem that the best way in which thermo regulation of receptor activity could be achieved would be by means of temperature changes induced in the intimate environs of individual receptors. In the case of the Meissner corpuscle, the primary and secondary dermal papillae provide a particularly favourable nidus from this point of view. The corpuscles themselves are each snugly embedded in sparse connective tissue close under a small dome like niche hollowed out from the undersurface of the living epidermis at the dermo-epidermal interface. They are thus quite near to the skin surface so that natural cooling is at its most efficient here. Only below are these papillae in continuity with the rest of the dermis and, in the case of the secondary papillae, this continuity is via a comparatively restricted 'neck' that allows the passage of only the nerves supplying the corpuscle above, other intrapapillary nerves and blood capillary loops. These structures are accompanied by an economical ration of connective tissue. Thus each Meissner corpuscle has what amounts to an individually tailored micro-environment virtually isolated (except for vital supply lines) from the remainder of the dermis. Consequently, localized vasodilation could raise the temperature of a corpuscle very quickly and this might increase its sensitivity whilst vaso-constriction would result in a somewhat slower temperature decline to the level of the general epidermal temperature (which is usually less than that of the deep body temperature) with a consequent reduction in corpuscular sensitivity.

ENVOI

Whilst recognizing the great importance of the role that will be played by neurophysiology and psychology in future experimental research into the complex mechanisms of digital sensation, it is evident that not only is further work on the structure, innervation and distribution of cutaneous sensory nerve endings in the skin highly desirable but it will also be essential to find out more about the micro-mechanical characteristics of skin - especially how it reacts and deals with applied physical stimuli.

ACKNOWLEDGEMENTS

I am grateful to Ian Barnes F.D.S. for scanning electron microscopy and to John V. Arnold for the preparation of the plates.

REFERENCES

(1)
Cauna, N. & Ross, L.L. The Fine Structure of Meissner's Touch Corpuscles of Human Fingers. Journal of Biophysical & Biochemical Cytology, 8, 467-482 (1960).

(2)
Griffin, C.J. The Fine Structure of Nerve Endings in Human Buccal Musosa. Archives of Oral Biology (in press).

(3)
Jänig, W. The Afferent Innervation of the Central Pad of the Cat's Hind Foot. Brain Research, 28, 203-216 (1971a).

(4)
Jänig, W. Morphology of Rapidly and Slowly Adapting Mechano receptors in the Hairless Skin of Cat's Hind Foot. Brain Research, 28, 217-231 (1971b).

(5)
Pease, D.C. & Quilliam, T.A. Electron Microscopy of the Pacinian Corpuscle. Journal of Biophysical & Biochemical Cytology, 3, 331-342 (1957)

(6)
Quilliam, T.A. Unit Design and Array Patterns in Receptor Organs. In: Touch, Heat and Pain (1966) pp. 86-116. Eds. A.V.S. Reuck and J. Knight London : Churchill.

(7)
Quilliam, T.A. Cutaneous Sensation Reconsidered. Medical & Biological Illustration, 21, 183-187 (1971).

(8)
Quilliam, T.A. Neuro-cutaneous Relationships in Fingerprint Skin. In: The Somatosensory System (1975a) pp. 193-199. Ed. H.H. Kornhuber. Stuttgart : George Thieme.

(9)
Quilliam, T.A. Some Hazards in the Interpretation of the Pattern of Structure in Lamellated Receptors. In: The Somatosensory System (1975b) pp. 200-204. Ed. H.H. Kornhuber. Stuttgart : George Thieme.

(10)
Quilliam, T.A. & Sato, M. The Distribution of Myelin on Nerve Fibres from Pacinian Corpuscles. Journal of Physiology, 129, 167-176 (1955).
(11)
Quilliam, T.A., Jayaraj, P. & Tilley, R. The Genesis of Free Nerve Terminals in the Epidermis. Proceedings of 3rd European Anatomical Congress, Manchester, 1, 78-79 (1973a).
(12)
Quilliam, T.A., Jayaraj, P. & Tilly, R. Epidermal Innervation in the Pig's Snout. Journal of Anatomy, 115, 156-158 (1973b).
(13)
Ridley, A. Silver Staining of the Innervation of Meissner Corpuscles in Peripheral Neuropathy. Brain, 91, 539-552 (1968).
(14)
Ridley, A. Silver Staining of the Nerve Endings in Human Digital Glabrous Skin. Journal of Anatomy, 104, 41-48 (1969).
(15)
Santini, M. A Theory of Sympathetic-Sensory Coupling. In: Sensory Functions of the Skin of Primates, pp. 15-35 (1976). Ed. Y. Zotterman. Oxford : Pergamon Press.
(16)
Wood-Jones, F. The Principles of Anatomy as Seen in the Hand (1941). 2nd Edition. London : Bailliere Tindall & Cox.

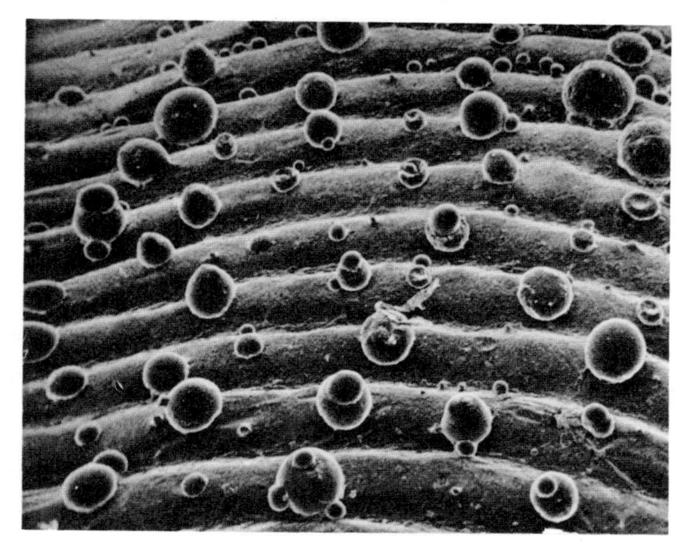

Fig. 1

Fig. 1 Scanning electron micrograph (S.E.M.) of finger tip skin. The papillary ridges and the intervening sulci recur at regular intervals. At short intervals, globules of sweat exude from the summits of the ridges. It is possible that the shape and the position of the sweat globules seemingly permanently poised on the ridges are artefacts of the method employed. (Human index finger; araldite replica; field of view 5 mm).

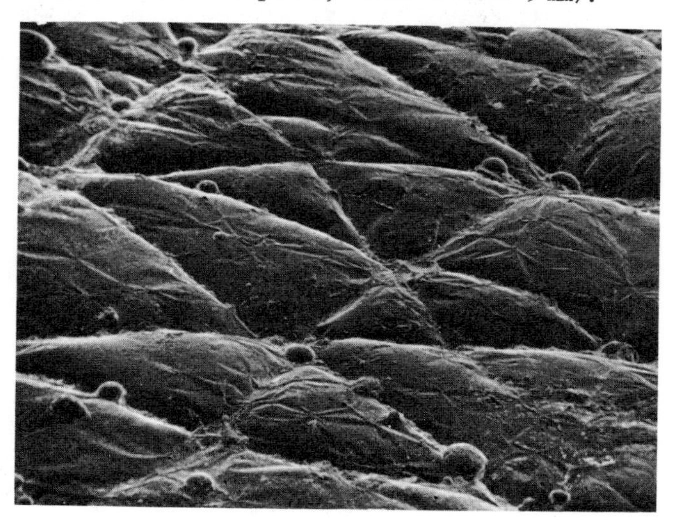

Fig. 2

Fig. 2 S.E.M. of non-hairy skin. The surface contours are irregularly fissured and occasional small globules of sweat are seen near the fissures. (Human lip ; araldite replica; field of view 200 μm).

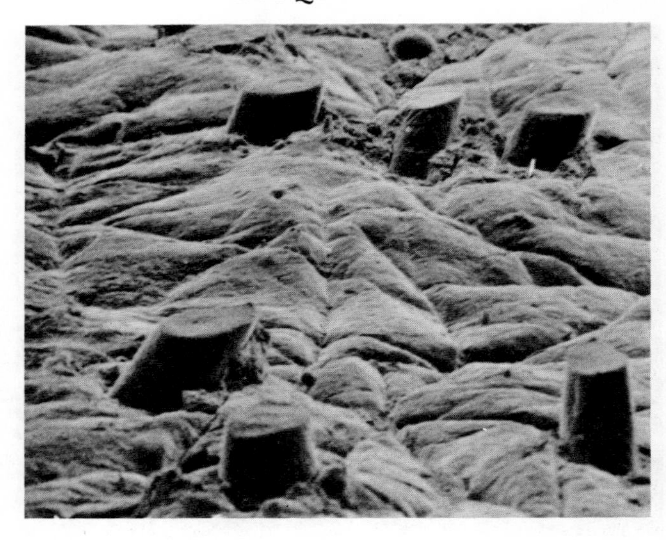

Fig. 3

Fig. 3 S.E.M. of hairy skin. Six shaven bristles are seen projecting from an irregularly fissured surface. A single globule of sweat is seen in the background. (Human chin ; araldite replica; field of view 200 µm).

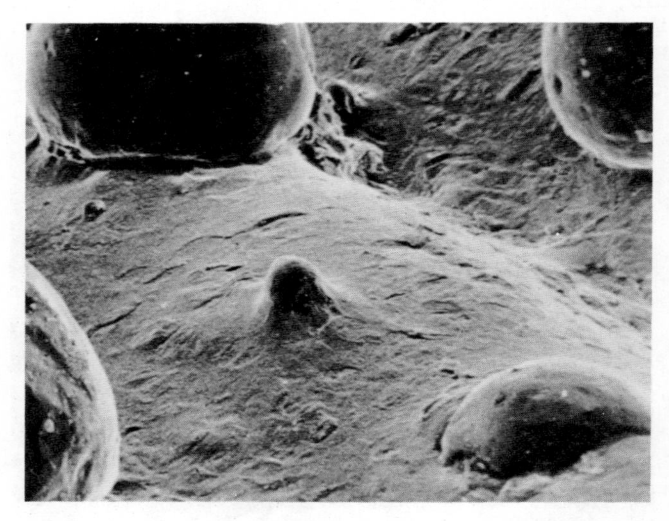

Fig. 4

Fig. 4 S.E.M. of part of a papillary ridge and a sulcus from finger tip skin. Globules of sweat in 3 different stages of production are seen. The contours of the ridges and sulci are gently rounded, flakes composed of several dead epidermal cells are beginning to be formed from the surface layers of the stratum disjunctum prior to their final desquamation. (Human index finger; araldite replica; field of view 500 µm).

SENSORY COUPLING FUNCTION AND THE MECHANICAL PROPERTIES OF THE SKIN

Julien Petit* and Yves Galifret**

Laboratoire de Psychophysiologie Sensorielle, Université Pierre et Marie Curie, 75230 Paris, Cedex 05, France

ABSTRACT

Using conditions similar to those employed in tactile physiology, the relation between the force and the resulting indentation was measured at different skin locations in the rat and in man. This relation is multiform and consequently the force sustained by a mechanoreceptor cannot be simply deduced from the cutaneous indentation. It is possible that the variety of coupling functions (input-output relation) which have been proposed for the slowly adapting mechanoreceptors can in part be explained by the fact that the input used was the cutaneous indentation.

INTRODUCTION

Curiously, Fechnerian psychophysics started with Weber's experiments on weight comparison; and it is in the realm of somesthesis that this psychophysics is most difficult to accept. It is true that even in the specific field of cutaneous mechanoreception it is more difficult to define the effective stimulus parameters than in hearing or in vision. Kruger and Kenton (1973) have stressed the difficulties of establishing the input-output relation in sensory physiology. Furthermore it may be asked whether the diversity of coupling functions (input-output relation) proposed in different studies (Lindblom, 1965; Werner and Mountcastle, 1965; Mountcastle, 1966; Chambers et al., 1972; Knibestöl, 1975) is due to the diversity of operation of various types of receptors or if it is not, at least in part, linked to the difficulties of understanding the adequate parameters of the stimulation. For the slowly adapting cutaneous mechanoreceptors the amplitude of cutaneous deformation is often taken as a measure of the stimulation. Is indentation a valid measure of the stimulation? In order to verify this, we have studied the behaviour of the skin under the stylus of the sensory physiologist, employing for indentation and force the range of values generally used in physiological experiments.

METHOD

The aim of the experiment was to measure simultaneously the force applied to the skin and the deformation thus provoked.

The system (Fig. 1A) devised and built by one of us (J.P.) consisted of a vibrator (V) whose motor element pushed against a spring (R). The spring deformed the skin by way of the probe (T). The movement (E1) of the motor

* J. Petit, Laboratoire de Neurophysiologie, College de France, 11 Place Marcelin Berthelot, 75231 Paris Cedex 05, France.

**Y. Galifret, Laboratoire de Psychologie Sensorielle, Universite Pierre and Marie Curie, 4 Place Jussieu, 75230 Paris Cedex 05, France.

element was measured by the differential transformer (M1), the movement of the probe (E2) was measured by the differential transformer (M2) and, using the spring deformation E = E1-E2, the force could be deduced: F = K E, this whole arrangement acting like a 'load cell'.

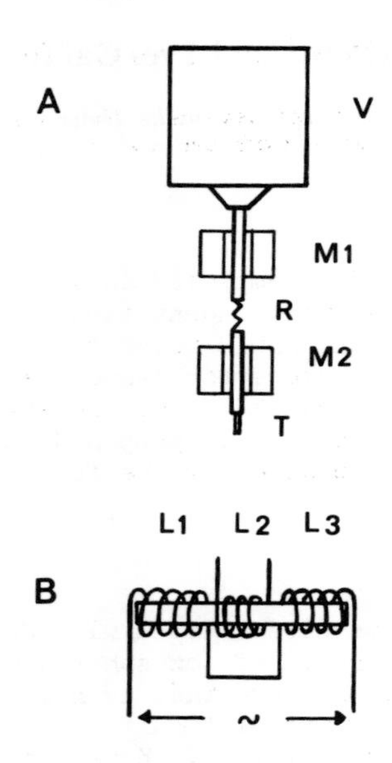

Fig. 1. A. Schema of the device. V : vibrator; M1 & M2: differential transformers; R:spring; T:probe
B. Differential transformer

Apparatus

Vibrator - A wave-form generator provided rectangular pulses. These pulses were fed into a power amplifier driving a coil which moved in the field of a permanent magnet. The form of the mechanical response given by the whole system was quasi identical to the form of the electrical signal up to about 1000 Hz. Thus rectangular mechanical pulses could be obtained.

Spring - It had an elasticity K = 5.5 g wt/mm. The minimum measurable force was approximately 27.5 mg wt., corresponding to a deformation of 5 μm.

Differential transformers (Fig. 1B). With an alternating current of 15,400 Hz in the opposing coils (L1 and L2) a band-pass of up to 1000 Hz could be obtained. The relation between the displacement and the voltage induced in the coil (L3) was linear up to 1600 μm with 2.5 V/mm. Oscilloscope readings were accurate to within 2% with a lower limit of 5 mV corresponding to a sensitivity of 2 μm. A calculation of the force, based upon measurement of two displacements, was thus found accurate to within 5%.

Probe - This is a cylinder of circular cross section; three diameters were used: 0.4, 1 and 1.5 mm. The probe was applied normal to the skin surface.

Mode of operation

The experiments were carried out on a vibration-proof table. The stimulation device was mounted on a carriage and could be moved by a micrometric drive (microelectrode driver) which permitted precise positioning. Measurements were performed on the skin of the rat and of human subjects.

Rat. The animal under deep anesthesia was placed in a containing apparatus. For stimulation of the hairy skin the involved regions were carefully shaved.

Man. The deformations were applied to the fingers. The forearm was fixed to the top of the table, the fingers were spread and separately fixed. Each joint was immobilized by being tightly bandaged yet not sufficiently to hamper the circulation.

In all cases, probe contact with the skin was checked under a dissection microscope as well as with the differential transformer, M2. It was thus noted that the skin was not fixed but moved up and down. Knibestöl (1970) has shown that this movement is synchronous with respiratory movements. Similarly, the proximity of a blood vessel produced pulsations synchronous with the cardiac rhythm. All these oscillations had amplitudes varying between 10 and 30 µm. Imposed indentations were always greater than 40 to 50 µm.

RESULTS

The determinations were made on the skin of the rat in different regions

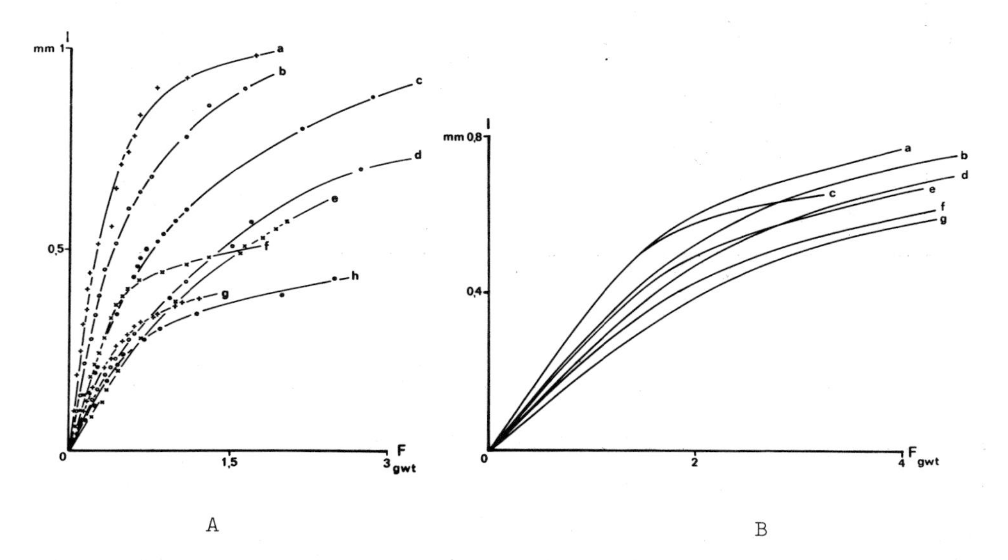

Fig. 2. Force-Indentation relation (probe diameter: 1.5mm)
A. For the skin of the rat : a and b, thigh; c, sole of hind paw; d and e, ankle; f and g, forepaw, dorsal side; h, second finger forepaw.
B. For the skin of a human subject; second phalanx (ventral side). a,c, first (index); b,d,e,f, third (ring) finger; g, second finger.

(fingers, palm, back of hand, wrist, thigh) and on the skin of human subjects on the ventral and dorsal surface of the second phalanx of different fingers employing a range of values generally used in neurophysiology: forces up to 3 or 4 gwt., indentations up to 1 mm.

Fig. 2A shows the Force-Indentation relation obtained with a probe of 1.5 mm diameter for 8 series of measurements in the rat (the location is indicated in the legend). All the curves show the same shape presenting an inflexion toward the higher values as the thickness of underlying tissues increases, particularly for the thigh (curves a and b). Fig. 2B shows the same relation for pressures exerted on the ventral face of the second phalanx in man. The curves show the same general shape; however the initial slope is not as steep and the inflexion is less pronounced. These data have been replotted (Fig. 3) using semi-logarithmic coordinates and it can be noted that the curves of Fig. 3B are thus almost superposable over the curves c,d,e of Fig. 3A.

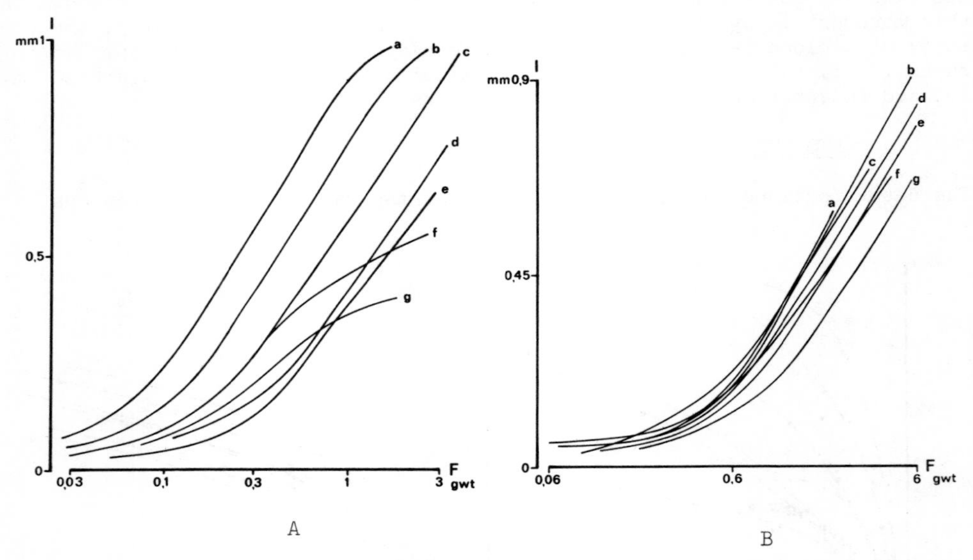

Fig. 3. Log (Force) Indentation relation same data as in
Fig. 2. (For clarity curve h in Fig. 2A has been discarded)

The four phases of the Force-Indentation relation

Looking at all the curves (with the exception of f,g,h of Fig. 2A and f,g of Fig. 3A) it can be noted that there exist two phases in which the relation between the force (F) and the indentation (E) is simple and obvious:
- a linear phase $E = kF$ at the origin, terminating at values of E of about 150 µm and
- a logarithmic phase $E = \gamma \log F$ starting at about 350-450 µm.
- between these two phases the transition can be approximated by a relation such as $E = \alpha \log (1+ \beta F)$.

For the curves f and g (Fig. 2A and 3A) it can be noted that the linear phase is reduced, particularly for g and that the log phase of both curves is equally very reduced in that a terminal inflexion intervenes almost immediately after the initial increase of slope. This contraction of the phases can be explained by the preponderant and early intervention of underlying bony structures when the subcutaneous cushion is lacking. In the curves a and b (Fig. 3A) the onset of a terminal inflexion at higher values is just noticeable. In man when pressure on the second phalanx is applied to the dorsal face the phenomenon is also observed and intervened earlier as the diameter of the probe was increased.

The role of the time factor

The previous relations were established for systems in equilibrium. Owing, however, to the viscoelasticity of the skin and underlying tissues this equilibrium state is reached only after a certain relaxation time.
As our force was applied through a deformable spring we were unable to impose a constant indentation and measure the evolution of the force during relaxation; in the seconds which followed the application of the probe the indentation

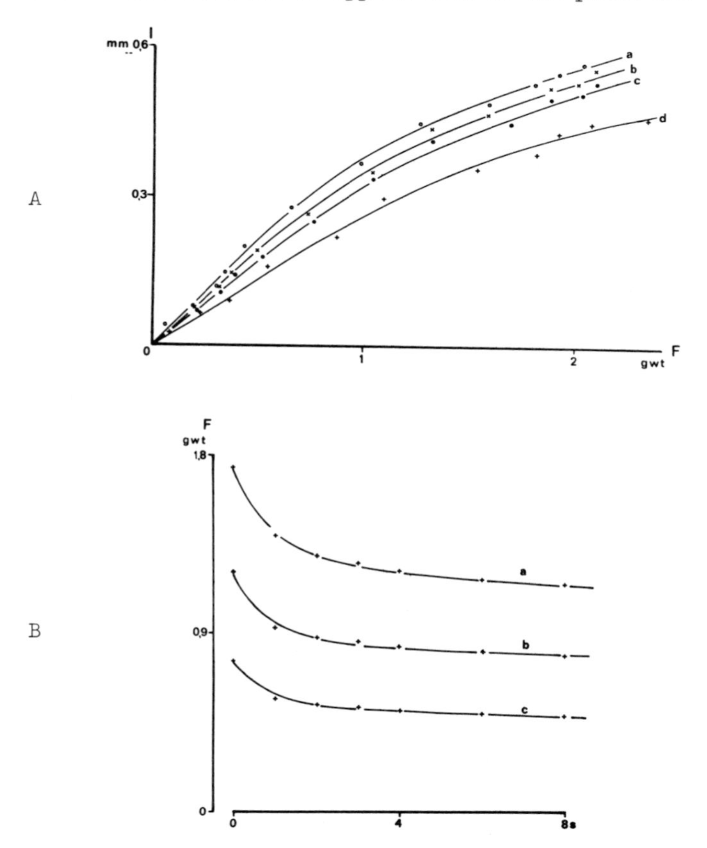

Fig. 4. A. Force-Indentation relation at different times during stimulation (rat, ankle) a,8s; b,4s; c,1s; d,0.1s. B. Variations of force with time for 3 different indentations. a,400 μm; b,300 μm; c,200 μm.

progressively increased and the force decreased. We therefore measured the force and indentation at different moments during stimulation; Fig. 4A shows an example of the curves obtained in this manner. From such curves we then determined for a given indentation, the force values at different instants; these values have been plotted in Fig. 4B. As the deformations remained small we can consider the curves of Fig. 4B as a satisfactory approximation of what would be obtained under isometric conditions. It can be noted that the relaxation time is in the order of 8 seconds.

Another phenomenon was observed: for strong indentations the deformation persisted following the removal of the probe. Several minutes (3 to 4) were required for the skin to regain its resting position. This phenomenon is related to what is referred to in mechanics as plasticity.

DISCUSSION

The mechanics of the skin

The different investigators who have been concerned with the mechanics of the skin and other biological tissues have often done this from the point of view of its application in surgical techniques. Large surfaces have been deformed by using either uniform pressure or simple traction. The forces employed have been as high as 1.5 kg, causing deformations in the order of centimeters. On the contrary our experiments involved a punching exerted over a reduced area, the force rarely surpassing 4 g wt. Nevertheless it should be noted that if we look at the pressures exerted, these are comparable and in the same order of 4 to 5 kg/cm2 at most. It is well worth noting that the general relation between the deformation and the force applied is in all cases essentially not linear except for very minor deformations. The latter case is classical in mechanics: the sag (indentation) is proportional to the strain, the proportionality being dependant on the degree of the initial tension. This state of tension differs according to the location and surgeons have established the map of the lines of maximal tension (lines of Langers). This fact can be related to our observation of variations in the initial slope of the curves (Fig. 2).

For a stronger punching, one can consider the situation as being analogous to the deformation of an envelope, and assuming that the area deformed is a plate, the effects caused by the punching can be described as being similar to those resulting from a simple traction. In this case, Ridge and Wright (1966) have shown that the stretching of the skin is proportional to the log of the force at least in a first phase corresponding to the alignment of the collagen fibres; in a second phase corresponding to the traction of these fibres the function becomes a power function. In our experiments the deformations are such as to be situated in the first phase.

The viscoelastic character of the skin imposing a certain relaxation time has also been investigated by Ridge and Wright (1966) and by Sanders (1973) the latter being equally interested in the phenomenon of plasticity. Thus the analysis of the behaviour of the skin under our probe reveals basically the same properties as those demonstrated on a large scale and intended for use in surgery.

Viscoelasticity and adaptation

Nakajima and Onodera (1969) have demonstrated that, for the crayfish stretch receptor, the receptor potential shows practically no adaptation if the load is clamped, whereas adaptation occurs when one uses constant stretch. Thus the viscoelasticity of the preparation can, for the most part, explain the adaptation phenomenon at this level.

The relaxation phenomenon observed in our study can be linked to the response adaptation of the slowly adapting type II receptor of the hairy skin of the cat as described by Chambers et al (1972). This receptor which is located in the deeper layers of the skin and oriented parallel to the surface is stimulated by variations in the curvature and above all by the tangential tractions sustained by the skin. It is not surprising that, at constant indentation, the force sustained decreasing, the neural response decreases as well.

What is the adequate parameter when establishing the coupling function for a slowly adapting mechanoreceptor?

Is it the indentation of the skin or the force applied?
One argument is formal : the indentation is already the result of the stimulation, the stimulus being the force exerted on the skin.

One argument is given by the mechanics: the response of the receptor depends on the deformation of its transducer site and this deformation is a function of the force sustained by the site. In a general manner this force is unknown - only the force exerted on the skin can be known - however at the end of the relaxation time the system (the tissues plus the stimulator) is in equilibrium and the force applied to the transducer site is proportional to that measured by the deformation of the spring. Hence, in the special case of the slowly adapting receptor, we have the rare opportunity to know the static response of the fiber and the value of the stimulus reaching the receptor. We have also seen that the relation between force and displacement is not at all a straightforward one and thus the cutaneous indentation is not a valid measure of the force.

Cases in the literature where the force exerted on the skin was used as a measure of the stimulus are rather rare. However when this was the case (Hensel and Boman, 1960; Gybels and Van Hees, 1972) the coupling function obtained was of the hyperbolic tangent log type as proposed by Naka and Rushton (1966) for the S-potentials of the fish retina. It has been shown that this coupling function is convenient for a great variety of sensory systems (Lipetz, 1971; Galifret and Petit, 1973); and Knibestöl (1973) has used it with success to express the relation between the indentation velocity and impulse frequency in the rapidly adapting mechanoreceptors of human glabrous skin.

In the case of the slowly adapting receptors let us assume that this same function is convenient for expressing the relation between the force applied on the skin and the steady discharge frequency. If now we take as the input not the force but the indentation, the indentation-frequency function will result from a combination of the log tan h function with the force-indentation function. In most cases it is a region of one function which combines with a region of the other. Under these conditions the results could vary considerably. Hence the log phase of the force-indentation relation combined with the mid-region of the log tan h function would give an approximately linear relation between the indentation and the frequency. If however the linear phase of the force-indentation relation is combined with the same mid-region a log relation will appear.

It is possible that even with regard to the slowly adapting mechanoreceptors, different types of coupling functions may exist; however this is difficult to affirm as long as such functions have not been determined using the real measure of the force applied.

REFERENCES

Chambers, M. R., Andres, K. H., Duering, M. V. & Iggo, A. The structure and function of the slowly adapting type II mechanoreceptor in hairy skin, Quart. J. exp. Physiol. 57, 417-445 (1972).

Galifret, Y. & Petit, J. La fonction de couplage des systèmes sensoriels. Arch. ital. Biol. 111, 352-371 (1973).

Gybels, J. & Van Hees, J. Unit activity from mechanoreceptors in human peripheral nerve during intensity discrimination of touch. Somjen, G.G. (1972) Neurophysiology studied in Man, Excerpta Medica, Congress series, Amsterdam.

Hensel, H. & Boman, K. Afferent impulses in cutaneous sensory nerves in human subjects. J. Neurophysiol. 23, 564-578 (1960).

Iggo, A. Relation of single receptor activity to parameters of stimuli. De Reuck, A.V.S. & Knight, J. (1966) In Touch Heat and Pain, Churchill; London.

Iggo, A. & Muir, A. R. The structure and function of a slowly adapting touch corpuscle in hairy skin. J. Physiol. 200, 763-796 (1969).

Knibestöl, M. Stimulus response functions of rapidly adapting mechanoreceptors in the human glabrous skin area. J. Physiol. 232, 427-452 (1973).

Knibestöl, M. Stimulus response functions of slowly adapting mechanoreceptors in the human glabrous skin area. J. Physiol. 245, 63-80 (1975).

Kruger, L. & Kenton, B. Quantitative neural and physiological data for cutaneous mechanoreceptor function. Brain Research. 49, 1-24 (1973).

Lindblom, U. Properties of touch receptors in distal glabrous skin of the monkey. J. Neurophysiol. 28, 966-985 (1965).

Lipetz, L. The relation of physiological and psychological aspects of sensory intensity. Loewenstein, W. R. (1971) In Handbook of Sensory Physiology, Springer; Berlin, Heidelberg.

Mountcastle, V. B., Talbot, W. H. & Kornhuber, H. H. The neural transformation of mechanical stimuli delivered to the monkey's hand. De Reuck, A. V. S. & Knight, J. (1966) In Touch Heat and Pain, Churchill; London.

Naka, K. I. & Rushton, W. A. H. S-potentials from colour units in the retina of fish (Cyprinidae). J. Physiol. 185, 536-555 (1966).

Nakajima, S. & Onodera, K. Membrane properties of the stretch receptor neurones of crayfish with particular reference to mechanisms of sensory adaptation. J. Physiol. 200, 161-185 (1969).

Nakajima, S. & Onodera, K. Adaptation of the generator potential in the crayfish stretch receptors under constant length and constant tension. J. Physiol. 200, 187-204 (1969).

Ridge, M. D. & Wright, V. Mechanical properties of skin: a bioengineering of skin structure. J. Appl. Physiol. 21, 1602-1606 (1966).

Sanders, R. Torsional elasticity of human skin in vivo. Pflügers Arch. 342, 255-260 (1973).

Werner, G. & Mountcastle, V. B. Neural activity in mechanoreceptive cutaneous afferents: stimulus response relations, Weber functions and information transmission. J. Neurophysiol. 28, 359-397 (1965).

THE TACTILE SENSORY INNERVATION OF THE GLABROUS SKIN OF THE HUMAN HAND

A. B. Vallbo and R. S. Johansson

Department of Physiology, University of Umea, S-901 87 Umea, Sweden

The concept of active touch is intimately linked with the function of the human hand (Ref. 16). Although many aspects of active touch involve very complex functions it is obvious that a basic element must be the input to the central nervous system provided by the first order afferents. The present report centres upon the tactile sensory innervation of the human hand.

Tactile sensory mechanisms may be defined as those which provide information on the presence and characteristics of mechanical events at the body surface. In humans, these mechanisms produce sensations which range from those resulting from passive touching to detailed and complex perceptive experiences of the shape, form, size, mechanical properties and surface structure of physical objects as they are actively explored. It seems likely that several types of peripheral sensory units contribute to sensations of this nature. One main group is obviously the sensory units which have terminals in the skin and the superficial subcutaneous tissues. Over the last decades evidence has been accumulating in support of the view that it is adequate to divide the cutaneous sensory units into three groups: mechanosensitive units, thermosensitive units and nociceptive or high threshold units (Ref. 4,23,44). The first two types are distinguished by their exquisite but not necessarily exclusive sensitivity to the particular type of physical energy indicated by the terminology. The nociceptive units, on the other hand, may respond to thermal and/or mechanical stimulus, but they require higher intensities which in typical cases approach damaging levels.

It is obvious that the cutaneous mechanosensitive units are of vital importance in regard to tactile sensibility, whereas it largely remains to be shown to what extent information from the thermosensitive and the nociceptive units also contributes to the finer points of tactile cutaneous sensations. Further, certain aspects of tactile sensibility are certainly also dependent upon information from non-cutaneous mechanosensitive units with end organs located in muscles, joint capsules, ligaments and other deep structures. It seems likely that these units are particularly significant for the complex tactile functions which fall under the heading of active touch. However, the present report is concerned exclusively with one of the four groups of units considered above: the cutaneous mechanosensitive units in the glabrous skin area of the human hand.

Available evidence indicates a fairly simple correspondance between unit type and nerve fibre diameter in the primate glabrous skin area. It seems that the cutaneous mechanosensitive units have large diameter myelinated nerve fibres (A α), whereas the small myelinated fibres (A δ) belong to cold sensitive as well as nociceptive units and finally the unmyelinated fibres (C) belong to heat sensitive as well as nociceptive units (Ref. 4,23,44).

METHODS

The neurophysiological findings are based on recordings of single unit impulses from afferents in forty waking human subjects. The subjects were mainly students; their ages largely ranged from eighteen to twenty-five years. Fine tungsten needle electrodes were inserted percutaneously in the median nerve on the upper arm (Ref. 61). The recording electrode was adjusted manually, in small steps, until single unit activity was discriminated from the background activity.

Skin indentations of controlled amplitudes were delivered with a feedback controlled moving coil stimulator. Its function has been described in detail in a previous study (Ref. 67) and the present account is limited to a schematic description. The stimulator probe started to move, at constant speed, at some distance from the skin. When the probe made contact with the skin an electric circuit was closed. The indentation which started at that moment was predetermined by a gate. The great advantage of this system is that small indentations of well controlled amplitudes can be delivered in spite of the fact that the skin surface of the human hand is constantly moving in phase with pulse and respiration. The amplitudes of these movements are in the range of 20 micrometers during quiet breathing,which is well above the threshold of many units. In some experiments simple stimulation devices were employed such as glass rods and von Frey hairs.

Psychophysical two point threshold tests were carried out on ten subjects. They had no previous experience of this type of test, although three of them had participated to a minor extent in psychophysical detection experiments. The stimuli were delivered manually with a modified metal sliding calliper. The points of the arms were ground to 0.5 mm diameter half-spherical endings. When the arms were in maximal apposition the distance between the points was 0.5 mm. The figures presented refer to the interpoint distance and not to the distance between the shafts. Blind stimuli were provided with a probe which had a tip of 0.5 by 1.0 mm with rounded edges. The indentation lasted for about 0.5 sec and the amplitude was about 0.5 mm, as estimated by eye. The stimulus was constantly delivered such that the two points were in the transverse direction of the hand. The threshold was estimated with the method of limits with an up-and-down procedure (Ref. 18). First, the approximate threshold was determined from two ascending and two descending series in which the distance between the stimulus points was changed in steps of 1.0 mm. Then the threshold was definitely assessed from six series in which the interpoint distance was changed in steps of 0.5 mm (Ref. 13,68). A small number of blind stimuli were semi-randomly intermixed. The subject was not told whether his responses were correct or not. The subjects were initially informed about the nature of the experiment while a few stimuli were applied to them. The instruction given was as follows. "When the points are sufficiently wide apart you will feel the stimulus as two points touching the skin as you may notice from the stimuli I am now applying to you. When the two points are close together, on the other hand, you will experience the stimulus as a single indentation, as you may notice from the stimuli I am now applying to you. After each stimulus you shall respond with 'one' or 'two'. Immediately before each stimulus I will say 'now'".

The median nerve specimen was prepared for histological examination and analysed as described by Behse, Buchthal, Carlsen and Knappeis (1).

RESULTS

Types of mechanosensitive units

It was first shown by Knibestöl and Vallbo (38) that there are four distinctly different types of mechanosensitive units in the glabrous skin area of the human hand. The four types are most readily differentiated on the basis of the units' receptive field properties and their responses to sustained indentations. A schematic presentation of the four unit types and their basic functional properties is given in Table I, which will be considered in more detail below. Two types of units responded to skin indentation exclusively while the stimulus was changing, whereas they exhibited no static discharge. They were conventionally denoted rapidly adapting. The other two types of units were slowly adapting in the sense that they exhibited a sustained discharge while a steady indentation was maintained.

TABLE 1 Types of Tactile Sensory Units in the Glabrous Skin of the Human Hand

	Receptive field characteristics	
Adaptation	distinct borders small size several sensitivity maxima	indistinct borders large size a single sensitivity maximum
rapid - no static response	RA	PC
slow - static response present	SA I	SA II

Among the rapidly adapting units the major group had small and quite distinct receptive fields. A few examples are shown in Fig. 1 A. The receptive fields were determined with von Frey hairs which produced a force five times the threshold force of the unit. The median value of 118 fields was 12.6 mm^2 which corresponds to a diameter of 4 mm. Similar units have been described in the glabrous skin of the monkey, the racoon and in the cat's footpad (Ref. 35,40, 54,55,59). They have been termed RA units, which stands for rapidly adapting,

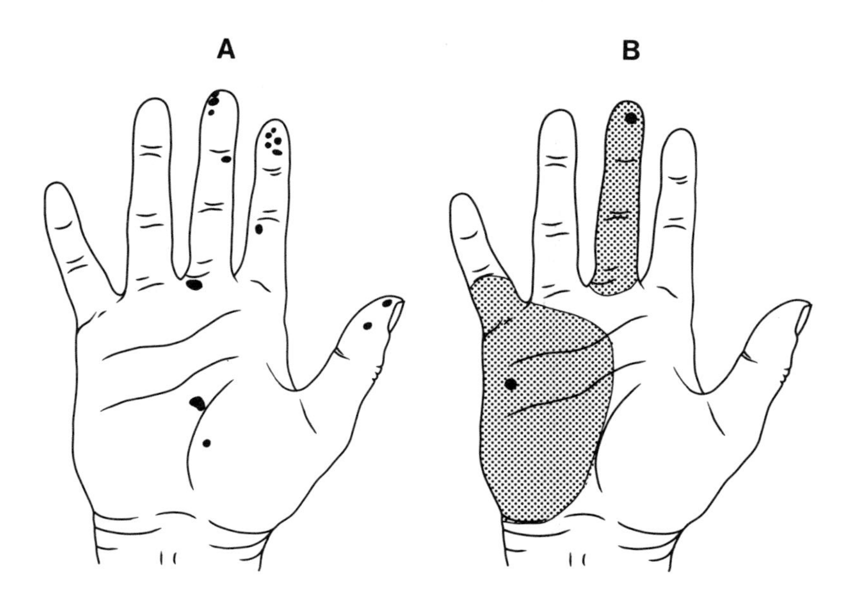

Fig. 1. A, receptive fields of fifteen RA units as measured with von Frey hairs providing a force five times the threshold force of the individual unit. B, receptive field characteristics of two PC units as determined with hand held instruments. The dots indicate points of maximal sensitivity to skin deformation whereas the dotted areas represent the regions from which activity was readily elicited with light taps with a pencil. One of these two units were recorded from the ulnar nerve.

or QA for quickly adapting. This terminology has been widely accepted although it is illogical because the RA units constitute but one of the two types of rapidly adapting mechanosensitive units in this skin area (see below). It has been tentatively suggested that the end organs of the primate RA units are the Meissner corpuscles (Ref. 36,40,51), whereas in the cat's footpad the terminals of the RA units have recently been identified as Krause's end organs by Iggo and Ogawa (27). The basic morphological structures of these two types of end organs are closely related and it seems likely that they are equivalents which are present in different areas and/or different species (Ref. 25,46,51,52).

In previous studies on primates it has been stated that the sensitivity is uniform within the receptive fields of RA units in the glabrous skin or that they have a single zone of maximal sensitivity (Ref. 22,33,36,37,38,40,49,54,59).

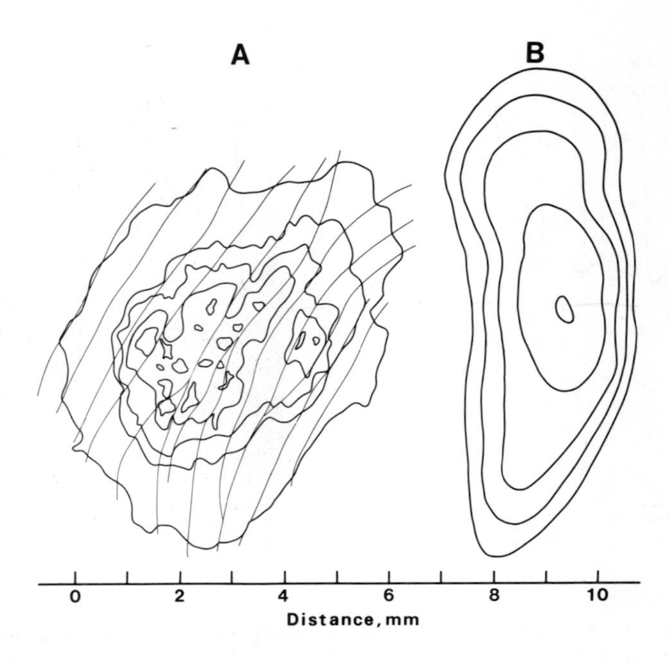

Fig. 2. Receptive field sensitivity maps of a RA unit (A) and a PC unit (B). The closed lines are iso-sensitivity lines whereas the thin lines indicate the grooves between the skin ridges. The iso-sensitivity lines refer to the following threshold amplitudes in A: 6 µm, 10 µm, 18 µm, 34 µm and 130 µm and the following amplitudes in B: 168 µm, 328 µm, 488 µm, 648 µm and 808 µm.

It was found that this was not the case in human subjects if the stimulus amplitude was adequately controlled (Ref. 28,29,30). To analyse this aspect systematically the receptive fields were mapped in detail, using triangular indentations with a constant velocity, and the area within which a response from the unit was obtained at a given stimulus intensity was indicated on a x-y plotter. This procedure was repeated with different indentation amplitudes. The probe had a rounded tip and a diameter of 0.4 mm.

An example of a RA field is shown in Fig. 2 A. The sensitivity is indicated as iso-sensitivity lines in the same way as the height of a mountain is indicated on a map. The continuous lines enclose areas from which responses were obtained with a given indentation amplitude as given in the legend. The thin lines represent the grooves between the skin ridges. At the peaks the threshold was only 6 μm and uniform among all of them. In the central area between the peaks the threshold was slightly higher, but not very much, whereas the threshold increased steeply towards the periphery away from the peaks, e.g. the outermost line indicates a threshold of 130 μm. It seems reasonable to assume that the locations of the peaks correspond to the locations of the end organs. Therefore, in this case, there would be at least 17 end organs, or more as there could be more than one end organ under each peak.

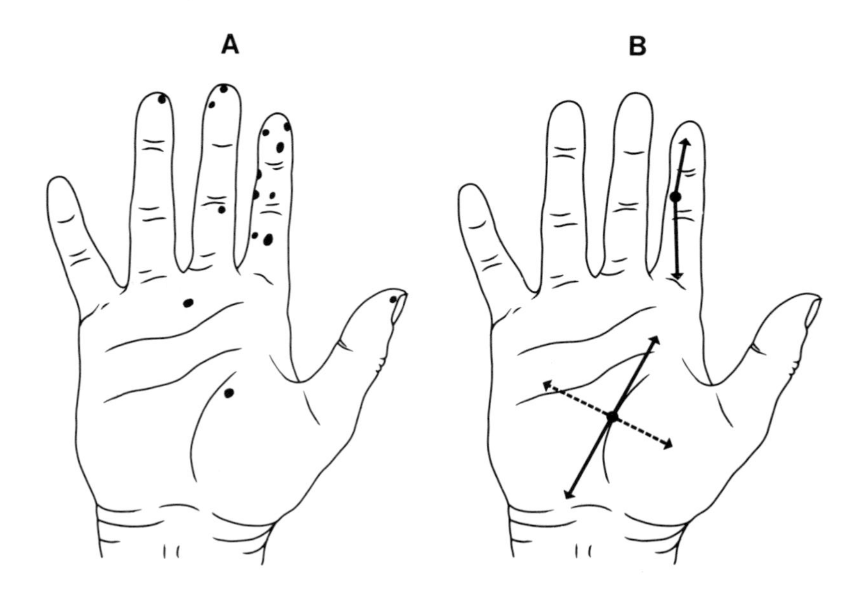

Fig. 3. A, receptive fields of fifteen SA I units as measured with von Frey hairs providing a force five times the threshold force of the individual unit. B, receptive field characteristics of two SA II units. The dots indicate points of maximal sensitivity to skin indentation. The lines indicate the directions and the approximate extents of skin stretch sensitivity. The heavy lines and the broken lines refer to an increase and a decrease of the afferent activity of the unit, respectively.

The other type of rapidly adapting unit has a completely different type of receptive field when tested with hand held instruments (Fig. 1 B). It is not small and well defined, instead there is a point or a small area of maximal sensitivity and a very wide region from which activity may be elicited by somewhat stronger stimuli. An analysis of the receptive field sensitivity with the same method as given above showed that there is a single point of maximal sensitivity. From this point the threshold rise is very gentle compared to that of the RA unit. For instance, in this case the outermost line represents a stimulus intensity which is only 4.8 times the minimal threshold for this unit which, in turn, was exceptionally high for this unit type (Ref. cf. 31,62). Other properties, such as the sensitivity to high frequency vibration and the liability to give an off-response when the indentation was removed, indicate that these units are very likely of the same nature as the PC units which have been described in many species and which have been shown to be connected to Pacinian corpuscles.

There were also two types of slowly adapting units. One had distinct receptive fields of small size, quite similar to those of the RA units, when tested with von Frey hairs (Fig. 3 A). Other properties were, in typical cases, similar to those of the SA I units of the hairy skin described in detail by Iggo and Muir (26). These properties were high dynamic sensitivity, irregular inter-spike interval during prolonged indentation and low sensitivity to skin stretch. In the study by Iggo and Muir the end organs have been identified as the Merkel cell neurite complexes. It is well known from earlier morphological studies that there are a large number of Merkel cell neurite complexes in the glabrous skin of the human hand (Ref. 6,9,45,46,47,56,69). It seemed therefore reasonable to denote the slowly adapting units with well defined receptive fields SA I units.

A detailed receptive field analysis of SA I units indicated a similar organization to the RA units: multiple points of maximal sensitivity located closely together and a steep rise of threshold with distance from these peaks (Fig. 4 A).

In addition to these three types another type of unit was found in the human glabrous skin. It is slowly adapting and may be continuously discharging in the absence of direct skin indentation. It is characterized by a high sensitivity to skin stretch. In most cases the direction of skin stretch is critical:

pulling in two opposite directions gives rise to an increase in the afferent activity whereas pulling at about right angles decreases previous activity, if any (Fig. 3 B). The discharge is very regular in contrast to that of the SA I

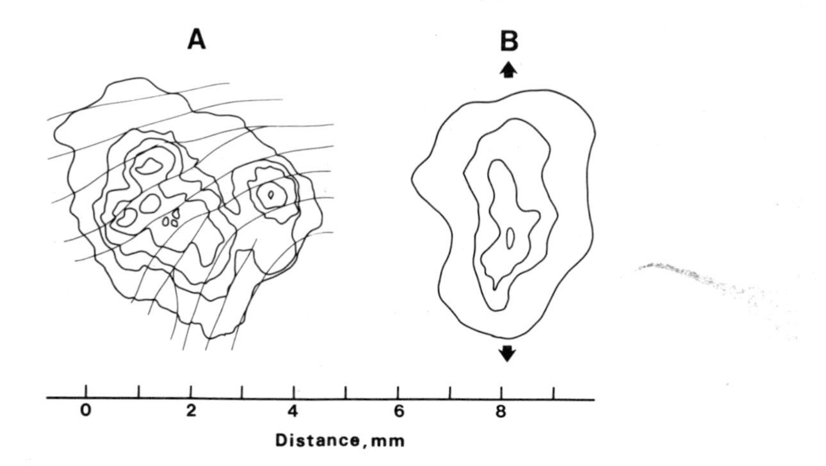

Fig. 4. Receptive field sensitivity maps of a SA I unit (A) and a SA II unit (B). The SA II unit was vigorously responding to stretching of the skin in the directions of the arrows. The closed lines are iso-sensitivity lines, whereas the thin lines indicate the grooves between the skin ridges. The iso-sensitivity lines refer to the following threshold amplitudes in A: 42 µm, 82 µm, 162 µm, 322 µm and 642 µm and the following amplitudes in B: 200 µm, 242 µm, 282 µm and 322 µm.

units. The functional properties of these units are similar to those of the SA II units described in the hairy skin of the cat by Chambers, Andres, Duering and Iggo (11), who also identified the end organs as the Ruffini endings. A receptive field analysis indicates similar basic field properties as for the PC units: a single point of maximal sensitivity and a gentle rise of the threshold from this point (Fig. 4 B).

No indication has been found of major functional differences between the human glabrous skin mechanoreceptor units and those described in other primates (Ref. 4,23,44). Certain properties have been studied in some detail, such as stimulus-response functions and absolute thresholds (Ref. 31,36,37,62). Others have been looked at more casually. However, from what has emerged so far it seems reasonable to extrapolate from the large amount of information available from subhuman primates when the tactile functions of the human hand are studied. The data concerning the receptive field sizes from the various species studied are clearly not uniform but differences in method as well as in size of the

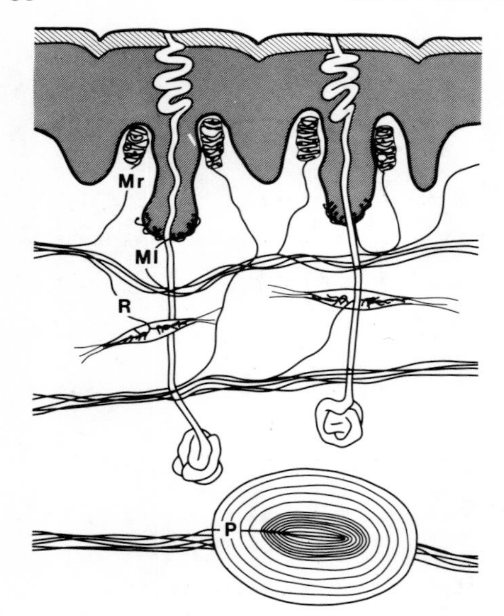

Fig. 5. A Schematic vertical section through the glabrous skin of the human hand demonstrating the locations of four types of organized nerve terminals tentatively suggested as the end organs of the four types of tactile sensory units found in this skin area. Types of end organs are indicated by abbreviations as follows: Mr - Meissner corpuscles, Ml - Merkel cell complexes, R - Ruffini ending and P - Pacinian corpuscles and paciniform endings. Modified after Miller, Ralston and Kasahara (Ref. 46).

species studied make a comparison complicated.

Table I summarizes some of the basic distinguishing properties of the four different types of cutaneous low threshold mechanoreceptive units in the glabrous skin area of the human hand.

Figure 5 shows schematically a section through the glabrous skin with four types of end organs (Ref. 46) which are the most likely terminals of the four types of tactile sensory units. The Meissner corpuscles which are located high up in the dermal papillae are probably the end organs of the RA units. At the tip of the intermediate epidermal ridge, which projects down into the dermis, is located the Merkel cell neurite complex which is with all likelihood the ending of the SA I units. In contrast to these two endings close to the epidermal-dermal border, the Ruffini ending is located more deeply within the dermis. Available evidence indicates that the Ruffini ending is the terminal of the SA II units. Finally the large Pacinian corpuscles represent the endings of the PC units. These endings are found in the subcutaneous tissues. Most of them are probably located more deeply below the skin surface than indicated in Fig. 5. The relation between structure and function for this unit type is obvious (Ref. 21,43). In addition to the huge Pacinian corpuscles there are

smaller lamellated endings of similar structure which are denoted paciniform endings and include the Golgi-Mazzoni bodies (Ref. 12,21,23,57,58).

It should be pointed out that there are no direct studies in human subjects on the relations between structure and function as considered above. On the other hand, the relations presented above seem to be the most reasonable ones on the basis of indirect evidence from separate morphological and physiological studies in man and combined studies of structure and functions in other species (Ref. 6,10,11,17,24,26,27,30,34,36,37,38,41,43,46,47,53,56,57,58).

The detailed receptive field analysis as presented in a previous section eliminates all ambiguity with regard to there being four distinctly different mechanosensitive unit types in this area. Further, this analysis supports the notion that the units in the human glabrous skin are analogous to those described in other species. Finally, the presumed relations between the structure of endings and the function of the sensory units, as considered above, are supported by the analysis. For instance, morphological evidence indicates that for Pacinian corpuscles and Ruffini endings the stem nerve fibre carries but a single end organ, whereas the nerve fibres innervating the Meissner corpuscles and the Merkel cell neurite complexes branch repeatedly to terminate in several endings (Ref. 7,8,10,43).

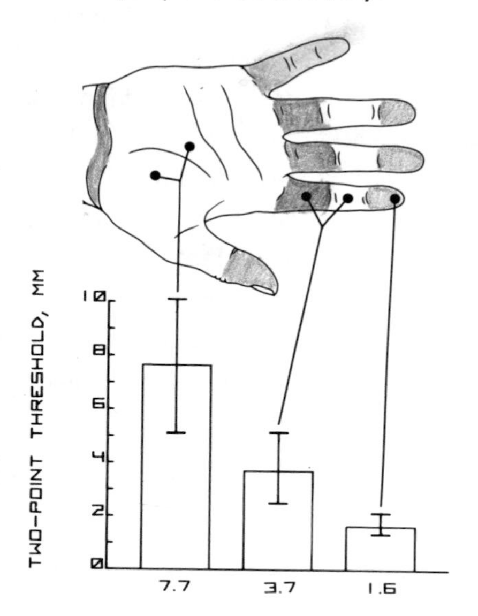

Fig. 6. Two point thresholds at three different skin regions as measured in ten young adults. The medians and the inter-quartile distances are indicated.

Densities and receptive field sizes of tactile sensory units

The functional properties as described in previous sections invite speculations on the differential functional role of the four types of sensory units in tactile sensibility. An important tactile function is the capacity for spatial analysis, which may be defined as the ability to locate skin indentations or components of skin indentations in relation to each other and in relation to the body surface whether the indentations vary temporally, or not, and whether they are caused by active or passive touch. It seems reasonable to suggest, just from the properties of the receptive fields, that the units with small and well defined receptive fields account for the spatial analysis. For instance when a minute object such as a small piece of shot is held and rotated between the thumb and the index, the input from the populations of RA and SA I units would consist of an intense activity in afferent fibres with end organs located underneath the skin area directly indented. The surrounding units of these two types would in contrast give a weak response. The activity in the populations of PC and SA II units, on the other hand, would originate from a much larger area and have a less clearly defined maximum due to their high sensitivity to remote stimuli.

A classical test of the capacity of tactile spatial analysis is the two point discrimination test. It has been repeatedly shown that there are pronounced regional variations in the two point threshold over the body surface. For instance, it is striking that the capacity increases in distal direction along the extremities (Ref. 20,39,63,65,66). Figure 6 shows the result of a two point threshold study on ten young adults. The points selected are of particular interest in this context as will be considered below: they were located on the finger tip, i.e. the distal half of the terminal phalanx, the main part of the fingers and the palm. The findings are essentially in agreement with those reported in earlier studies(Ref.20,39,63,64,65,66)in showing that the two point threshold varies by a factor of about five (4.8) between the palm and the finger tip. The median values for the three regions were 7.7 mm, 3.7 mm and 1.6 mm which are also in reasonable agreement with those obtained in previous studies.

A reasonable hypothesis, which goes back to Weber's work (Ref. 2,3,64), is that the two point threshold is closely related to the sizes of the receptive fields. The underlying assumption is that there must be one relatively silent

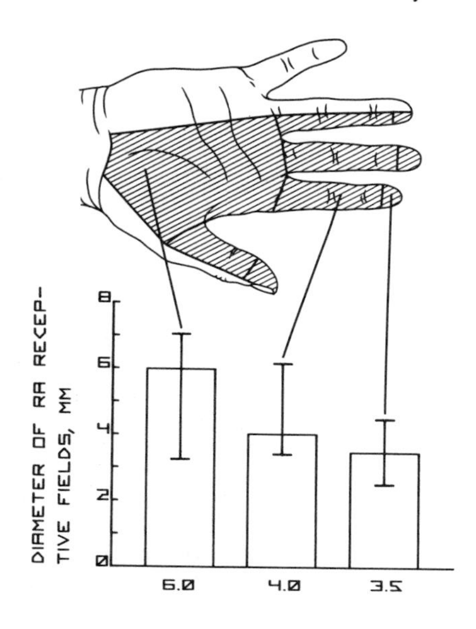

Fig. 7. Sizes of RA receptive fields in three different skin regions. The diameters given in the figure are those corresponding to round fields even though about three fourths of the fields were oval in shape. The medians and the inter-quartile distance are indicated.

receptive field between the two points of the stimulus. It seems that a theory of this nature is based upon two assumptions with regard to the properties of the single sensory units. One is that the sensitivity of the separate endings belonging to the same unit is uniform to preclude the possibility that the central nervous system can deduce information regarding the location of a stimulus within the receptive field from the relative intensity of the activity of the individual unit. The receptive field analyses as described above indicate that this assumption is justified. The other assumption is that occlusion is pronounced between the effects of two stimuli within the same receptive field. Available evidence from recent neurophysiological studies indicates that also this assumption is justified (Ref. 26,42). The above hypothesis predicts that the field diameters would be about five times larger in the palm than at the finger tip if the two point threshold is determined by the size of receptive fields. We determined the field sizes with von Frey hairs which gave a force of five times the threshold force of the unit. Separate analysis showed that the field as determined in this way corresponds fairly well to the area where the end organs of the RA and SA I units are located (Ref. 30). In a sample of 255 units it was found that the sizes of the fields were not significantly different for the SA I, the SA II, nor the PC units between the finger tip, the main part of the finger and the palm. For the RA units there was a regional difference. The medians of the RA field diameters are given in Fig. 7. They

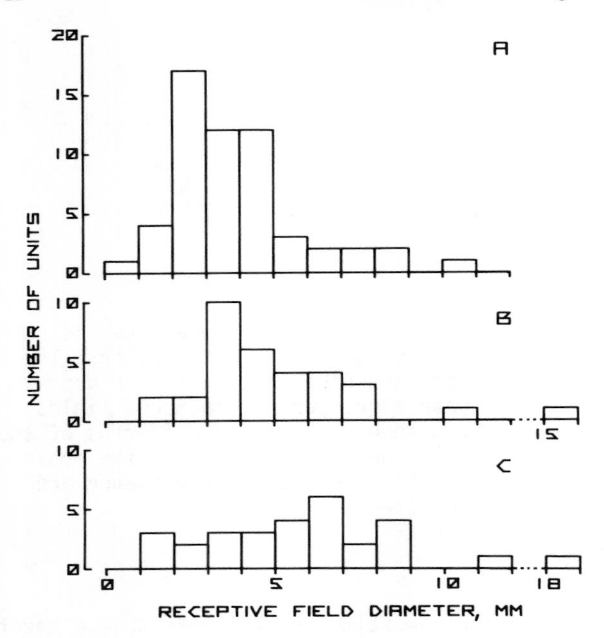

Fig. 8. Distribution of the sizes of the RA receptive field within three skin regions as indicated in Fig. 7: A the finger tip, B the main part of the finger and C the palm. The diameters given in the figure are those corresponding to round fields even though about three fourths of the fields were oval in shape.

varied by a factor of only 1.7 from the finger tip to the palm. The findings imply that there is a difference in the field size of the RA units which qualitatively matches the regional variation in the two point threshold.

However, it may be argued that it is an oversimplification to study the relation between the two point threshold and the medians of the receptive field diameters, unless the field sizes are uniform. Figure 8 shows that the RA fields are not all that uniform in size. The fields in the palm, in particular, are almost evenly spread within a relatively wide range. It may also be seen that the sizes of the smallest RA fields are not much different in the three regions.

It seems reasonable to assume that not only the sizes of the receptive fields but also the densities of units may be of significance for the two point threshold. One aspect is that the density of fields which are small in relation to the two point threshold must not be too low. To elucidate this point an estimate was made of the absolute unit density in the various regions of the glabrous skin. This estimate was based upon two sets of data. The first was a sample of 354 skin mechanoreceptor units collected from the median nerve. The

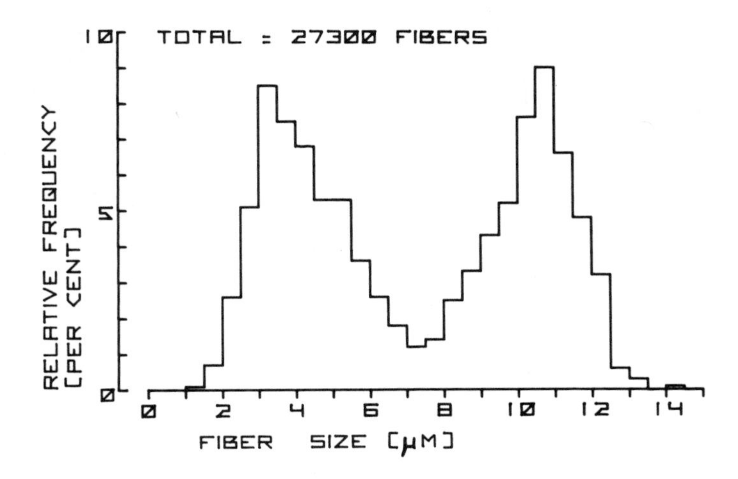

Fig. 9. Distribution of nerve fibre diameters measured from a section of a median nerve at the wrist. The calibre spectrum is based upon measurements of one thousand fibres.

units were analysed with regard to unit type and receptive field locations on the skin surface. The other set of data was an estimate of the total number of myelinated nerve fibres in the median nerve at the wrist. This was made from a specimen taken from a 23 year old woman. The histological analysis was made by dr Buchthal and dr Behse in Copenhagen. The total number of myelinated fibres was estimated to be 27 300. Figure 9 shows the caliber spectrum of the nerve based upon measurements of 1 000 fibres. Fifty per cent of the fibres were large. It can be inferred that the vast majority of the large fibres belong to the four types of tactile sensory units described above (Ref. 32). The separate steps involved in making an estimation of the unit density and the assumptions involved are given in another report (Ref. 32).

The analysis of the absolute unit density refers to an intermediate band of the median nerve territory as indicated in Figs. 7,10,12. The purpose of this restriction was to avoid errors due to edge effects in the unit sampling procedure. There were some indications of such effects in the material. The results are presented in Fig. 10, where each point represents a receptive field. The figure illustrates the density of units exclusively and not the

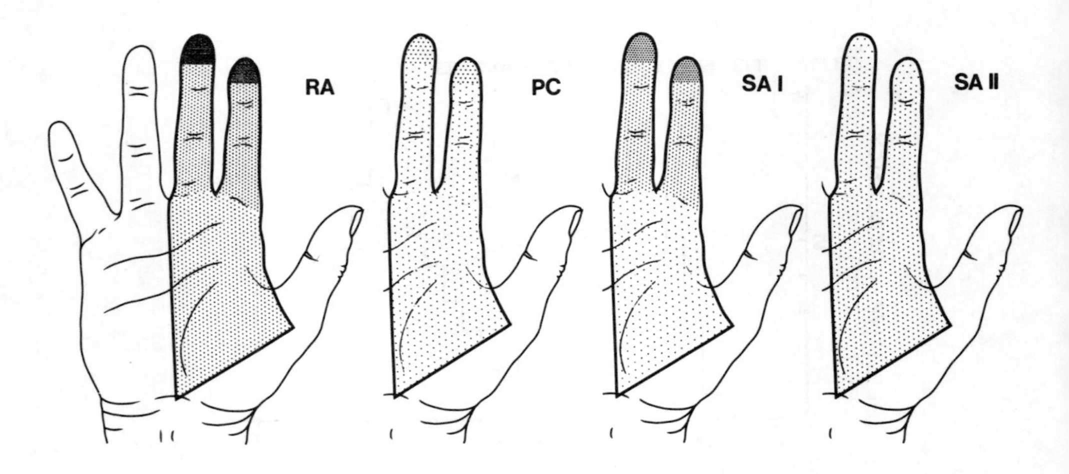

Fig. 10. Estimated absolute densities of the four types of tactile sensory units in the glabrous skin. Each dot represents a single sensory unit innervating the skin area. It should be emphasized that the figure illustrates only the average unit density within the region whereas the dot size and the exact location of the individual dot in relation to the neighbours has no relevance with regard to receptive field size and spatial distribution of fields.

sizes, nor the relative locations, of the fields. It was found that there was an enormous accumulation of RA and SA I units at the finger tip, i.e. the area distal to the whorl of the skin ridges. The densities of these units were also higher on the main part of the finger than in the palm. The densities of PC and SA II units on the other hand were more even over the entire glabrous skin area. No support was obtained for the idea of a continuous rise in total unit density in the distal direction, but rather for a two step rise as indicated in Fig. 10. It is not claimed, however, that the borders are as sharp as illustrated. The estimated density of RA units at the fingertip was 141 units per square cm whereas in the palm it was only 25 units per square cm. It may be seen that the PC units were much fewer than the RA units: at the finger tip there were 21 and in the palm 9 units per square cm. The gradient of SA I unit density was, in principle, similar to that of the RA units but the densities were generally lower: at the finger tip 70 per square cm and in the palm only 8 units per square cm. Finally the density of the SA II units was 9 units per square cm at the finger tip and 16 units per square cm in the palm.

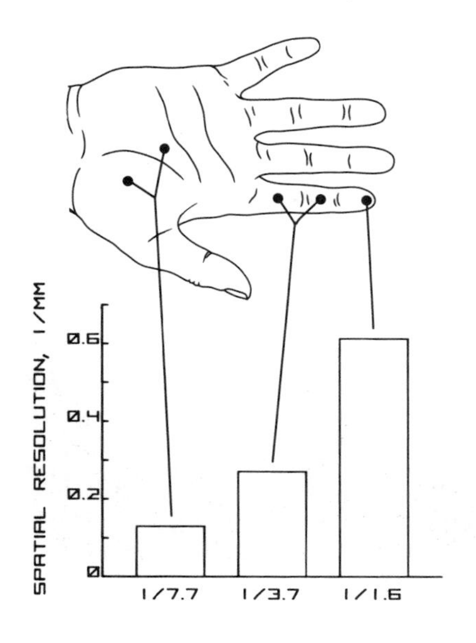

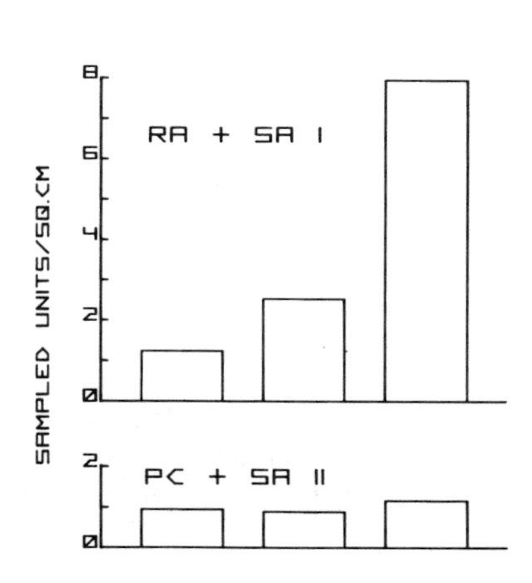

Fig. 11. Spatial resolution in psychophysical two point threshold tests (A) and density of tactile sensory units (B) in three regions of the glabrous skin. The bars in the left hand diagram give the inverse of the two point threshold in units of mm^{-1}. The same data as presented in Fig. 6. The diagrams to the right give the sampled densities of two groups of tactile sensory units within the same skin regions; above, the units characterized by small and well defined receptive fields; below, the units characterized by receptive fields with indistinct borders and relatively high sensitivity to remote skin stimuli.

A comparison between the regional variations in the two point threshold and the variations in unit density indicates a striking relation. Figure 11 A shows the inverse of the two point threshold, which has been denoted spatial resolution (mm^{-1}). Figure 11 B shows the relative densities of the units separated into two groups: above, the RA and SA I units which are the unit types with small and well defined receptive fields and, below, the PC and SA II units which are the unit types which have a high sensitivity to remote stimuli. It is striking that the relation is similar between the finger tip, the main parts of the finger and the palm with regard to spatial resolution and relative density of the RA and SA I units summated, whereas there is no such correspondance for the PC and SA II units. It seems reasonable to assume that the density of units, as well as the sizes of the receptive fields, are of significance for the minimal

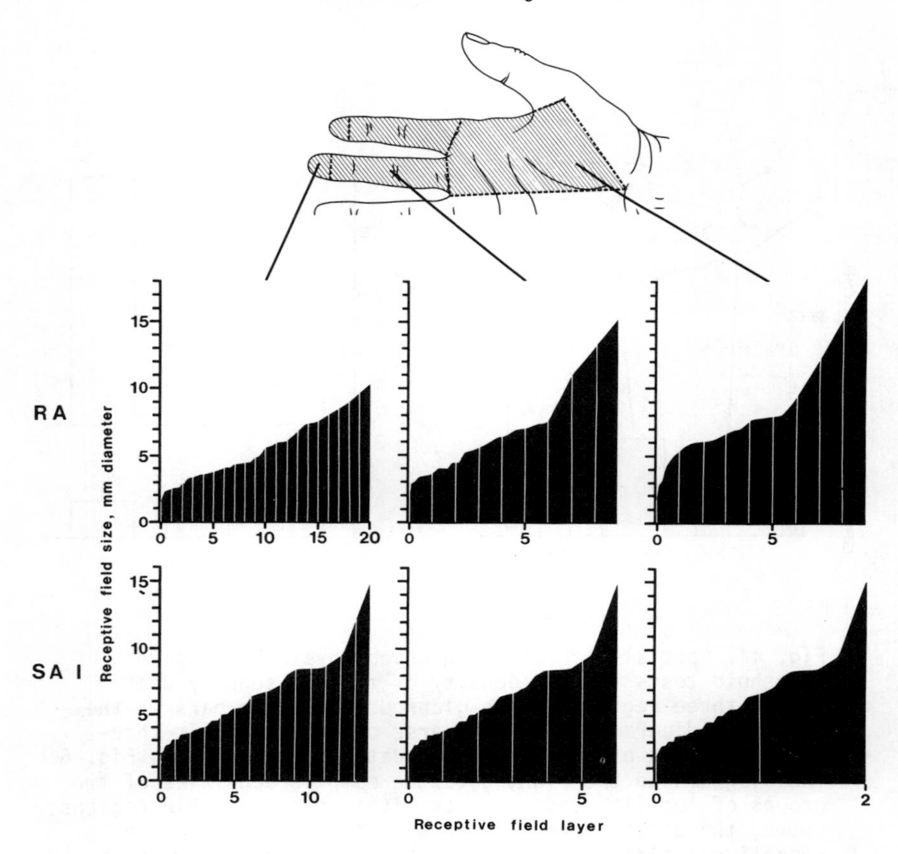

Fig. 12. Distribution of the sizes of the receptive fields of the RA and SA I units in three different skin regions. The abscissae give the sum of the receptive field areas in units of total skin area from which the sample was collected. Each column represents a set of units, the receptive fields of which would cover the whole skin area. The term receptive field layer was used for such a set of receptive fields.

distance between two stimuli which gives rise to a trough of activity between two peaks in a two point threshold test. The higher the unit density, the higher would be the probability for end organs belonging to a silent unit being intercalated between two excited ones, even though the stimuli are separated by a smaller distance than the field diameter.

As pointed out above it seems of interest to estimate the density of fields which are small in relation to the two point threshold. Figure 12 gives the outcome of a calculation along this line. The abscissae give the sum of the receptive field areas in units of total skin area from which the sample was collected and the ordinates give the cumulative distribution of field diameters. Hence, the individual column represents a set of units, the receptive fields of which would cover the whole skin area if the fields had been layed out as a mosaic. Schematically, such a set of fields may be seen as a single layer covering the skin area, and the term receptive field layer was therefore adopted. It may be seen, for instance, that the number of such layers of RA fields are as many as twenty at the finger tip and nine in the palm. The corresponding figures for the SA I units are fourteen and two, respectively.

Of particular interest in relation to the capacity for spatial analysis is the density of small fields. Related quantities are the diameters of the smallest fields which would account for a single receptive field layer. The first columns in the diagrams give these diameters. It may be seen that RA fields with diameters up to 2.4 mm would cover the whole skin area at the finger tip. The corresponding figures at the main part of the finger and in the palm are 3.7 mm and 5.9 mm, respectively. This gives a factor of 2.5 between the palm and the finger tip, which implies a longitudinal gradient in the same direction as that of the two point threshold, although the latter gradient is higher. The corresponding factor for the SA I units is 2.7.

Discussion

Some quantitative data on the tactile sensory innervation of the glabrous skin of the human hand have been discussed in relation to the two point threshold in this area. Emphasis has been put upon gradients in the proximo-distal direction of the innervation parameters matching the gradient in spatial discrimination capacity. On the other hand, a quantitative model of the afferent input provided by two point stimuli has not been developed. It seems that a more sophisticated test of spatial discrimination capacity is needed to make a more quantitative approach meaningful. The two point threshold is a much debated measure of the spatial discrimination capacity (Ref. 3,15,20,60). It has been pointed out in several investigations that the criterion used by a subject to decide whether the stimulus is one or two points may vary, e.g. the subject may base his decision on the perceived size of the stimulus rather than on the perceived unity or duality of the stimulus (Ref. 3,15). This factor might, for instance, account for the effect of training (Ref. 3,5,14,63). There seems to be no doubt, however, that there is a proximo-distal gradient in the two point threshold. There is also a reasonable agreement between many investigations with regard to the magnitude of this gradient, as well as the absolute values of the two point thresholds. On the basis of the above considerations, it seems reasonable to look for gradients in the peripheral organization of the mechanoreceptive units to provide evidence for the functional role of the different types of tactile sensory units.

It was found that there was a gradient in the sum of the RA and SA I unit density similar to the gradient in the two point threshold. A gradient was also found in the medians of the receptive field size of one type of sensory unit, although this gradient was smaller than the two point threshold gradient. In an analysis of the absolute number of tactile sensory units, it was found that there is an abundance of RA and SA I units with diameters smaller than the two point threshold except at the finger tip. It should be stressed, however, that the receptive field size as given in the present study is not an unequivocal quantity. First, the size was given as the diameter, although many fields were not round but oval in shape with the long axis in the proximo-distal direction. Second, the size of a receptive field is clearly dependent upon the intensity of the stimulus as is clearly shown in the first section of this report. Finally, the method of measuring the field size used in the present study may entail a relative overestimation of the sizes of the smaller fields.

The fact that there are proximo-distal gradients in the number of units, as well as the receptive field size of the RA and SA I units seems to strongly support the notion that these units are responsible for spatial discriminative capacity in this skin area. Particularly striking is the very high density of these two unit types at the finger tips, indicating that this is a skin area with outstanding peripheral equipment for tactile spatial analysis.

The considerations presented above concerning the measurements of the receptive field size, as well as the two point thresholds, emphasize that the exact figures should not be stressed too much. On the other hand, the findings clearly indicate a general agreement between psychophysical and neurophysiological data supporting the notion that the RA and SA I units account for the tactile spatial analysis as defined above.

No direct experimental evidence has been presented for the role of the other two types of mechanosensitive units in this area: the PC and the SA II units. The role of the PC units has been discussed in many other investigations (Ref. 19,48,50,59) and it seems reasonable to emphasize their capacity to inform about high frequency transients and vibrations of objects which are held in the hand or against which the hand is leaning. It seems clear that the other types of units are considerably less sensitive to these types of stimuli, indicating that our ability to detect and characterize such stimuli are dependent upon the PC units. The significance of the ability of the PC units to detect transients generated within the body may also be considered.

The role of the SA II units has also been discussed in previous studies (Ref. 11,29,30,37,38). Their most striking functional characteristic is their high sensitivity to tangential forces in the skin. Information about the amount of skin stretch may be of significance for evaluating the positions of joints, as well as the shearing forces between the skin and a hand held object. However, these questions remain to be investigated.

ACKNOWLEDGEMENTS

This study was supported by grants from the Swedish Medical Research Council (project no. 04X-3548), Gunvor och Josef Anērs Stiftelse and the University of Umeå (Reservationsanslaget för främjande av ograduerade forskares vetenskapliga verksamhet). We are greatly indebted to drs. F. Buchthal and F. Behse for the analysis of the median nerve specimen.

REFERENCES

(1) Behse, F., Buchthal, F., Carlsen, F. and Knappeis, G. G. (1974). Endoneurial space and its consituents in the sural nerve of patients with neuropathy, Brain 97, 773-784.

(2) Boring, E. G. (1930). The two-point limen and the error of localization, Am. J. Psychol. 42, 446-449.

(3) Boring, E. G. (1942). Tactual sensibility. In Sensation and perception in the history of experimental psychology, pp. 463-522, Appleton, New York.

(4) Burgess, P. R. and Perl, E. R. (1973). Cutaneous mechanoreceptors and nociceptors. In Handbook of Sensory Physiology, vol. II, ed. Iggo, A., pp. 29-78, Springer, Berlin.

(5) Camerer, W. (1883). Versuche über den Raumsinn der Haut nach der Methode der richtigen und falschen Fälle, Z. Biol. 19, 280-300.

(6) Cauna, N. (1954). Nature and function of the papillary ridges of the digital skin, Anat. Rec. 119, 449-468.

(7) Cauna, N. (1956). Nerve supply and nerve endings in Meissner's corpuscles, Am. J. Anat. 99, 315-350.

(8) Cauna, N. (1959). The mode of termination of the sensory nerves and its significance, J. comp. Neurol. 113, 169-199.

(9) Cauna, N. (1961). Cholinesterase activity in cutaneous receptors of man and some quadrupeds, Bibl. anat. (Basel) 2, 128-138.

(10) Cauna, N. and Mannan, G. (1958). The structure of human digital pacinian corpuscles (Corpuscula lamellosa) and its functional significance, J. Anat. 92, 1-20.

(11) Chambers, M. R., Andres, K. H., v. Duering, M. and Iggo, A. (1972). The structure and function of the slowly adapting type II mechanoreceptor in hairy skin, Q.Jl.exp. Physiol. 57, 417-445.

(12) Chovuchkov, Ch. N. (1973). The fine structure of small encapsulated receptors in human digital glabrous skin, J. Anat. 114, 25-33.

(13) Dixon, W. J. and Mood, A. M. (1948). A method for obtaining and analyzing sensitivity data, J. Am. Stat. Ass. 43, 109-126.

(14) Dresslar, G. (1894). Studies in the psychology of touch, Am. J. Psychol. 6, 313-368.

(15) Friedline, C. L. (1918). The discrimination of cutaneous patterns below the two-point limen, Am. J. Psychol. 29, 400-419.

(16) Gibson, J. J. (1962). Observations on active touch, Psychol. Rev. 69, 477-491.

(17) Gottschaldt, K. M., Iggo, A. and Young, B. W. (1973). Functional characteristics of mechanoreceptors in sinus hair follicles of the cat, J. Physiol. 235, 287-315.

(18) Guilford, J. P. (1954). Psychometric methods, 2nd ed., 597 pp. McGraw-Hill, New York.

(19) Harrington, T. and Merzenich, M. M. (1970). Neural coding in the sense of touch: Human sensation of skin indentation compared with the responses of slowly adapting mechanoreceptive afferents innervating the hairy skin of monkeys, Exp.Brain Res. 10, 251-264.

(20) Henri, V. (1898). Über die raumwharnehmungen des Tastsinnes, pp. 1-228, Reuther and Reichard, Berlin.

(21) Hunt, C. C. (1974). The pacinian corpuscle. In The Peripheral Nervous System, ed. Hubbard, J. I., pp. 405-419, Plenum Press, New York.

(22) Iggo, A. (1963). An electrophysiological analysis of afferent fibres in primate skin, Acta neuroveg. 24 225-240.

(23) Iggo, A. (1974). Cutaneous receptors. In The Peripheral Nervous System, ed. Hubbard, J. I., pp. 347-404, Plenum Press, New York.

(24) Iggo, A. (1976). Is the physiology of cutaneous receptors determined by morphology? In Progress in Brain Research, vol. 43, eds. Iggo, A. and Ilyinsky O. B., pp. 15-31, Elsvier, Amsterdam.

(25) Iggo, A. (1977). Cutaneous and subcutaneous sense organs, Br. med. Bull. 33, 97-102.

(26) Iggo, A. and Muir, A. R. (1969), The structure and function of a slowly adapting touch corpuscle in hairy skin, J. Physiol. 200, 703-796.

(27) Iggo, A. and Ogawa, H. (1977). Correlative physiological and morphological studies of rapidly adapting mechanoreceptors in cat´s glabrous skin, J. Physiol. 266, 275-296.

(28) Johansson, R. S. (1976). Receptive field sensitivity profile of mechanosensitive units innervating the glabrous skin of the human hand, Brain Res. 104, 330-334.

(29) Johansson, R. S. (1976). Skin mechanoreceptors in the human hand: Receptive field characteristics. In Sensory Functions of the Skin in Primates, ed. Zotterman, Y., pp. 159-170, Pergamon Press, Oxford.

(30) Johansson, R. S. Tactile sensibility in the human hand: Receptive field characteristics of mechanoreceptor units in the glabrous skin. To be published.

(31) Johansson, R. S. and Vallbo, Å. B. (1976). Skin mechanoreceptors in the human hand: An inference of some population properties. In Sensory Functions of the Skin in Primates, ed., Zotterman, Y., pp. 171-184, Pergamon Press, Oxford.

(32) Johansson, R. S. and Vallbo, Å. B. Tactile sensibility in the human hand: Spatial distribution of four types of mechanoreceptive units in the glabrous skin. To be published.

(33) Johnson, K. O. (1974). Reconstruction of population response to a vibratory stimulus in quickly adapting mechanoreceptive afferent fibre population innervating glabrous skin of the monkey, J. Neurophysiol. 37, 48-72.

(34) Jänig, W. (1971). Morphology of rapidly and slowly adapting mechanoreceptors in the hairless skin of the cat´s hind foot, Brain Res. 28, 217-231.

(35) Jänig, W., Schmidt, R. F. and Zimmerman, M. (1968). Single unit responses and the total afferent outflow from the cat´s foot pad upon mechanical stimulation, Exp. Brain Res. 6, 100-115.

(36) Knibestöl, M. (1973). Stimulus-response functions of rapidly adapting mechanoreceptors in the human glabrous skin area, J. Physiol. 232, 427-452.

(37) Knibestöl, M. (1975). Stimulus response functions of slowly adapting mechanoreceptors in the human glabrous skin area, J. Physiol. 245, 63-80.

(38) Knibestöl, M. and Vallbo, Å. B. (1970). Single unit analysis of mechanoreceptor activity from the human glabrous skin, Acta physiol. scand. 80, 178-195.

(39) Kottenkampf, R. and Ullrich, H. (1870). Versuche über den Raumsinn der Haut der oberen Extremität, Z. Biol. 6, 37-52.

(40) Lindblom, U. (1965). Properties of touch receptors in distal glabrous skin of the monkey, J. Neurophysiol. 28, 966-985.

(41) Lindblom, U. and Lund, L. (1966). The discharge from vibration-sensitive receptors in the monkey foot, Expl Neurol. 15, 401-417.

(42) Lindblom, U. and Tapper, D. N. (1966). Integration of impulse activity in a peripheral sensory unit, Expl Neurol. 15, 63-69.

(43) Lynn, B. (1969). The nature and location of certain phasic mechanoreceptors in cat´s foot pad, J. Physiol. 201, 768-773.

(44) Lynn, B. (1975). Somatosensory receptors and their CNS connections, Ann. Rev. Physiol. 37, 105-127.

(45) McGavran, M. (1964). Chromaffin cell: electron microscopic identification in the human dermis, Science 145, 275-276.

(46) Miller, M. R., Ralston, H. J. and Kasahara, M. (1958). The pattern of cutaneous innervation of the human hand, Am. J. Anat. 102, 183-197.

(47) Miller, M. R., Ralston III, H. J. and Kasahara, M. (1960). The pattern of cutaneous innervation of the human hand, foot and breast. In Advances in biology of skin, vol. I, ed. Montagne, W., pp. 1-47, Pergamon Press, Oxford.

(48) Merzenich, M. M. and Harrington, T. (1969). The sense of flutter-vibration evoked by stimulation of the hairy skin of primates: Comparison of human sensory capacity with the responses of mechanoreceptive afferents innervating the hairy skin of monkeys, Exp. Brain Res. 9, 236-260.

(49) Mountcastle, V. B., Talbot, W. H. and Kornhuber H. H. (1966). The neural transformation of mechanical stimuli delivered to the monkey´s hand. In Touch, Heat and Pain, ed. de Reuck, A. V. S. and Knight J.

pp. 325-345, Ciba Foundation, Churchill, London.

(50) Mountcastle, V. B., LaMotte, R. H. and Carli, G. (1972). Detection thresholds for stimuli in humans and monkeys: Comparison with threshold events in mechanoreceptive afferent nerve fibres innervating the monkey hand, J. Neurophysiol. 35, 122-136.

(51) Munger, B. L. (1971). Patterns of organization of peripheral sensory receptors. In Handbook of Sensory Physiology, vol. 1, ed. Loewenstein, W. R., pp. 523-556, Springer, Berlin.

(52) Munger, B. L. (1975). Cytology of mechanoreceptors in oral mucosa and facial skin of the Rhesus monkey. In The Nervous stystem, vol. 1, ed. Tower, D. B., New York.

(53) Munger, B. L., Pubols, L. M. and Pubols, B. H. (1971). The Merkel rete papilla - a slowly adapting sensory receptor in mammalian glabrous skin, Brain Res. 29, 47-61.

(54) Pubols, B. H. and Pubols, L. M. (1976). Coding of mechanical stimulus velocity and indentation depth by squirrel monkey and raccoon glabrous skin mechanoreceptors, J. Neurophysiol. 39, 773-787.

(55) Pubols, L. M., Pubols, B. H. and Munger, B. L. (1971). Functional properties of mechanoreceptors in glabrous skin of the raccoon´s forepaw, Expl Neurol. 31, 165-182.

(56) Quilliam, T. A. and Ridley, A. (1971). The receptor community in the finger tip. J. Physiol. 216, 15-16P.

(57) Sakada, S. (1971). Response of Golgi-Mazzoni corpuscles in the cat periostea to mechanical stimuli. In Oral-Facial Sensory and Motor Mechanisms, eds. Dubner R. and Kawamura, Y. pp. 105-122, Appelton, New York.

(58) Stilwell Jr., D. L. (1957). The innervation of deep structures of the hand, Am. J. Anat. 101, 75-92.

(59) Talbot, W. H., Darian-Smith, I., Kornhuber, H. H., Mountcastle, V. B. (1968). The sense of flutter-vibration: Comparison of the human capacity with response patterns of mechanoreceptive afferents from the monkey hand, J. Neurophysiol. 31, 301-334.

(60) Titchener, E. B. (1916). On ethnological tests of sensation and perception with special reference to tests of colour vision and tactile discrimination described in the reports of the Cambridge anthropological expedition to Torres Straits. Proc. Am. phil. Soc. 60, 204-236.

(61) Vallbo, Å. B. and Hagbarth, K-E. (1968). Activity from skin mechanoreceptors recorded percutaneously in awake human subjects, Expl Neurol. 21, 270-289.

(62) Vallbo, Å. B. and Johansson, R. S. (1976). Skin mechanoreceptors in the human hand: Neural and psychophysical thresholds. In Sensory Functions

of the Skin in Primates, ed. Zotterman, Y., pp. 185-199, Pergamon Press, Oxford.

(63) Volkmann, A. (1858). Über den einfluss der übung auf das erkennen raümlicher distanzen, Ber. D. Sächs, Ges. d. Wiss. 38-69.

(64) Weber, E. H. (1834). De pulsu, resorptione, auditu et tactu. Annotationes anatomicae et physiologicae, Fasciculi tres, 1-175.

(65) Weber, E. H. (1835). Über den Tastsinn, Arch. Anat. Physiol. 152-159.

(66) Weinstein, S. (1968). Intensive and extensive aspects of tactile sensitivity as a function of body part, sex and laterality. In The skin senses, ed. Kenshalo, D. R., pp. 193-218, Springfield, Illinois.

(67) Westling, G., Johansson, R. S. and Vallbo, Å. B. (1976). A method for mechanical stimulation of skin receptors. In Sensory Functions of the Skin in Primates, ed. Zotterman, Y., pp. 151-158, Pergamon Press, Oxford.

(68) Wetherhill, G. B. (1963). Sequential estimation of quantal response curves, Jl R. statist. Soc. Series B, 25, 1-48.

(69) Winkelmann, R. K. and Breathnach, A. S. (1973). The Merkel cell, J. invest. Derm. 60, 2-15.

CORTICAL PROCESSING OF TACTILE INFORMATION IN THE FIRST SOMATOSENSORY AND PARIETAL ASSOCIATION AREAS IN THE MONKEY

H. Sakata and Y. Iwamura

Tokyo Metropolitan Institute for Neurosciences, Fuchu, Tokyo, and Department of Physiology, Toho University School of Medicine, Ohta-ku, Tokyo, Japan

INTRODUCTION

In early neurophysiological studies of the somatosensory system,the classical ideas of "epicritic" and "protopathic" sensations were correlated with the lemniscal and spinothalamic systems respectively (Mountcastle,1959).

Now, the lemniscal system is considered to have been specialized by evolution for "active touch" in contrast to "passive touch", attributed to the spinothalamic system (Werner,1974). This idea was based on neuropsychological studies of tactile perception. Lesions of the first somatosensory area (SI) in the monkey are known to impair shape discrimination by palpatory exploration of the tactile objects (Orbach and Chow,1959), yet tactile threshold is not altered (Schwartzman and Semmes,1971). Furthermore, lesions of area 5 in the parietal lobe that receives a heavy projection from SI (Jones and Powell, 1969; Pandya and Kuypers,1969) also affect active tactile discrimination selectively (Ettlinger et al.,1966; Ettlinger and Kalsbeck,1962). On the other hand, discrimination of passively received cutaneous stimuli involves primarily the second somatic sensory area (SII), which is considered to belong to the spinothalamic system (Glassman,1970; Schwartzman and Semmes,1971). The recent advances in microelectrode recordings in either unanesthetized, immobilized animals or behaving animals (Evarts,1966) have opened up great possibilities in studying the neuronal activity in any part of the brain in the condition of full awareness.

In this review, we are going to concentrate on such types of single unit analysis in the lemniscal system so as to follow the sensory information processing up to the level of the association cortex in the hope of finding some clue as to the neural mechanisms of tactile perception.

PROCESSING IN THE FIRST SOMATOSENSORY AREA (SI)

In the initial single unit studies of the somatosensory cortex in cats and monkeys, the most prominent fact was the clustering of cells with the same submodality and somatotopic location in vertical columns (Mountcastle,1957; Powell and Mountcastle,1959). This led to the idea of modality segregation and point-to-point representation of the peripheral receptor sheets in the cortical neurons. Indeed it was confirmed in more recent investigations in unanesthetized monkeys (Mountcastle et al.,1969) that the sensory information encoded in the peripheral afferents is transmitted and reproduced in SI with remarkable fidelity. Many neurons of SI were correlated with one receptor type such as Meissner's corpuscle, Merkel's disk, pacinian corpuscle etc. They responded to relatively simple and punctate stimuli of the skin. This is in sharp contrast to the visual cortex where drastic modifications of sensory messages occur at the initial stage of cortical processing in the primary projection area, and most of the neurons are insensitive to diffuse light or spot-like stimuli (Hubel and Wiesel,1962). However, those simple and sensitive neurons of SI were mainly localized in area 3b which receives the heaviest projection from the ventrobasal complex of the thalamus, whereas neurons of area 1 and 2 in the more posterior part of SI which receive cortico-cortical projection from area 3b were more complicated in their functional properties. Therefore, there seems to be a hierarchical order of processing within SI corresponding to the step-wise projection from area 3b to areas 1 and 2 (Hyvärinen, 1976). We are going to present some evidence to support this hypothesis.

Feature Extraction in SI

It is now widely accepted that the function of the simple cells and complex cells of the visual cortex is concerned with the process of feature extraction in the machine of pattern recognition. The majority of them are concerned with the extraction of contour out of the retinal image of the object.

Apparently this is not the case in SI, and the first examples of feature extraction were the directionally selective neurons sensitive to motion on the skin surface (Whitsel et al.,1972; Schwarz and Frederickson,1971). This is, of course, quite useful for active touch in rubbing the surfaces and edges of objects to recognize them. Figure 1 shows an example of such a neuron in area 2 with the receptive field in the glabrous skin of the hand. This neuron was insensitive to punctate indentation applied with a glass rod, but vigorous discharges were elicited when the rod was moved on the skin surface in the proximal to distal direction. Almost no response was obtained by the same stimulus movement in the opposite direction. The receptive field was fairly wide in this case, including most of the postaxial side of the palm, but receptive fields may also be narrower and confined, for example, to one segment of the phalanx.

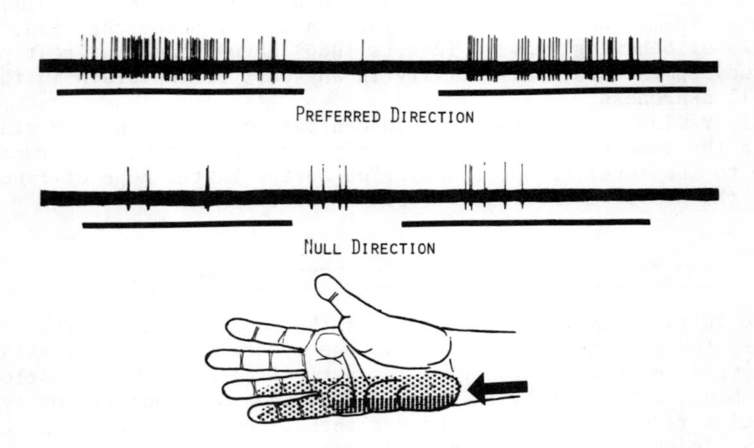

Fig.1 Directionally selective neurons of area 2.

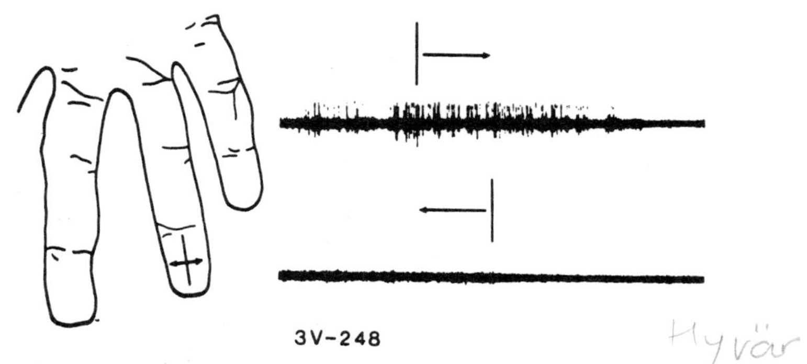

Fig.2. Responses of an area 2 neuron to movement of a metal edge; receptive field in the distal phalanx of the contralateral 4th finger (Hyvärinen,1976).

This type of neuron is quite common in area 1 and 2 and mainly localized in layer III where most of the cells are pyramidal cells sending association fibers to the other areas of the cortex (Whitsel et al.,1972).

There are other specialized types of movement-sensitive neurons in area 2 as shown in Fig.2 (Hyvärinen,1976). This neuron was sensitive to the movement of a metal edge in a particular direction orthogonal to the preferred direction of the movement, but unresponsive to skin indentation or movement of a small probe tip.

More recently, a new type of feature detector neuron was found in the palm area of the racoon's SI (Pubols and Leroy,1977). It responded preferentially to linear or elongated tactile stimuli of a particular orientation. In contrast to the previous ones, these neurons were sensitive to static indentation rather than moving stimuli. Similar types of feature detectors are also likely to be found in the monkey SI in further investigations. So far, most of the feature detectors were found preferentially in areas 1 and 2, at least in the monkey, suggesting a difference in the level of processing between area 3b and areas 1 and 2. However, these feature extractions do not constitute the whole story of cortical processing in SI.

Integration of Receptive Fields in SI

A systematic study of sizes and shapes of receptive fields (RF) of the SI neurons in unanesthetized cats by Iwamura and Tanaka (1977,a,b,c) demonstrated that cutaneous RFs were integrated into functionally significant blocks of skin surface. They compared the RFs between the coronal region (area 3b) and the ansate region (area 1 or 2). In the former, most of the RFs were small, focal, and confined to single digits or pads (Fig.3A), whereas in the latter, most of the unit RFs covered two or three digits or the ulnar or radial side of the dorsal forepaw surface including the digit tips (Fig.3B).

RFs of area 3b were similar to the RFs of the thalamic ventrobasal (VB) neurons. If the RFs of VB neurons and posterior SI neurons were compared in the same region of the body, it was clear that the RFs of the cortical neurons are not only larger in size, but also more uniform in shape. The results suggest the presence of an integratory mechanism for moulding and sculpturing the RF configuration within SI. Moreover, some SI neurons had inhibitory receptive fields. These inhibitory areas were found to be adjacent to the excitatory ones, and their configurations resembled those of excitatory fields of neigh-

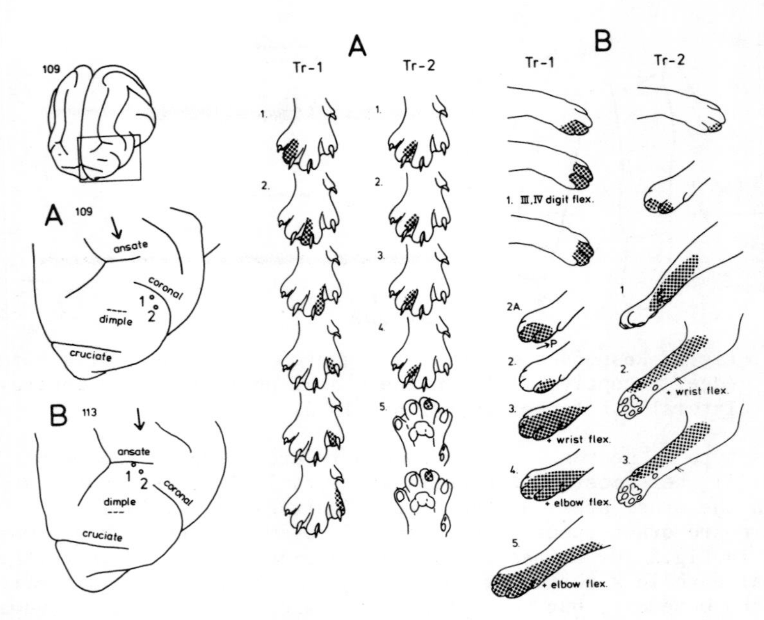

Fig.3. Receptive fields of neurons recorded along penetrations in the coronal region (A) and the ansate region (B) (Adopted from Iwamura and Tanaka, 1977, b, c).

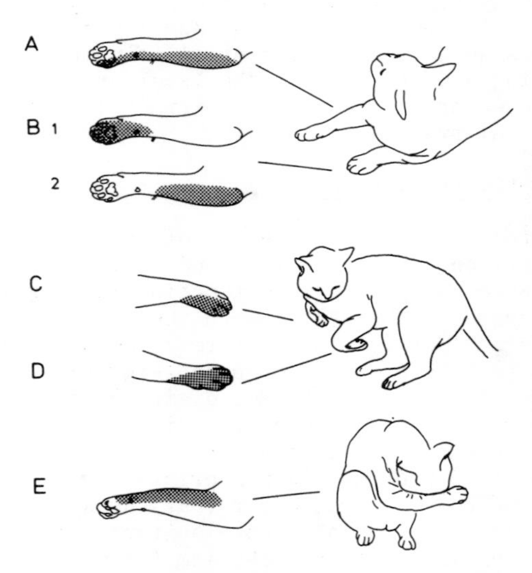

Fig.4. Correspondence of receptive fields of posterior SI neurons to various poses of cats (Iwamura and Tanaka, 1977, c).

boring cortical units. These inhibitory RFs also suggest a specific processing for discrimination of particular skin regions from neighboring ones. The functional significance of this mode of integration becomes apparent through observation of the natural behavior of the cat.

The shapes of the receptive fields resembled corresponding skin areas which come into contact with objects or other parts of the body when the animal sits or assumes some characteristic pose. Examples of such correspondence are shown in Fig.4. Large RFs on the ulnar and ventral sides of the forearm cover the area which comes into contact with the ground (A and B) when the cat lies down as illustrated, and each corresponds to a subtle difference in pose. RFs in the dorsal forepaw surface (C and D) come into contact with the ground when the cat assumes the characteristically feline crouching position. Typical long RFs on the radial side of the forearm (E) may be correlated with the grooming behavior in cats when they wipe their face with their forearms. Thus each of these integrated types of RFs found are considered to represent functional surfaces.

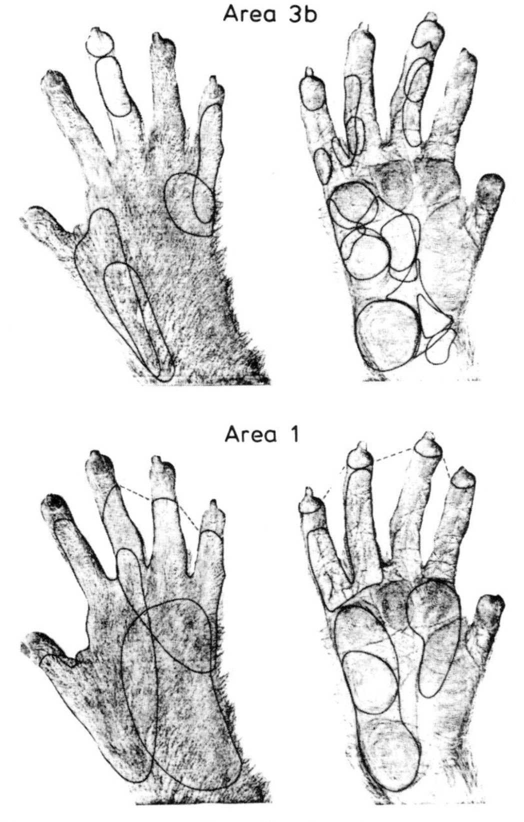

Fig.5. Cutaneous receptive fields of area 3b and area 1 neurons of the hand region of the monkey's SI. Note that the RF in the finger tip in the lower right-hand figure of each digit belongs to the same neuron as indicated by the dotted line (Iwamura et al.,unpublished observations).

This leads to an important conclusion that the increase in the size of RFs within SI is also a process of data reduction to extract meaningful information about somatotopic locations of stimuli. This may be related to the fact that "surface" is as important as "contour" for tactile perception.

A similar type of RF integration was also found in the monkey SI (Iwamura et al.,unpublished observations). Figure 5 illustrates the receptive fields of SI neurons collected in area 3b and area 1 of a rhesus monkey. The receptive fields were smaller and more fragmentary in area 3b in both the volar and the dorsal surfaces of the hand. The RFs of area 1 neurons were larger in size and involved several RFs of area 3b neurons. Furthermore, it may be noted that in the monkey also, the increase of RF size is not the result of random convergence but occurs in a way as to cover functionally significant regions often independent of dermatomal order. For example, a unit in area 1 which covered the tips of all fingers except for the thumb may be significant for recognition of a single surface with four fingers. These clear-cut differences of RF size and configuration between area 3b and area 1 support the hypothesis of hierarchical processing within SI.

Form Discrimination by Grasping

During single unit analysis in a behaving monkey, Iwamura and others (in unpublished observations), found some peculiar neurons which can probably be related to discrimination of shapes of tactile objects. They were found in the posterior part of the hand area of SI, at the boundary between areas 2 and 5, that is, the lateral part of the anterior bank of the intraparietal sulcus. Many neurons were undrivable here by way of ordinary skin stimulation or by simple mechanical stimulation of wrist and finger joints. Some of these undrivable units were driven almost by chance when the monkey grasped objects of particular shapes as illustrated in Figs.6 and 7. The unit shown in Fig.6 was driven when the monkey grasped a rectangular block of wood or a straight-edged ruler in its hand. However, no response was obtained when a small cylindrical bottle was grasped instead. In contrast, the unit shown in Fig.7 was activated by the grasping of round objects such as a cylindrical bottle or an apple but not by a rectangular block. The tests were performed out of the monkeys' sight and the same units were not driven by mere contact of palmar skin to a flat surface, or only by bending the fingers in ways similar to grasping.

Therefore, a particular combination of finger joints and cutaneous stimulation were considered to be necessary to drive these units. It looks as if "the covariance of cutaneous and articular motion is information in its own right" (Gibson,1966) for these neurons.

CORTICAL PROCESSING IN AREA 5

Tactile perception has long been attributed to the function of the posterior parietal association cortex including both area 5 and 7 (Ruch et al., 1938; Blum et al.,1950; Ettlinger and Kalsbeck,1962). However, this was only part of its various functions revealed by the "parietal lobe syndrome", including body image, visual space perception, and control of exploratory movements (Critchley,1953; Denny-Brown and Chambers,1958).

Area 5 was not established as the somatosensory association area until recently when the studies of cortico-cortical connections demonstrated a heavy projection from SI to area 5, although not directly to area 7 (Jones and Powell,1969; Pandya and Kuypers,1969). On the basis of this finding, Jones and Powell suggested that area 5 is the site of higher-order processing of somatosensory information, receiving processed input from SI. This was exactly

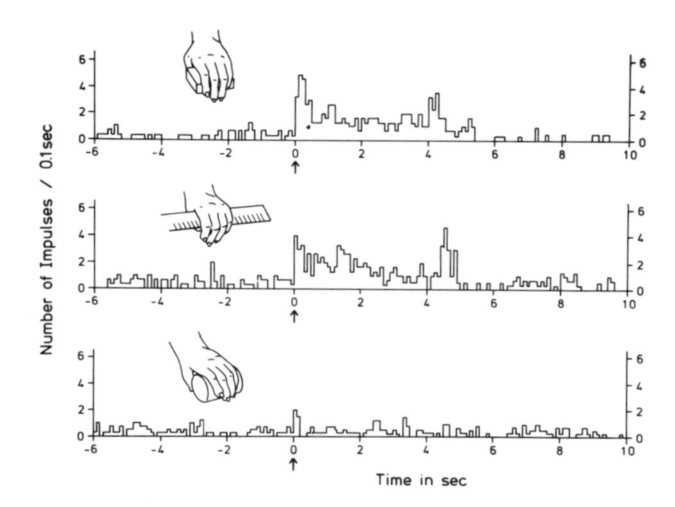

Fig.6. Response histograms of a posterior SI neuron during grasping of different objects (4 to 5 trials were averaged). Top: rectangular block; middle: straight-edged ruler; bottom: cylindrical bottle. Arrows indicate the onset of grasping which continued for about five seconds (Iwamura et al., unpublished observations).

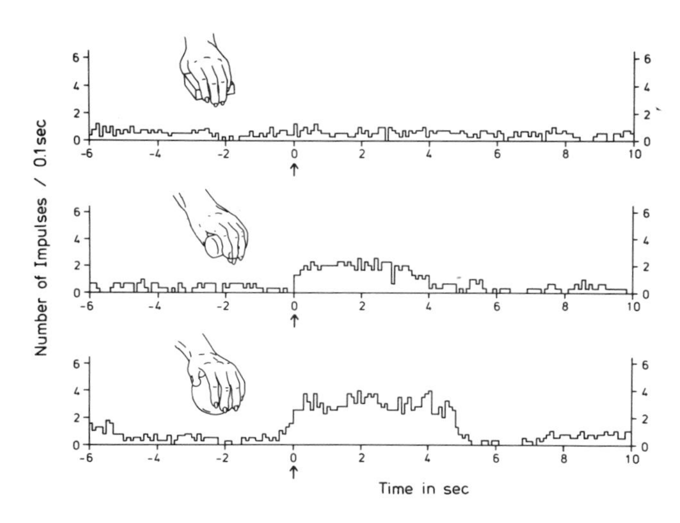

Fig.7. Response histograms of a posterior SI neuron during grasping of different objects. Top: rectangular block; middle: cylindrical bottle; bottom: apple (Iwamura et al., unpublished observations).

what we observed in the single unit study of area 5 in unanesthetized immobilized monkeys (Sakata et al.,1973; Duffy and Burchfiel,1971).

The general response characteristics of area 5 neurons are summarized in Table 1. The most salient feature was the convergence of various parts of the body and different submodalities. It should be emphasized that this convergence was not diffuse and non-specific, but quite selective in terms of position, pose of the body, etc.

Topographic Convergence

There is a considerable overlap of incoming fibers in area 5 from different somatotopic subdivisions of SI, although gross somatotopical order is preserved (Fig.8). The anatomical picture of this topographic convergence agrees well with our neurophysiological observations of the topographic properties of area 5 neurons summarized in a map in Fig.9. The distribution of neurons with receptive fields covering both forelimb and hindlimb or trunk regions coincided remarkably in the same region with the overlapping of projection from both the hand area and the trunk and hindlimb area. Another pathway to facilitate topographic convergence is the interhemispheric connection from the contralateral SI to area 5 (Jones and Powell,1969; Pandya and Vignolo,1969). The consequences of this input are the ipsilateral and bilateral receptive fields of a considerable number of area 5 neurons (Table 1).

Another piece of evidence of topographic convergence was the general increase of the size of the receptive fields of skin neurons, which often covered one side of one whole extremity or bilateral chest, abdomen, etc.

Joint Combination Neurons

This is another example of topographic convergence which has special meaning as Gibson pointed out, "The angle of one joint is meaningless (for tactile space perception) if it is not related to the angles of all the joints above it in the postural hierarchy".

Various types of interaction of multiple joints were observed in approximately one-fourth of area 5 neurons that were recorded. This is in sharp contrast to SI where most of the joint neurons are related to one single joint and a single direction.

TABLE 1 Classification of Area 5 Neurons

Category	Contralat.	Ipsilat.	Bilateral	Total	%
Skin(directional)	13	1	8	22	9.0
Skin(non-directional)	10	2	2	14	5.7
Joint(single)	29	13	-	42	17.2
Joint(combination)	35	10	18	63	25.7
Deep-others	5	0	10	15	6.1
Joint and Skin	50	12	27	89	36.3
Total	142	38	65	245	
%	58.0	15.5	26.5		

(Sakata et al.,1973)

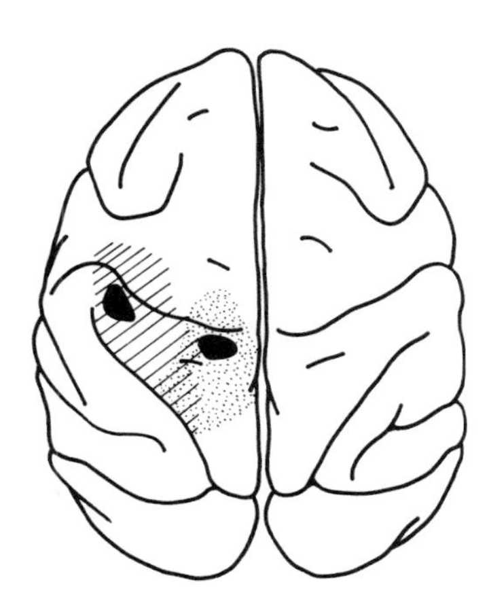

Fig.8. Two Nauta experiments reconstructed on a tracing of the same monkey brain, demonstrating overlap in area 5 of cortico-cortical connections from the trunk and hindlimb (medial) and the forelimb (lateral) area of SI (Jones,1974).

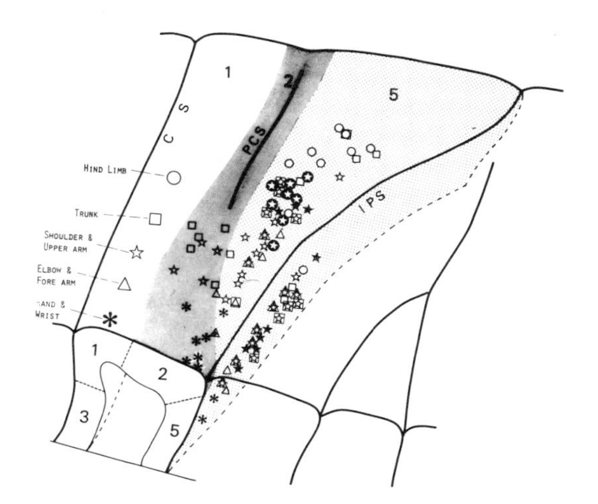

Fig.9. Somatotopical properties of penetrations in area 5 (and 2). Major subdivisions of SI are indicated in area 1 with large symbols which also serve as keys. Superimposed symbols indicate the convergence of subdivisions. The stars on filled circles indicate forelimb-hindlimb convergence (Sakata,1975).

Figure 10 illustrates a simple example of this class of neuron, which is related to both the wrist and the elbow of the ipsilateral side. In this example, either the wrist or the elbow was moved while the other joint was held in a fixed position. Elbow flexion elicited a sustained response when the wrist was fixed in the flexed position (Fig.10A). On the other hand, very little response was evoked when the wrist was kept extended (Fig.10B). Likewise wrist flexion elicited a good response if the elbow was flexed (Fig.10C), but almost no response was obtained with a combination of wrist flexion and elbow flexion. The most plausible neuronal circuit of this unit is that the inhibitory inputs from antagonistic joint neurons (elbow extension and wrist extension) converge upon the excitatory input from synergistic joint neurons (wrist flexion and elbow flexion).

On the other hand, Duffy and Burchfiel (1971) reported a summating type of interaction of the finger, wrist, and elbow joints which suggested simple convergence of excitatory inputs from multiple joints. The combination of elbow and shoulder was also common, through which the approximate position of the hand was to be determined (refer to Fig.14). Furthermore, bilateral interaction was also fairly common, in which proximal joints were involved, such as shoulder and hip joints. One typical unit of bilateral interaction was driven by a parallel movement of both arms from right to left while a separate movement of the right or left arm in the same direction was almost ineffective; apparently, the main components were right shoulder pronation plus left shoulder supination. A reciprocal type of neuron with optimal response to the parallel arm rotation from left to right was recorded in the neighborhood of this neuron.

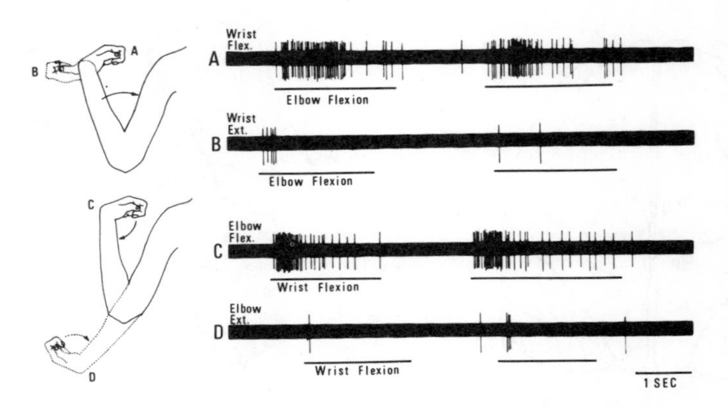

Fig.10. Joint combination neuron of area 5 with optimal combination of wrist flexion and elbow flexion. (Sakata et al.,1973).

Interaction of Joint and Skin

The most common type of submodality convergence in area 5 is the combination of joint and skin inputs in single neurons. We have already suggested that the neurons showing selectivity for the shapes of tactile objects may have this type of convergence, although this has yet to be examined. Joint and skin neurons were far more common in area 5 than in SI (more than a third as shown in Table 1), although the same type of submodality convergence has been observed in SI (Hyvärinen,1976). The degree of complexity of the maximally effective stimulus varied greatly among different neurons.

One of the simplest examples of this type of unit is shown in Fig.11. Maximum response was elicited by rubbing the skin in combination with flexing the elbow (first trace); the receptive field was in the skin covering the palmar side of the hand and the ventral side of the forearm. The response was minimal and transient if elbow flexion was applied alone, while holding the forearm outside of the receptive field, i.e., on the dorsolateral side (second trace). Almost no response was recorded through skin rubbing alone when the elbow was half extended and fixed (third trace). The effective tactile stimulation consisted of surface movement rather than a steady touch, and directionality was from the preaxial to the postaxial side opposite to the movement of the hand. This type of intermodal interaction may be considered to be excitatory, and different from the inhibitory skin to joint interaction found in SI (Mountcastle and Powell,1959).

The effects of joint rotation were sometimes marked even in the steady position. For example, in a joint and skin neuron, a cutaneous moving stimulus was effective when the shoulder was abducted. A possible underlying mechanism of this response is an inhibition from joint to skin. This type of neuron is likely to be essential for the discrimination of the spatial positions of tactile objects.

The most complicated type of joint and skin neuron was that of what we called "matching neurons". As an example, there was a neuron which had a vigorous discharge when the monkey's hand rubbed the foot dorsum from the proximal to the distal direction. This unit did not respond well when the skin of the palm or the foot dorsum were rubbed separately. The "best" stimulus for

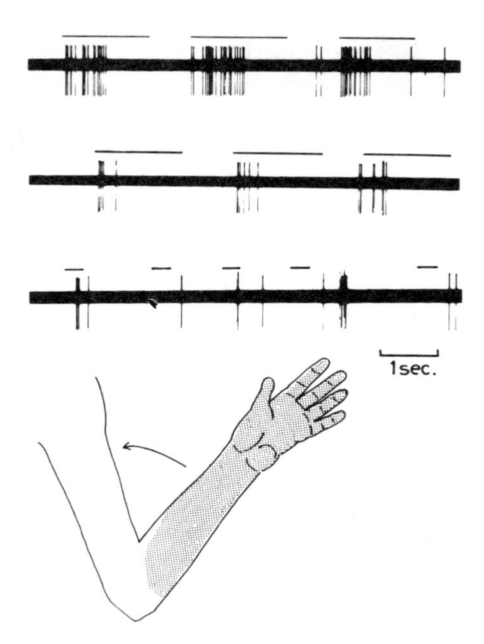

Fig.11 Joint and skin neurons of area 5 with optimal combination of elbow flexion and skin rubbing on the ventral side of hand and forearm (Sakata,1975).

this type of unit was to bring a part of a limb in contact with a part of another limb or the trunk, as if to match two body regions to each other. Although it may be puzzling that such a complex stimulus proved to be optimal for a single neuron, it is quite plausible since these are poses which the monkey often assumes in its daily behavior.

Neurons of Manual Reaching and Hand Manipulation

In addition to the various types of somatosensory neurons, a new type of neuron was found in area 5 in behaving monkeys, which is not activated by any form of passive discharge, but rather, discharge at high rates during exploratory movements of the hands and arms (Hyvärinen and Poranen,1974; Mountcastle et al.,1975). One group of this type of neuron is related to manual reaching and the other group is active during manipulative movements of the hand. These were named projection neurons and hand-manipulation neurons respectively (Mountcastle et al.,1975). Figure 12 shows the activity of such a neuron during the task of releasing a key, and reaching for a lever and then pulling it (Mountcastle,1975). It illustrates the response pattern of a projection neuron which discharges during the act of reaching for a target. As shown in the PST histogram to the left, the discharge rate starts to increase just a little before the monkey releases the key, and declines before it pulls the lever. On the other hand, the hand-manipulation neuron starts to increase its discharge rate as the hand approaches the lever, and reaches its peak just when the lever is pulled.

The same type of neurons were also observed in area 7. Hyvärinen and Poranen (1974) reported in their study of area 7 that the activating directions

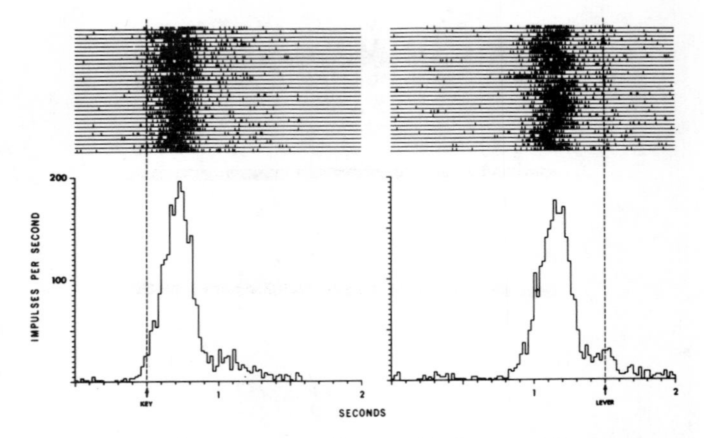

Fig.12. Projection neuron of area 5. Raster displays and PST histograms of responses during the tasks of releasing the key and pulling the lever switch when a signal light comes on. Left records are synchronized at the time of the key release and the right records are synchronized at the time of the lever pull (adapted from Mountcastle et al.,1975).

in space were quite specific for each group although Mountcastle *et al.* (1975) did not observe spatial selectivity of these neurons.

These neurons may be command cells sending control signals for exploratory movement as suggested by Mountcastle and others (1975). However, there is another possibility that they represent corollary discharge of forelimb movement which is necessary to discriminate active manipulation from passive movement of the hand and arm (Hyvärinen,1976).

SOME CHARACTERISTICS OF AREA 7 NEURONS

Recently, intensive studies of area 7 neurons in behaving monkeys have been carried out in several laboratories (Hyvärinen and Poranen,1974; Mountcastle et al.,1975; Lynch et al.,1977). Their functional properties are summarized in Table 2. Most of them are "visual neurons except for the projection and hand-manipulation neurons mentioned above together with some exceptional units. We would like to pick up some interesting observations which may have relevance to the present theme.

TABLE 2 Classifications of Neurons of Cortical Area 7

	No.	%
Projection and Hand Manipulation	124	32.9
Visual Space	49	13.0
Visual Tracking	49	13.0
Visual Fixation	129	34.2
Special	19	5.0
Joint Neurons, Passively Activated	7	1.9
Total	377	100.0

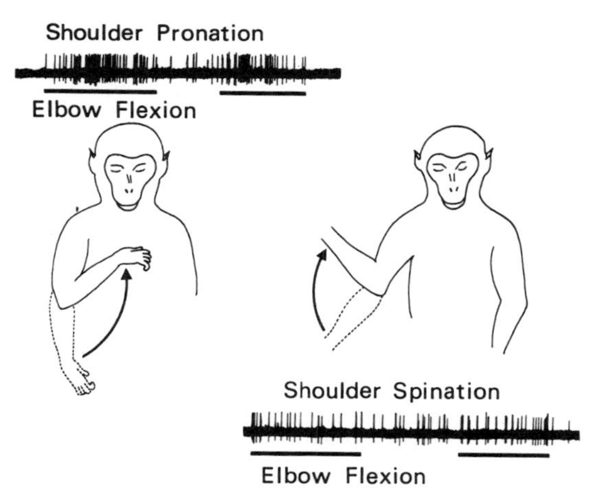

Fig.13. Responses of polymodal neuron of area 7 to combined joint stimulation. Note the joint combination to the left is approximately equivalent to approaching the mouth with the hand (Sakata and Kawarasaki,unpublished observations).

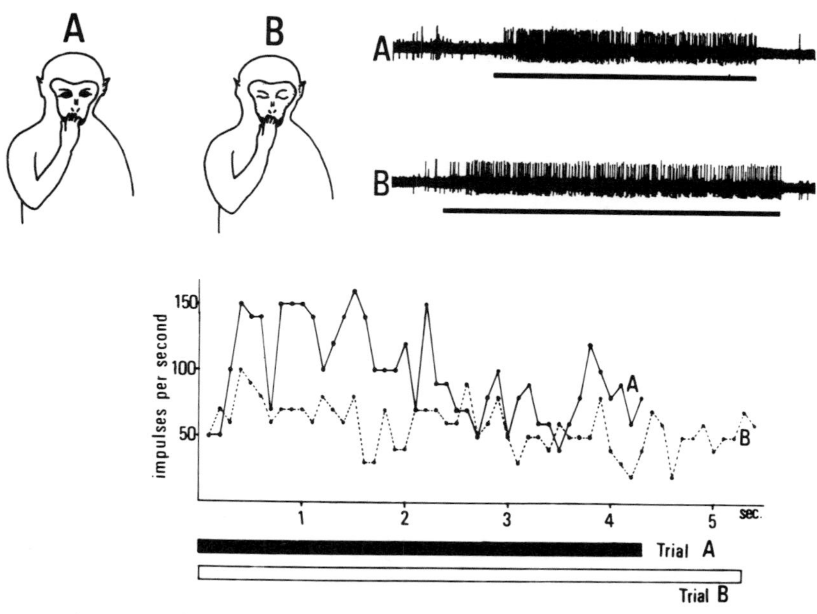

Fig.14. Enhancement of response area 7 neuron to joint combination with visual input. Upper right: impulse record. Bottom graph of firing rate plotted at every 100 msec. (Sakata and Kawarasaki,unpublished observations).

Interaction of Visual and Somatosensory Input

The association cortex has long been considered to have polymodal convergence of auditory, visual, and somatosensory inputs (e.g., Thompson et al., 1963). Therefore it is not surprising to find some neurons with convergence of visual and somatosensory inputs. However, it is interesting that many of such neurons found in areas 5 and 7 are specifically related to particular patterns of natural behavior of the monkey. Figures 13 and 14 illustrate a peculiar neuron of area 7 which was activated maximally when the monkey brought something to its mouth with its right hand. Somatosensory components were analyzed with the eyes closed, and they turned out to be a joint combination as described above.The "best" stimulus was the combination of elbow flexion and shoulder pronation so as to result in the hand approaching the mouth. In contrast, shoulder supination suppressed the response to elbow flexion (Fig.13). Next the effects of the visual component were examined by comparing open eye and closed eye conditions during the same joint movements applied passively. There was a marked enhancement of response in the open eye condition (Fig.14A) when compared to the closed eye condition (Fig.14B).

The most likely source of visual stimulus was the approaching image of the hand. Indeed, there are some neurons in area 7 which are sensitive to movement in depth, and respond to an approaching visual object, while being suppressed by receding movement (Fig.15). If such neurons send excitatory input to the same type of neuron as shown in Fig.14, that is enough to enhance the response in the open eye condition.

Visual Tracking Neurons and Perception of Visual Movements

In the experiment shown in Fig.15, the monkey was trained to fixate on a target light until it dimmed, at which point it had to release the key to get

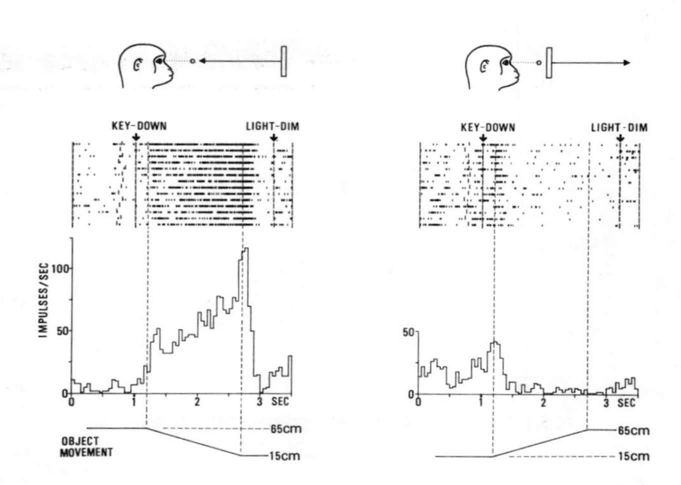

Fig.15. Visual space neuron of area 7, responding to the approaching object and suppressed by the receding movement while the monkey is fixating on the target light in the front (Sakata et al.,1977).

a drop of juice. Similar types of experiments were conducted in our laboratory to study the spatial selectivities of "visual" neurons in area 7 (Sakata et al.,1977). We would like to briefly mention one experiment which may give some hints as to the neural mechanisms of movement perception in general.

This was an experiment on a visual tracking neuron activated during smooth pursuit of an object in a particular direction. As shown in Fig.16, the unit was activated during tracking of a spot of light emitting diode from up to down, although it was suppressed by movement from down to up (left column). There was a remarkable reduction of response when tracking was conducted in the dark (middle column). Since we considered it possibly due to the lack of a retinal image of the background in the dark, as a way of simulating the background movement, we moved a luminous frame around a spot, while letting the monkey fixate on a stationary spot in the center. There was a clear-cut response of the neuron when the frame was moved slowly upward, opposite to the optimal direction of tracking, and a slight suppression occurred during downward movement of the frame (right column). This set-up is quite similar to the one used by Duncker (1929) for the psychological experiment of "induced motion". Actually we perceive an illusive movement of the spot in the opposite direction of the movement of the frame. It looked as if the neuron was perceiving this illusive movement. It is clear that this neuron is not a command neuron to send control signals for tracking eye movement, but it is likely to be a perceptual neuron responding both to real and apparent movement of the visual target.

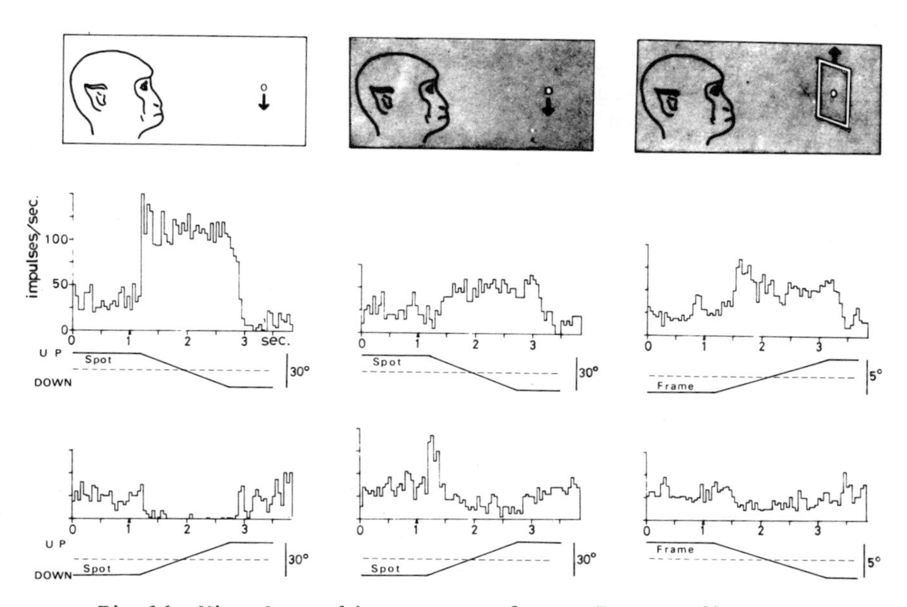

Fig.16. Visual tracking neuron of area 7 responding to both the real movement of the spot and the "induced" motion of the spot by frame movement. Left: tracking in the light; middle: tracking in the dark; right: movement of the frame during fixation on the spot. Movements are indicated under each PST histogram (Sakata et al.,1977).

During tracking in the light, both the retinal signal of image movement in the exrafoveal region, and the eye movement signal were summated to form a vigorous discharge. In contrast, during tracking in the dark, the only cue of target movement is the extraretinal signal of eye movement. The latter is most likely to be attributed to the "corollary discharge". Thus this observation may give concrete evidence for the corollary discharge hypothesis of movement perception. It suggests that signals of active movement are an important, though not the only cue for movement perception of an object when an animal is actively tracking it.

SUMMARY

In this article, we have reviewed recent single unit studies of SI and posterior parietal association areas in unanesthetized monkeys including related investigations in some other animals. During the course of these studies it has become increasingly clear that various kinds of integrative processes are going on in these areas of the cortex so as to pick up functionally significant information out of the sensory signals sent from the peripheral receptor sheet.

Some of the feature detectors found in SI such as directionally selective neurons and orientation (or edge) detectors are quite similar to those found in the visual cortex. However, a more common type of integration of receptive fields in SI is size increase. We pointed out the fact that this happens in such a way so as to be useful for discriminating different ways of contact of the body to the surroundings and the tactile objects. These integrative processes are more prominent in areas 1 and 2 than in area 3b, the typical koniocortex comparable to the striate cortex. Therefore, it is suggested that there is a hierarchical order of processing within SI. We also presented some recent evidence suggesting that the posterior part of the hand area of SI may be concerned with form discrimination of an object grasped in the hand, presumably as the result of submodality convergence of skin and joint inputs.

Submodality convergence is much more common and obvious in area 5, together with a remarkable topographic convergence. As a result, more complicated synthetic processings are going on in area 5. There are "joint combination" neurons for signalling positions and movements of hands and feet in three dimensional space, and "joint and skin" neurons for discriminating the shapes and positions of tactile objects. Some of them were so complicated as to represent certain poses of the body, suggesting a mode of formation of the body image in the parietal association areas.

There are also neurons which discharge during active movement in reaching and in hand-manipulation in areas 5 and 7. We suggested the possibilities that some of these neurons may be activated by "corollary" discharge of forelimb movements. Furthermore, there are some polymodal neurons in areas 5 and 7 which respond both to somatosensory and visual stimulation. An example was taken from a peculiar neuron of area 7, which indicates a specific correlation between the spatial patterns of two modalities of stimulation.

Finally, we described a group of area 7 neurons which are directly related to the perception of visual movement during active tracking, suggesting the synthesis of retinal signals of relative movement and extraretinal signals of eye movement. A similar type of synthesis of signals of active movement and passive sensory stimuli might well be utilized for tactile sensory perception during active manipulation of objects.

REFERENCES

1. Blum,J.S., K.L.Chow, and K.H.Pribram, A behavioral analysis of the organization of the parieto-temporal-occipital cortex. J. comp. Neurol., 93: 53-100, (1950).
2. Critchley,M., The Parietal Lobes, Edward Arnold, London, 1953.
3. Denny-Brown,D. and R.A.Chambers, The parietal lobe and behavior. Res. Pub. Ass. nerv. ment. Dis., 36: 35-117, (1958).
4. Duffy, F.H. and J.L.Burchfiel, Somatosensory system: organizational hierarchy from single units in monkey area 5. Science, 172: 273-275, (1971).
5. Duncker,K., Über induzierte Bewegung. Psychol. Forsch., 12: 180-259,(1929).
6. Ettlinger,G. and J.L.Kalsbeck, Changes in tactile discrimination and visual reaching after successive and simultaneous bilateral posterior parietal ablations in the monkey. J. Neurol. Neurosurg. Psychiat., 25: 256-268 (1962).
7. Ettlinger,G., H.B.Morton, and A.Moffett, Tactile discrimination performance in the monkey: the effect of bilateral posterior parietal and lateral frontal ablations and of callosal section. Cortex, 2: 6-29, (1966).
8. Evarts,E.V., Methods for recording activity of individual neurons in moving animals. In Methods of Medical Research, R.F.Rushmer (ed.), Chicago, Yearbook vol.11, (1966).
9. Gibson,J.J., The Senses Considered as Perceptual Systems, Houghton Mifflin Co., Boston, 1966.
10. Gilman,S. and Denny-Brown,D., Disorders of movement and behavior following dorsal column lesions. Brain, 89: 397-518, (1966).
11. Glassman,R., Cutaneous discrimination and motor control following somatosensory cortical ablations. Physiol. Behav., 5: 1009-1019, (1970).
12. Hubel,D.H. and T.N.Wiesel, Receptive fields, binocular interaction, and functional architecture in the cat's visual cortex. J. Physiol., 160: 106-154, (1962).
13. Hyvärinen,J., Cellular mechanisms in the parietal cortex in alert monkey. In Sensory Functions of the Skin in Primates with Special Reference to Man, Y.Zotterman (ed.), Pergamon Press, Oxford, 1976.
14. Hyvärinen,J., and A.Poranen, Function of the associative area 7 as revealed from cellular discharges in alert monkeys. Brain, 97: 673-692, (1974).
15. Iwamura,Y. and M.Tanaka, Multiple representation of the forepaw skin area in the cat somatosensory cortex (SI), Abstract for the 27th Int'l Congress of Physiol. Sci., (1977a).
16. Iwamura,Y. and M.Tanaka, Functional organization of receptive fields in the cat somatosensory cortex I: integration within the coronal region. Brain Research (submitted for publication), (1977b).
17. Iwamura,Y. and M.Tanaka, Functional organization of receptive fields in the cat somatosensory cortex II: second representation of the forepaw in the ansate region. Brain Research (submitted for publication), (1977c).
18. Jones,E.G., The anatomy of extrageniculostriate visual mechanisms. In The Neurosciences 3rd Study Program, F.O.Schmitt and F.G.Worden (eds.), MIT Press, Cambridge,Mass., 1974.
19. Jones,E.G. and T.P.S.Powell, Connections of the somatic sensory cortex of the rhesus monkey I, ipsilateral cortical connections. Brain, 92: 477-502 (1969a).
20. Jones,E.G. and T.P.S.Powell, Connections of the somatic sensory cortex of the rhesus monkey II, contralateral cortical connections. Brain, 92: 717-730, (1969b).
21. Jones,E.G. and T.P.S.Powell, An anatomical study of converging sensory pathways within the cerebral cortex of the monkey. Brain, 93: 789-821,

(1970).

22. Lynch,J.C.,V.B.Mountcastle, W.H.Talbot, and T.C.T.Yin, Parietal lobe mech-smisms for directed visual attention. J.Neurophysiol., 40: 362-389, (1977).
23. Mountcastle,V.B., Modality and topographic properties of single neurons of cat's somatic sensory cortex, J. Neurophysiol., 20: 408-434, (1957).
24. Mountcastle,V.B., Some functional properties of the somatic afferent system. In Sensory Communication, W.A.Rosenblith (ed.), NIT Press, Cambridge,Mass, 1961).
25. Mountcastle,V.B., J.C.Lynch, A.Georgopoulos, H.Sakata, and C.Acuna, Posterior parietal association cortex of the monkey: command functions for operations within extrapersonal space. J. Neurophysiol., 38:871-908, (1975).
26. Mountcastle,V.B. and T.P.S.Powell, Central nervous mechanisms subserving position sense and kinesthesis. Bull. Johns Hopkins Hosp., 105: 173-200, (1969).
27. Mountcastle,V.B., W.H.Talbot, H.Sakata, and J.Hyvärinen, Cortical neuronal mechanisms in flutter-vibration studied in unanesthetized monkeys, neuronal periodicity and frequency discrimination. J. Neurophysiol., 32: 452-484, (1969).
28. Orbach,J. and K.Chow, Differential effects of resection of somatic areas I and II in monkeys, J. Neurophysiol., 22: 195-203, (1959).
29. Pandya,D.N. and H.G.J.M.Kuypers, Cortico-cortical connections in rhesus monkey. Brain Research, 13: 13-36, (1969).
30. Pandya,D.N. and L.A.Vignolo, Interhemispheric projections of the parietal lobe in the rhesus monkey. Brain Research, 15: 49-65, (1969).
31. Powell,T.P.S. and V.B.Mountcastle, Some aspects of the cortex of the post-central gyrus of the monkey, a correlation of findings obtained in a single unit analysis with cytoarchitecture. Bull. Johns Hopkins Hosp., 105: 133-162, (1959).
32. Pubols,L.M. and R.F.Leroy, Orientation detectors in the primary somatosensory neocortex of the racoon. Brain Res. 129: 61-74, (1977).
33. Ruch,T.C., J.F.Fulton, and W.J.German, Sensory discrimination in monkey, chimpanzee, and man after lesions of the parietal lobe. Arch. Neurol. Psychiat., 39: 919-937, (1938).
34. Sakata,H., Somatosensory responses of neurons in the parietal association area (area 5) of monkeys, In The Somatosensory System, H.H.Kornhuber (ed.), George Thieme, Stuttgart, 1975.
35. Sakata,H., Y.Takaoka, A.Kawarasaki, and H.Shibutani, Somatosensory properties of neurons in the superior parietal cortex (area 5) of the rhesus monkey. Brain Res. 64: 85-102, (1973).
36. Sakata,H., H.Shibutani, and K.Kawano, Spatial selectivity of "visual" neurons in the posterior parietal association cortex of the monkey. Abstract for the 27th Int'l Congress of Physiol. Sci., (1977).
37. Schwartzman,R.J. and J.Semmes. The sensory cortex and tactile sensitivity. Expl. Neurol., 33: 147-158, (1977).
38. Schwarz,D.W.F. and J.M.Frederickson, Tactile direction sensitivity of area 2 oral neurons in the rhesus monkey cortex. Brain Res., 27: 397-401, (1971).
39. Thompson,R.F., H.E.Smith, and D.Bliss, Auditory, somatic sensory, and visual response interactions in association and primary sensory cortical fields of the cat. J. Neurophysiol., 26: 365-378, (1963).
40. Werner,G., Neural information processing with stimulus feature extractors. In The Neurosciences, 3rd Study Program, F.O.Schmitt and F.G.Worden (eds)., MIT Press, Cambridge,Mass.,1974.
41. Whitsel,B.L.,J.R.Roppolo, and G.Werner, Cortical information processing of stimulus motion on primate skin. J. Neuropjysiol., 35:619-711, (1972).

NEURAL PROCESSING OF TEMPORALLY-ORDERED SOMESTHETIC INPUT: REMAINING CAPACITY IN MONKEYS FOLLOWING LESIONS OF THE PARIETAL LOBE

Robert H. LaMotte* and Vernon B. Mountcastle

Department of Physiology, The Johns Hopkins University School of Medicine, 725 N. Wolfe Street, Baltimore, Maryland 21205, U.S.A.

The lemniscal somatic afferent system and the primary somatic area of the cerebral cortex (SI) are important for the processing of temporally ordered afferent input (1,2,3,4,5). A temporally-ordered pattern of neuronal signals may be produced by laterally displacing the surface of the skin across the textured surface of an object. Alternatively, mechanical sinusoids can be delivered to one locus on the restrained hand making it possible to isolate for study the physical parameters of intensity and frequency of skin indentations from the more complex spatio-temporal mechanical stimulations that occur during active touch. The present series of experiments demonstrate that cortical lesions involving SI impair the capacity of monkeys to detect and discriminate between mechanical sinusoids delivered to the contralateral hand. Preliminary results also suggest that vibratory patterns -similar in certain physical characteristics to those evoked by mechanical sinusoids- are important for spatial discriminations between textured surfaces during active touch.

Previous studies have shown that mechanical sinusoids in the frequency range of 5 to 40 Hz delivered to the glabrous skin of the hand in humans evoke sensations of flutter while sinusoids of higher frequencies elicit the sense of vibration (5). The capacities of monkeys and humans to detect the presence of sinusoidal stimulation of the glabrous skin of the hand or to discriminate between sinusoids of different frequencies were compared with the responses to these same stimuli of mechanoreceptive afferents innervating the monkey hand (2,4). Results supported the hypothesis that the necessary primary afferent input to the central nervous system required for detection of oscillating mechanical stimuli was the appearance of minimal activity in two populations of quickly-adapting mechanoreceptive afferents: those terminating in Meissner corpuscles within the dermal papillae for frequencies within the flutter range (5 to 40 Hz) and those terminating in Pacinian corpuscles beneath the skin for higher frequencies (40 to 400 Hz). The results also suggested that the capacity to discriminate between sinusoids of different frequencies within the flutter range depended upon the presence of a neural representation of the periodicity or cycle length of each stimulus in the "Meissner" afferents and in the cortical neurons to which they were linked. For example, both monkeys and humans were able to detect the presence of a 30 Hz stimulus at an amplitude that evoked minimal, untuned activity in a small number of Meissner afferents innervating the monkey hand. However, frequency discrimination was not possible for either monkeys or

*Present address: Department of Anesthesiology, Yale University School of Medicine, 333 Cedar Street, New Haven, Connecticut 06510

humans until the sinewave amplitude was raised to about 8db above this level to a point sufficient to evoke tuned activity in these afferents (2). The 8db difference in minimal amplitude required for detection and frequency discrimination was called the "atonal interval".

It has also been shown that sinewave amplitude influenced the variability of responses of those cortical neurons linked to Meissner afferents when sinusoids were delivered to their receptive fields on the hand (3). The variability in the periodic responses of these cortical neurons to sinusoids was highest near detection threshold and reached a minimal value about 7db above threshold near the top of the atonal interval. It was inferred from these and other results that in order for frequency discrimination to occur, a cortical processing mechanism within SI must operate upon differences in the dominant period of the neuronal signals evoked by the frequencies to be discriminated (3).

In the present series of experiments we determined the effects of lesions in the parietal cortex on the capacity of monkeys to perform the following sensory tasks: (1) detection of and discrimination between mechanical sinusoids which differed in intensity when frequency was held constant and for which the first order neural coding mechanism was probably the total amount of activity within the Meissner or Pacinian afferents (2); (2) discrimination between sinusoids of different frequency within the flutter range -a task believed to require central measurement of differences in cycle length between neuronal signals set forth by the Meissner afferents.

Sixteen monkeys were trained in a variable delay, reaction-time task (4) to detect the presence of sinusoids delivered to the glabrous skin of the restrained hand. Test frequencies were 10, 30 and 200 Hz. Three animals were further trained to discriminate between sinusoids of the same frequency but different amplitude and 6 were trained to discriminate between sinusoids of the same subjective intensity but different frequency. Stimuli in the discrimination tasks were presented as a two-alternative, forced-choice. On each trial a standard sinewave was followed, after a brief period, by a comparison sinewave after which the animal made one of two responses to indicate whether the comparison was greater or lesser than the standard in amplitude (or frequency, depending upon the task). In the amplitude discrimination task, discrimination thresholds (DLs) were determined for sinewave standards of 102 or 154 μm at 30 Hz or 83 μm at 40 Hz. Frequency DLs were usually obtained with a 30 Hz standard at 48 and 129 μm (comparison frequencies: 24 to 36 Hz in 2 Hz steps). Test stimuli in both the detection and discrimination tasks were delivered in random order with each presented an equal number of times. Detection and discrimination thresholds were obtained with confidence intervals. Preoperative discrimination thresholds averaged about 10% of the standard stimulus (see refs. 2 and 4 for detailed discussion of methods and preoperative performance).

Each animal was tested before and after each lesion and most were tested on the hand ipsilateral as well contralateral to the lesion. At the end of the postoperative test period, many animals received a second lesion in the intact hemisphere and this was followed by another period of postoperative testing. The most frequent unilateral lesion was one of the following: (1) removal of the postcentral gyrus (Brodmann's areas 1,2,3,5 and the second somatic area, SII), (2) complete parietal lobectomy (areas 1-3,5,7 and SII), (3) parietal lobectomy plus motor cortex removal (areas 1-7 and SII).

Postoperative Sinewave Detection

A unilateral parietal lobectomy, or in some cases partial removals which included either SI alone or all of the postcentral gyrus, resulted in elevated detection thresholds on the contralateral hand of 2 to 7 times preoperative values. Removal of motor cortex along with the parietal lobectomy did not, in most cases, raise detection threshold higher than that resulting from parietal removal alone. Following parietal lobectomy, with continued testing over a 2 to 8 week postoperative period, detection threshold returned either to normal or to a residual level of about 2 times preoperative value. Unilateral removal of SII alone or the posterior parietal cortex (areas 5 and 7) had no significant effect on detection threshold. No defects were observed for stimuli delivered to the hand ipsilateral to any lesion.

Postoperative Amplitude Discrimination

Complete unilateral removal of parietal and motor cortex or parietal lobectomy alone resulted in an elevation of amplitude DLs 2 to 5 times the preoperative values for stimuli delivered to the contralateral hand. These effects were probably due to removal of postcentral gyrus since removal of the latter alone produced an elevation in the DL of 2 times, while removal of the posterior parietal cortex without the postcentral gyrus had no significant effect upon amplitude discrimination. No defects were observed on the hand ipsilateral to any lesion.

Postoperative Frequency Discrimination

Following unilateral removal of the postcentral gyrus contralateral to the tested hand, all 3 animals tested failed to discriminate during the first 6 test sessions. During subsequent tests, one animal recovered to 1.6 times his preoperative DL while the other two animals never recovered. In other animals, parietal lobectomy either with or without removal of motor cortex resulted in a permanent loss in the capacity to make frequency discriminations on the contralateral hand. No defects were observed on the hand ipsilateral to any unilateral lesion.

At the end of the postoperative test period each animal with a complete parietal removal was tested for the capacity to discriminate between very large differences in frequency (a standard of 30 Hz and comparison frequencies of 10 to 50 Hz in 5 Hz steps). These stimuli, when delivered to the human hand, were reported as requiring only a gross discrimination of "flutter" from "vibration". DLs were elevated for this task to about 4 to 5 times preoperative values for the hand contralateral to the lesion. This defect represented a large sensory loss in the capacity to process neural information in the temporal mode. Thus, parietal removal eliminated the capacity to make discriminations on the contralateral hand based upon differences in cycle length for stimulus frequencies within the flutter range; it also greatly reduced the number of frequencies which could be identified across a much broader frequency range. This was further illustrated by the postoperative performance of an animal trained to identify each of 4 categories of frequency within the 20 to 60 Hz range — with 90% accuracy — by selecting one of 4 response keys for each of 4 frequencies. The preoperative performance of this animal was matched only by well-trained human observers. Following a unilateral parietal lobectomy contralateral to the tested hand, the animal was capable of identifying only the extremes of this frequency range (two categories instead of 4). This residual capacity was possibly based upon a primitive recognition of neural activity in a single labeled line -for example, the identification of activity in the Pacinian primary afferent population and the central neurons to which it projects.

Neural Processing in the Spatio-temporal Mode

The results of preliminary studies using spatially textured stimuli suggest that certain spatial discriminations between textured surfaces may require recognition of differences in vibratory patterns set up by lateral movement of the surfaces across the skin. In one experiment, discriminations between "spatial frequencies" (wire-wound cylinders of 9,10,11,12 and 13 wire turns/cm) required the presence of vibratory patterns on the surface of the skin. The standard stimulus was 11 turns/cm and subjects were required to state whether each comparison stimulus was greater or lesser in frequency (number of wire turns) than the standard. Movement of the finger tip pad back and forth along the length of each cylinder placed the transversely organized epidermal ridges parallel to the wire striations so that both a vibratory pattern and spatial texture were felt. Under these conditions, discrimination was possible and DLs averaged 1.2 turns/cm. Conversely, movement of the finger tip in a direction perpendicular to the length of the cylinder created neither a vibratory effect nor a sensation of texture. Under these conditions, subjects could not discriminate between spatial frequencies.

In other experiments, the responses of Meissner afferents in monkeys were recorded when textured surfaces were displaced laterally across their receptive fields on the tip of the finger (1). Nylon monofilament fabrics of different yarn counts (number of weft plus the number of warp per inch) were each fastened to a small plate which was displaced laterally back and forth over a distance of $\pm$ 2 mm at 4 Hz. Pressure exerted by the plate against the skin was varied and optically calibrated in microns of skin displacement. In parallel experiments, human observers numerically ranked the roughness of the same fabric surfaces. The cumulative nerve impulse count in the Meissner afferent during each stimulus (1 sec duration) increased with increases in skin displacement. Also, for any level of skin displacement, impulse count increased with decreasing yarn count and this relationship correlated with the numerical ranking of apparent roughness. That is, lower yarn counts were judged as rougher than higher yarn counts (1).

The results of these experiments suggest that those neural mechanisms important for discrimination between mechanical sinusoids are also important for processing spatio-temporal input.

REFERENCES

1. R.H. LaMotte, Psychophysical and neurophysical studies of tactile sensibility, In: Clothing Comfort: Interaction of thermal, ventilation, construction and assessment factors, Edited by N. Hollies and R. Goldman, Ann Arbor Science, Ann Arbor, 1977.

2. LaMotte, R.H. and Mountcastle, V.B., Capacities of humans and monkeys to discriminate between vibratory stimuli of different frequency and amplitude: A correlation between neural events and psychophysical measurements, J. Neurophysiol. 38, 539-559, 1975.

3. Mountcastle, V.B., Talbot, W.H., Sakata, H., and Hyvarinen, J., Cortical neuronal mechanisms in flutter-vibration studied in unanesthetized monkeys. Neuronal periodicity and frequency discrimination, J. Neurophysiol. 32, 452, 1969.

4. Mountcastle, V.B., LaMotte, R.H. and Carli, G., Detection thresholds for vibratory stimuli in humans and monkeys: comparison with threshold events in mechanoreceptive afferent nerve fibers innervating the monkey hand, J. Neurophysiol. 35, 122-136, 1972.

5. Talbot, W.H., Darian-Smith, I., Kornhuber, H.H. and Mountcastle, V.B., The sense of flutter-vibration: comparison of the human capacity with response patterns of mechanoreceptive afferents from the monkey hand, J. Neurophysiol. 31, 301-334, 1968.

EFFECTS OF PARIETAL LOBE COOLING ON MANIPULATIVE BEHAVIOUR IN THE CONSCIOUS MONKEY

John Stein

University Laboratory of Physiology, Oxford, England

SUMMARY

1. Cooling plates were implanted over parietal lobe areas 5 and 7 in monkeys trained in visuomotor and manipulative tasks.

2. Cooling the area 7 plate to 25°C led to inaccurate tracking of a visual target by the contralateral limb in the contralateral visual field; to inaccurate reaching in that half field; a prolonged reaction time to hit buttons, and visual inattention on that side.

3. Cooling the area 7 plate to 0°C caused, in addition to the visuomotor disturbance of the contralateral limb in the contralateral half field, delayed and inaccurate responses of the ipsilateral limb in the contralateral half field.

4. Cooling the area 7 plate did not cause a visual field defect, as it had no effect on the monkey's ability to detect lights in either visual hemifield, and respond by hitting buttons straight in front of him.

5. Cooling the area 5 plate in the monkeys trained to perform visuomotor tasks caused clumsiness of the contralateral limb; but no particular difficulty with visuomotor coordination in either visual field.

6. Cooling the area 5 plate in a monkey trained to perform a rough/smooth discrimination by touch alone, prevented him making the discrimination with the contralateral hand. The ipsilateral hand was unaffected however. Cooling the area 7 plate had no effect on his ability to make the discrimination with either hand.

7. It is concluded that area 5 is concerned in controlling contralateral movements in relation to cutaneous and proprioceptive feedback (active touch); whilst area 7 is important in controlling the movements of both contralateral and ipsilateral limbs in relation to visual targets in the contralateral visual field.

Active movements of the hand are said to confer a new dimension to the sense of touch. A single object pressed passively on to two fingers feels like two objects; actively palpated it becomes one. The hardness of an object can be appreciated by grasping it more tightly. Similar pressures applied passively to the skin are perceived as just that, increasing pressure on the skin. There is no sensation of hardness (Gibson, 1962). Bentley (1900) slowly raised a beaker containing various liquids (water, treacle, mercury) at identical temperatures over a subject's immobile finger. The subject was quite unable to say what the liquids were. However, when he was allowed to immerse his finger in the liquid himself, the subject had no difficulty in discriminating. Clearly active movement is essential to a full tactile experience.

If this is so, there is presumably some region of the brain where the relevant sensory and motor signals converge and interact. This region is probably within the cerebral cortex, since sensations are part of our conscious experience, and proper functioning of the cortex is the necessary and sufficient condition for consciousness. The obvious place to look for somaesthetic and motor interactions is in the postcentral gyrus. This receives a topographically organised record of cutaneous impressions, together with some information from the motor cortex in front about motor activity. However the elaboration of active touch sensations probably takes place at a 'higher' level than this. For it requires interaction of different modalities of somaesthesia and proprioception from different regions of the hand and arm with information about the activity of motor centres; whereas neurones in the primary sensory area are place and modality specific, and not highly sensitive to coincident motor activity (Mountcastle, 1975). Of course lesions in SI interfere with active touch because they render the contralateral limb anaesthetic; but this only implies that SI is on a pathway to the higher level, not that it is the higher level.

Clinical experience suggests that higher sensory analysis occurs further back in the parietal lobe. Lesions in its posterior part cause a complex of disturbances, given a confusing array of lengthy classical neologisms by neurologists whose favourite sport this is. The clinical literature is additionally confusing about the extent to which symptoms are confined to the contralateral side of the body following lesions of the dominant or non dominant hemispheres. This is partly due to difficulties with the clinical definition of dominance, partly to the disparate tests and neologisms used to describe the results of the tests by different clinicians, and partly due to a lack of precise anatomical knowledge about the site and extent of lesions. However there now appears to be a sort of consensus about the 'parietal' syndrome (Critchley, 1953). Small lesions on either side lead to neglect of stimuli impinging upon the opposite half of the body and in the opposite half of visual space (Brain, 1941); they lead to an inability to determine by touch alone the shape of an object (astereognosis);to an inability to appreciate the spatial relationships of the patient's contralateral side or of contralateral visual space (amorphosynthesis), (Denny-Brown & Chambers, 1958); to inaccurate reaching using either tactile or visual control; and an unwillingness or inability to use the contralateral arm for controlled voluntary movements, though automatic movements may be relatively unimpaired (apraxia). Larger lesions of the dominant hemisphere lead to the additional symptoms of Gerstmann's syndrome (bilateral finger agnosia, acalculia, agraphia and inability to distinguish left from right). Still larger lesions involve the speech centres, giving rise to the various aphasias. Large lesions of the non-dominant hemisphere cause, in addition to the contralateral deficits, global symptoms of disordered spatial sensation. The patient may exhibit difficulties with spatial memory and spatial thinking. He may not be able to find his way out of his own house, or he may not be able to copy drawings with their correct spatial relationships (constructional apraxia).

What do these symptoms reveal about the functions of the parietal lobes, and how do they relate to the problem of active touch? First we may place on one side the symptoms in humans which depend upon the specifically human attributes of language, calculation and spatial thinking; i.e. those global effects of lesions affecting dominant and non-dominant hemispheres differently. This leaves the simpler sensory deficits. They are confined for the most part to the contralateral half of the body. They may be divided into those relating to the patient's own body image, somaesthetically sensed, which include tactile neglect, astereognosis, amorphosynthesis, tactile apraxia; and a second category which consists of [illegible] in which visual disturbances predominate: visual neglect, misreaching to [illegible] stimulus, visual apraxia and so on. The somaesthetic deficiencies [illegible] interfere with active touch, whilst the visual disturbances interfere [illegible] visuomotor coordination.

This classification invites the speculation that there might be two separate regions in the posterior part of the parietal lobe responsible for these two types of sensory-motor association. As the experiments of nature in humans have no respect for anatomical boundaries; and more precisely delineated surgical excisions of cerebral lesions for epilepsy (Hecaen et al, 1956) suffer from the fact that the brain tissue removed must already have been abnormal, we have to turn to animal experiments to throw light on this. There have been rather few animal studies in which the inferior (area 7) and the superior (area 5) parietal lobules have been removed separately. Peele (1944) noted that area 5 ablations in the monkey led to marked ataxia of both the arm and leg, which could however be brought under control if the animal was allowed to perform the movement under visual control. Area 7 lesions led to similar clumsiness which was not so easily corrected by close visual attention. All disturbances were much more marked in the period immediately following the operation than after several weeks. Of course how far the acute effects were the temporary consequence of oedema, haemorrhage and so on, and how far the later recovery was the result of other regions assuming the normal role of areas 5 and 7 it was impossible to say. However the studies of Ettlinger and co-workers (Ettlinger & Kalsbeck, 1962; Hartje & Ettlinger, 1974; Ratcliffe et al, 1977) also suggest that there are some differences between the functions of areas 5 and 7. Their monkeys with anterior lesions were more inaccurate than those with posterior lesions at discriminating shapes by feel, whereas the latter were worse than the former at reaching under visual control. These hints in the literature suggest that area 5 may be important for tactile control of movement, whilst area 7 may deal with visual guidance. Recent recordings in conscious monkeys (Hyvärinen & Poranen, 1974; Mountcastle et al, 1975; Lynch et al, 1977) go further to support this idea. They demonstrate a preponderance of neurones in area 5 related to active rather than passive movements of limb joints; and in area 7, of neurones related to visual targets. Mountcastle et al (1975) propose that their visual fixation and visual tracking neurones constitute a class of 'command' neurones which initiate movements of the limbs into visual space. Likewise the joint rotation and projection and hand manipulation neurones, found most commonly in area 5, could constitute a class of command cells for active touch. The neurones in area 7 could associate the initial position of a limb, signalled partly by a motor outflow copy of where motor systems last 'intended' to place it, and partly by feedback from the periphery about where it actually was, with a visual signal from a desired object. This sort of association must be an essential first step in the generation of a command to move a limb from its initial position to grasp an object in the external world. Similarly the association of signals about the present position of the hand with information about the cutaneous impressions of an object being palpated, would be an essential first step in building up a motor command for the next stage of the palpation.

There is now some evidence, therefore, that there is a division of labour in the posterior part of the parietal lobe; area 7 dealing with visually controlled movements, area 5 with active hand manipulations. However the evidence is fragmentary. The results from single unit recording suffer from sampling problems; both, can one be sure that the discharges observed were representative of the whole population of neurones; and also, was the selection of tests performed on the units able to reveal the full range of their functions, or only that part of it which happened to occur to the experimenter at the time? And the results from ablation studies are even more difficult to evaluate. The acute effects are confused by the immediate sequelae of surgery whilst the longer term effects are confused by remaining regions assuming the normal function of the ablated area. Permanent defects may therefore be more an indication of what other regions cannot do, than what the ablated area normally does do.

We have developed a new technique for cooling selected areas of cerebral cortex in trained conscious monkeys, in order to study the visual control of movement (Wattam-Bell & Stein, 1975; Stein, 1976; Stein, 1977; Caan & Stein, 1977). The advantages of cooling over ablation as a technique for studying the function of nervous structures are threefold. In the first place the effects are readily reversible, so that each monkey's performance before cooling can serve as the normal control to compare with changes during cooling. Secondly, as the duration of cooling is short, it is unlikely that other areas, normally less important for the function under study, can take over the role of the cooled region. This means that cooling is more likely to reveal the usual function of the area. Thirdly the degree of cooling can be graded so that subtle deficits hidden by the grosser results of total ablation may be revealed. So we have used this technique to examine the role of the posterior part of the parietal lobe in the visual and tactile control of movement in rhesus monkeys.

The cooling plate is made of silver and shaped to fit over the surface of the superior (area 5) or inferior (area 7) parietal lobules. (Fig. 1). Separate copper coils are soldered to each plate and brought out to manifolds to which the cooling apparatus can be attached. The coolant is absolute alcohol raised to a pressure of 60 lbs/sq inch in a steel pressure chamber, and then passed through a copper heat exchanger immersed in alcohol containing dry ice (solid CO_2); thence it is routed via a two way tap to either the anterior (area 5) or posterior (area 7) plate. The temperature of the plates is monitored by means of thermistors fixed to their upper surfaces. The plates are inserted under the dura over areas 5 and 7 as shown in the figure and fixed in place with self tapping bone screws and dental acrylic, in an operation under nembutal anaesthesia observing full aseptic precautions. Cooling experiments can be performed for about 6 weeks afterwards, before connective tissue grows under the plate and renders it ineffective.

The temperature profile shown in Fig. 2 was recorded in one of the monkeys in a final, acute experiment under anaesthetic. We used the opposite parietal lobe which had not been used for the chronic cooling experiments. The plate was cooled to $0^{\circ}C$, and the figure shows isotherms plotted around the edge of the plate. Note that even when the upper surface of the plate was at $0^{\circ}C$ the surface of the cortex was $10^{\circ}C$ higher due to its efficient circulation.

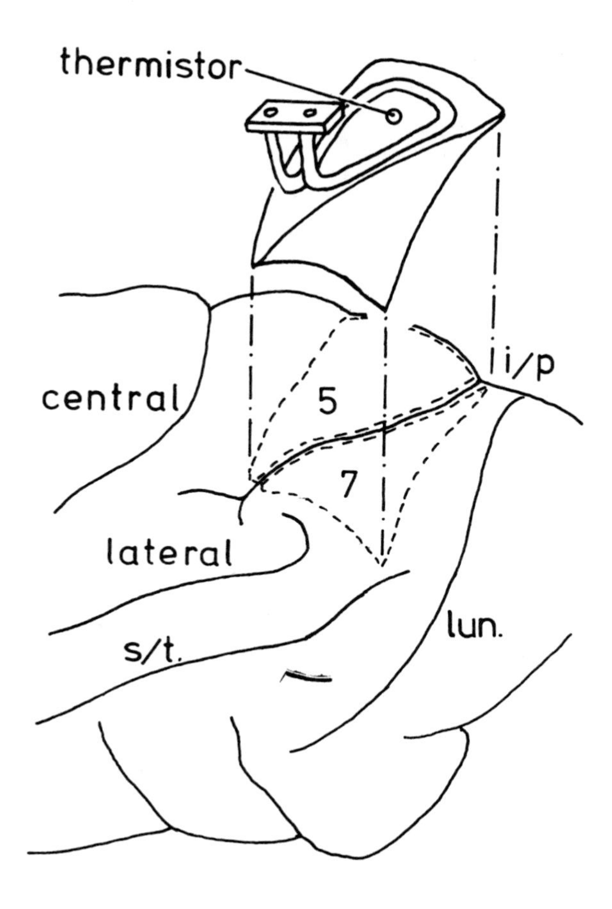

Figure 1. Silver plates shaped to fit over surface of areas 5 and 7. Coolant circulates through tubes soldered to the plates (only that on area 5 plate shown for clarity). i/t = intraparietal, s/t = sup. temporal sulci.

The dotted line in Fig. 2 shows the approximate depth of cortex at this point. So when the cooling plate was at $0^{\circ}C$ all the grey matter was below $25^{\circ}C$. In some other experiments in which we recorded potentials associated with movement when cooling the motor cortex, the first changes in either the monkey's performance or in associated motor potentials, occurred at a temperature of $25^{\circ}C$. (Caan & Stein, 1977). Unfortunately of course the cooling spreads horizontally as well. This means that it is impossible to cool area 7 or 5 alone without somewhat cooling the other. In addition much of parietal cortex lies not on the surface but in the depths of the intraparietal sulcus (approximately 10 mm deep). Much of this would not have been effectively cooled by our method.

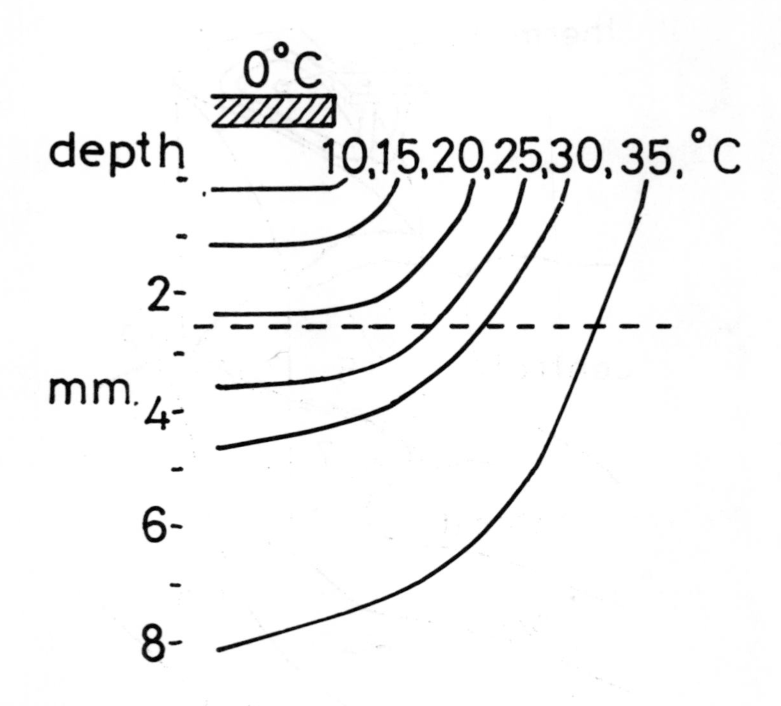

Figure 2. Temperature isotherms at edge of plate cooled to 0°C. Note surface of cortex is still at 10°C and regions 3 mm laterally and 3 mm deep are below 25°C. Approx depth of cortex given by dotted line.

The results were however striking. In the first monkey a plate covering both areas 5 and 7 was inserted on the left hand side. The presence of the plate caused no noticeable changes in his manipulative or visuomotor abilities using either hand. However during cooling to a plate temperature of 25°C slight clumsiness of the contralateral hand was observed. This was particularly noticeable for movements in the contralateral visual field. Cooling both regions to a lower temperature rendered the contralateral arm apparently anaesthetic, so that the monkey was entirely unaware if a raisin was placed in it. He was also incapable of making precise movements, though certainly not paralysed, as he was able to make clumsy lunges in the vague direction of visual targets. His eye movements appeared unaffected. At this lower temperature the ipsilateral limb was also disturbed however. The most striking disability was that he was unable to guide this limb (his left) into the contralateral visual field accurately and tended to ignore the contralateral hemifield completely, in a manner reminiscent of humans with parietal lobe lesions (Brain, 1941).

In the first monkey however it was impossible to separate the effects of cooling area 5 from area 7; nor were the effects quantifiable. So in two further monkeys separate cooling plates were placed over areas 5 and 7 and the animals were trained in reproducible tasks in order to measure the deficits at the various stages of cooling. One monkey was trained in visual tasks - a visual tracking game and a button pushing one: the other was trained in a tactile discrimination task.

The 'visual' monkey was first trained to track a circular target displayed in front of him on an oscilloscope, with a lever held in his hand. The lever's position was signalled by a spot on the screen. If he held the spot inside the target circle by matching his movements of the lever to those of the target for about 2 seconds he received a reward of apple puree. So by appropriate movements of the target we could get the monkey to move either contralateral or ipsilateral limb anywhere we liked at any speed. After he was fully trained, the insertion of separate cooling plates over areas 5 and 7 of the left parietal lobe had no noticeable effect on his ability to perform the task. Mild cooling of area 5 (to 25°C) had no effects on his ability to track with either arm. Mild cooling of area 7 at first sight appeared to have no effects either. However it soon became clear that the monkey was having difficulty making rapid movements into the visual field contralateral to the cooling plate, as in Fig. 3. Movements into this field were slower, jerkier and more inaccurate than into the ipsilateral field.

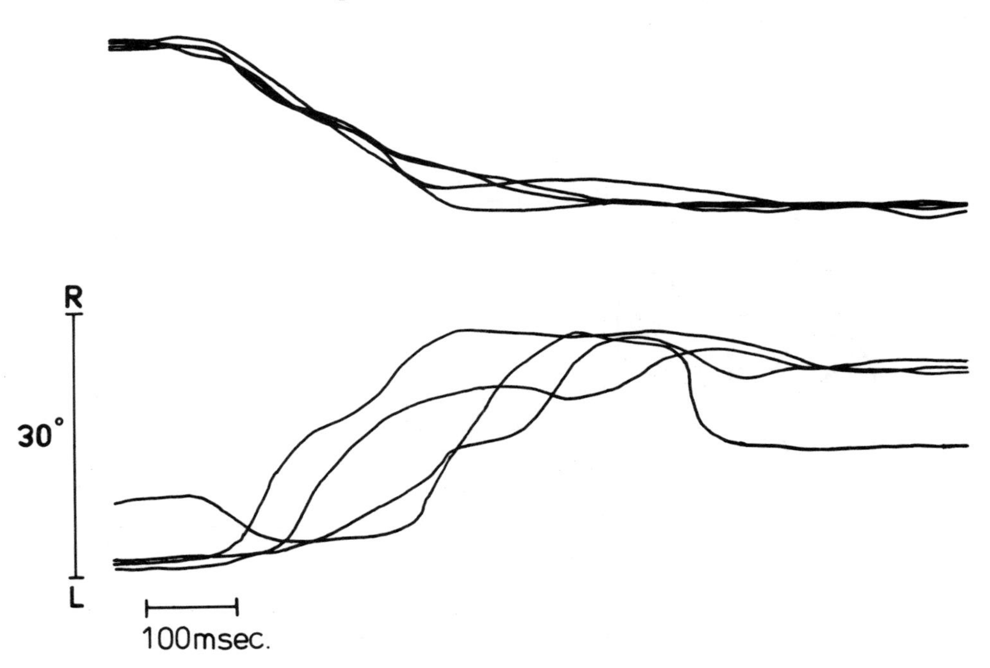

Figure 3. Cooling L. area 7 to 25°C. Upper trace : normal movements following target which jumped rapidly from right to left side of visual field at beginning of time record. Lower trace : movements following target jump from left to right, slower and more inaccurate.

Slow movements were unaffected. This was probably because when the monkey was following a slowly moving target he was able to keep it on his fovea by appropriate eye movements; whereas when the target jumped rapidly into the contralateral hemifield he had to generate an appropriate motor command for the limb to move into the contralateral hemifield before the eyes moved.

In order to check the hypothesis that visuomotor coordination in the contralateral hemifield was disrupted by cooling area 7, we quickly trained this monkey in a rather different task, designed to test his ability to control movements in either the contralateral or ipsilateral visual fields. The monkey initiated each trial by pressing a lighted button in the centre of a display of 7 (1 in the centre, 3 to the left at 10°, 20° and 30° from the centre, and 3 to the right (see Fig. 4)). So he was always looking straight ahead at the beginning of each trial. As soon as he pressed the centre button it went out and any one of the 6 buttons in the periphery, selected randomly, would light. He then had to press that button within 1.5 seconds in order to receive a reward. The average time taken to press each button and the number of times a given button was ignored were measured before, during and after cooling area 7 and area 5. During cooling area 7 to 25°C the average time taken by the monkey to hit buttons with his contralateral hand in the contralateral half field was very much longer than in the ipsilateral field. This difference was particularly marked for the outer lights (Fig. 4).

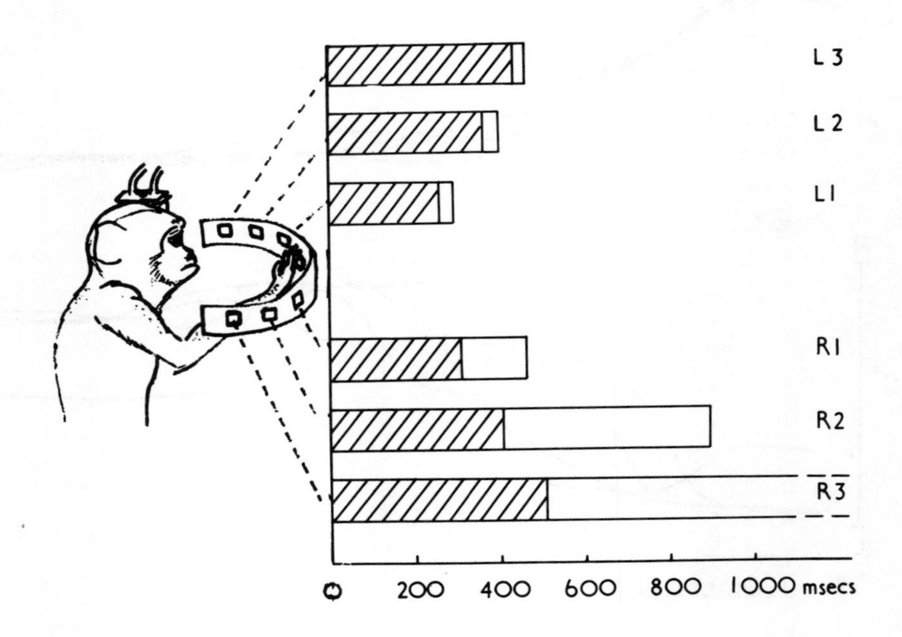

Figure 4. Cooling L. area 7 to 25°C. Average time taken to hit lighted buttons. Shaded area - before cooling. Note greatly increased reaction time to lights in right visual field.

Not only was the monkey slower on this side, but in many of the trials he ignored the light on the contralateral side altogether. Further, when he did respond on the contralateral side he very consistently first misreached medial to the target, and then readjusted by feeling the appropriate button. Again this is very reminiscent of some of the clinical reports for posterior parietal lobe lesions.

Cooling area 5 however led to general clumsiness of the contralateral limb and longer reaction times in both visual fields - not confined to the contralateral field. The animal gave the impression of not knowing where his hand was, knocking into objects and making inappropriate corrections having done so, as did the parietal monkeys reported by Ratcliffe et al (1977) when feeling in the dark.

More profound cooling of area 7 (to a plate surface temperature of $0^{\circ}C$; probable cortical surface temperature around $10^{\circ}C$) extended the visuomotor incoordination to the ipsilateral limb. So the left limb was now inaccurately directed into contralateral visual space as well. However at this temperature it was very difficult to get the monkey to work with the right limb at all in order to compare their disabilities.

The fact that the monkey sometimes ignored the lighted button altogether raised the possibility that he simply could not see it; for instance that the cooling had spread to involve the optic radiations. So this same monkey was trained in a third task. He had to initiate a trial by pressing the central button as before; but now he had only quickly to press a button below and slightly to the left if he saw a light in the left visual field and a button below and to the right if he detected a light in the right visual field. So he had only to register that he had seen the light; but did not have to reach into and control limb movements in the contralateral visual fields. Under these circumstances cooling area 7 did not prolong his reaction time for either left or right sided lights. So we can conclude that the deficit caused by cooling area 7 was not one of blindness in the contralateral field, but of the visual control of movement in that half field. This involved the contralateral arm mainly, but affected the ipsilateral arm also during severe cooling. Cooling area 5 reduced the dexterity of the contralateral limb but this was not confined to movements in the contralateral visual field.

These results therefore support the idea that area 7 is responsible for visuomotor control, perhaps by an association of the animal's awareness of the position of his limb with visual information about the position of a desired target. However the evidence about the role of area 5 was less strong. During area 5 cooling the monkeys seemed altogether clumsier, as though they did not know precisely where their contralateral arm was; and, they could not tell by touch alone the nature of objects placed in that hand. But these observations remained rather anecdotal and qualitative.

We have therefore trained a third animal in a tactile discrimination task, to obtain more quantitative data. The monkey was presented with two boxes, side by side, which he could not see. One had a rough knob with 1 mm deep grooves with which to open it; the other a smooth knob. He was trained to feel the knobs and to open only the box with the rough one in order to receive a reward. The "rough" and "smooth" boxes were presented on the left or right side in random sequence, and there were no other cues to tell him which was which. Within a few days the monkey selected the rough knob on 95% of trials. Area 5 and 7 coolers were then implanted, and the monkey retested. His correct response rate had not changed. During cooling of the area 5 plate to 25°C however, though he appeared to be still trying to make the discrimination, his success rate dropped to only c. 60% and his manipulations were clearly more clumsy. In addition if given the chance he would bring his ipsilateral hand to the aid of the other. If he was now allowed to see the boxes his ability to retrieve the food was much improved. During cooling of the area 5 plate to 0°C he became quite unable to perform the task at all with the contralateral hand, both because he could not make the discrimination (his hand wandered all over the boxes without any apparent guidance towards the knobs) but also because he was unable to perform the manipulations necessary to get the boxes open even with the aid of vision. The ipsilateral hand was unaffected.

Cooling area 7 however led to none of these disturbances. It appeared therefore that activity of area 5 is necessary for the manipulations and tactile discriminations of the contralateral hand. It might be suspected however that the contralateral hand was simply anaesthetised perhaps by spread of cooling to SI, a somewhat uninteresting result. However this is unlikely to be the full explanation for the effects for three reasons. First the hands were not completely anaesthetised, as the animal would always withdraw his hand if pressed or pinched. Second, mild cooling to 25°C would not have cooled cortex further than 2 mm in front to any appreciable extent. Yet it hindered the monkey's discriminations. Third, the monkey's deficits did not suggest SI ablation which causes at first almost as dramatic an effect on movement as ablation of the motor cortex. Rather area 5 cooling caused a restricted disturbance of cutaneously guided movements. The limb was as active as before, but no longer controlled by somaesthetic sensation.

If the reader accepts these rather different roles for areas 5 and 7, the next questions are anatomical. What are the inputs to these regions? First how does the information from cutaneous proprioceptive sources reach areas 5 and 7? Second how does information about the activity of motor centres reach them? Third where does the information about visual targets come from? The answer to the first question is relatively simple. SI projects convergent signals to area 5 (Jones & Powell, 1970) and area 5 passes this on to area 7. The answer to the second question is conjectural. One source of motor signals may be by connexions from motor cortex via sensory cortex and thence to area 5. Another may be from premotor areas. An extremely important one is from the cerebellum (Sasaki et al, 1972). The third question about the visual input to area 7 is also difficult to answer. It is uncertain whether there are significant projections from striate and prestriate areas to area 7; and in any case if there are they must be "gated" in some way by other inputs, since the activity of neurones in area 7 depends not just on the position of an object in the visual field but upon the significance of that visual target to the monkey (Mountcastle et al, 1975). There is also an extensive input to area 7 from the tectopulvinar system whose significance is unclear. Again the cerebellar input must be considered.

It is even more important to know what the outputs from these areas are. The major ones appear to be to the premotor cortical fields, but not to the motor cortex itself; to the basal ganglia; to the cerebellum; and into the pyramidal tract. The corticocortical pathway is not essential for visually controlled limb movements (Myers & Sperry, 1962) though it may be for visual control of individual finger movements (Haaxma & Kuypers, 1975). The route to the basal ganglia may not be important for visually controlled movements either, since Parkinsonian patients are often noted for their ability to compensate for their motor disturbances by making use of visual control (Denny Brown, 1962).

So probably the parietal outputs to the cerebellum and directly to the spinal cord are essential for the visual control of limb movements (Stein, 1977). However, we do not know if similar arguments apply to the somaesthetic control of movements. This invites further examination.

ACKNOWLEDGEMENTS

My thanks are due to Guy Goodwin with whom these experiments were first discussed and to Woody Caan and Stephen Gilbey who assisted with them.

REFERENCES

W. Bentley, The synthetic experiment, Am. J. Psychol. 11, 405-425 (1900).

W. R. Brain, Visual disorientation with special reference to lesions of the right cerebral hemisphere, Brain 64, 244-212 (1941).

A. W. Caan & J. F. Stein, The effect of cooling the motor cortex of a rhesus monkey upon visually controlled movement and associated motor potentials, J. Physiol. 263, 142-143 (1977).

Critchley, M. (1952) The Parietal Lobes, Arnold, London.

Denny Brown, D. (1962) The basal ganglia, Oxford University Press, London.

D. Denny Brown & R. A. Chambers, The parietal lobe and behaviour, Res. Publs. Ass. Res. nerv. ment. Dis. 36, 35-117 (1958).

G. Ettlinger & J. E. Kalsbeck, Changes in tactile discrimination and in visual reaching after successive and simultaneous bilateral posterior parietal ablations in the monkey, J. Neurol. Neurosurg. Psychiat. 25, 256-268 (1974).

J. J. Gibson, Active Touch, Psychol. Rev. 69, 477-491 (1962).

R. Haaxma & H. G. J. M. Kuypers, Intrahemispheric cortical connexions and visual guidance of hand and finger movements in the rhesus monkey, Brain, 98, 239-260 (1975).

W. Hartje & G. Ettlinger, Reaching in the light and dark after unilateral posterior parietal ablations in the monkey, Cortex 9, 346-349 (1974).

H. Hecaen, W. Penfield, C. Bertrand, & R. Malmo, The syndrome of apractognosia due to lesions of the minor cerebral hemisphere, Arch. Neurol. Psych. 75, 400-434 (1956).

J. Hyvärinen & A. Poranen, Function of parietal association area 7 as revealed from cellular discharges in alert monkeys, Brain 97, 673-692 (1974).

E. G. Jones & T. P. S. Powell, An anatomical study of converging sensory pathways within the cerebral cortex of the monkey, Brain 93, 793-820 (1970).

J. C. Lynch, V. B. Mountcastle, W. H. Talbot & T. C. T. Yin, Parietal lobe mechanisms for directed visual attention, J. Neurophysiol. 40, 362-389 (1977).

V. B. Mountcastle, J. C. Lynch, A. Georgopoulos, H. Sakata & C. Acuna, Posterior parietal association cortex of the monkey: Command functions for operations within extra personal space, J. Neurophysiol. 38, 871-908 (1975).

R. E. Myers & R. W. Sperry, Neural mechanisms in visual guidance of limb movement, Arch. Neurol. Psych.7, 41 (1962).

T. L. Peele, Acute and chronic effects of parietal lobe ablation in monkeys, J. Neurophysiol. 7, 269-286 (1944).

G. Ratcliffe, G. Ridley & G. Ettlinger, Spatial disorientation in monkeys, Cortex 13, 62-65 (1977).

K. Sasaki, Y. Matsuda, S. Kawaguchi & N. Mizumo, On the cerebello-thalamo-cerebral pathways for the parietal cortex, Exp. Br. Res. 16, 89-103 (1972).

J. F. Stein, The effect of cooling parietal lobe areas 5 and 7 upon voluntary movement in awake rhesus monkeys, J. Physiol. 258, 62-63P (1976).

J. F. Stein, The effects of transient cooling of parietal cortex and cerebellar nuclei during tracking tasks, in Prog. Clin.Neurophysiol. Vol. 4, Ed. J. Desmedt, Karger, Basel. (1977).

J. Wattam-Bell & J. F. Stein, The effect of cooling N. interpositus in rhesus monkeys on the tracking of a visual target, J. Physiol. 252, 47-48P (1975).

SHORT-LATENCY PERIPHERAL AFFERENT INPUTS TO PYRAMIDAL AND OTHER NEURONES IN THE PRECENTRAL CORTEX OF CONSCIOUS MONKEYS

R. N. Lemon* and R. Porter

Department of Physiology, University of Sheffield, U.K. and Department of Physiology, Monash University, Australia

SUMMARY

Pyramidal tract and other neurones in area 4 of the conscious relaxed monkey were found to have small, stable input zones on the contralateral limb. These neurones were found to be particularly sensitive to joint movement. Short-latency (<20 msec) responses for neurones with cutaneous and muscle input zones suggest a fast pathway to the motor cortex from the periphery. Preliminary observations on the responsiveness of these neurones during voluntary movements suggest that transmission in this pathway is considerably modified during such movements.

INTRODUCTION

There has been considerable interest in the nature of the afferent inputs to the motor cortex ever since the suggestion by Phillips in 1969 that these inputs might be important for control of fine voluntary movements. In the anaesthetised and sedated animal, neurones in area 4 of the cortex were relatively difficult to influence from the periphery. In the conscious monkey, however, the majority of neurones (over 80%) were shown to have clear afferent input zones on the contralateral limb from which stable and reproducible responses could be obtained (Fetz and Baker, 1969; Lemon and Porter, 1976a,b). Within the arm region of area 4, both pyramidal tract neurones (PTN) and unidentified neurones, whose discharge was clearly related to some aspect of voluntary movement performed by the monkey, received afferent inputs from the contralateral arm and hand and these inputs usually came from parts of the limb involved in the voluntary movement with which the neurone's change in discharge frequency was associated (Lemon, Hanby and Porter, 1976).

The pathways providing this specific afferent input to the motor cortex are as yet undetermined. One helpful clue would be the response latency of these neurones. Unfortunately, joint movement is the most effective adequate stimulus for these neurones, and it is difficult to time the onset of this stimulus accurately, particularly if the neurone only responds over a limited range of joint movement (Lemon and Porter, 1976a). Evidence, is presented, however, for a short-latency input to neurones with cutaneous or muscle input zones.

Since many neurones in area 4 send their axons to the thalamus, dorsal column nuclei and dorsal horn of the spinal grey matter, it is quite probable that the motor cortex plays a role in the control of sensory input, and this would

*R.N. Lemon, Afdeling Anatomie, Erasmus Universiteit Rotterdam, Faculteit der Geneeskunde, Postbus 1738, Rotterdam, The Netherlands.

be of particular importance during tactile exploration. This paper presents some preliminary evidence on the effects of voluntary movements on the responsiveness of neurones in area 4 to peripheral inputs.

METHODS

In 5 conscious monkeys, neurones were recorded with tungsten microelectrodes in the arm region of area 4 in the left cortex. The techniques have been fully described previously (Lemon and Porter, 1976a). The monkeys were trained to perform a stereotyped movement task (accurate pulling of a lever) with the right arm. They were also trained to accept passive manipulation and stimulation of the limbs without struggling. Electromyographic (EMG) recordings were taken from selected arm muscles, and in 3 monkeys stimulating electrodes were successfully implanted in the medullary pyramidal tract for antidromic identification of PTNs. In 2 monkeys, a silastic cuff containing an array of electrodes was chronically implanted on the right ulnar nerve at the wrist.

Response latencies to natural stimulation were determined using a touch-sensitive probe which gave an accurate indication of the time of contact with the input zone. Responses to at least 50 stimuli were recorded, and statistical analysis carried out on peri-stimulus histograms. Response latencies were taken at the first post-stimulus bin containing sufficient discharges to make it significantly different (at the $P < 0.001$ level) from the mean pre-stimulus discharge activity.

RESULTS

Afferent Input to Area 4 Neurones

The total number of fully-tested area 4 neurones was 390, including 69 identified PTNs. All these neurones showed a marked change in discharge frequency during the performance of the voluntary task. This was established by constructing an on-line peri-response histogram of the cell's activity. An example is shown in Fig. 5.

Once the pattern of the neurone's activity in relation to voluntary movement was established, its response to peripheral stimulation was tested in the relaxed, passive monkey. EMG recordings were continuously checked for quiescence during the testing period. 324/390 neurones (83%) responded to natural stimulation. Most of these neurones had small stable input zones on the contralateral limb. Many neurones, including PTNs could be powerfully influenced by such stimulation. Fig. 1 shows the activity of a PTN with a consistent discharge just prior to the monkey grasping the lever and pulling it, and a second discharge several hundred msec later, after lever release and just before the collection of a food reward in a position which forced the monkey to flex his wrist. The peak discharge frequency of this neurone was about 100 Hz. Similar frequencies were observed in the fully relaxed monkey when the wrist was flexed by the experimenter.

The majority of responsive neurones (227/324 or 70%) were influenced by joint movement. 171 neurones responded to movement of a single joint, and

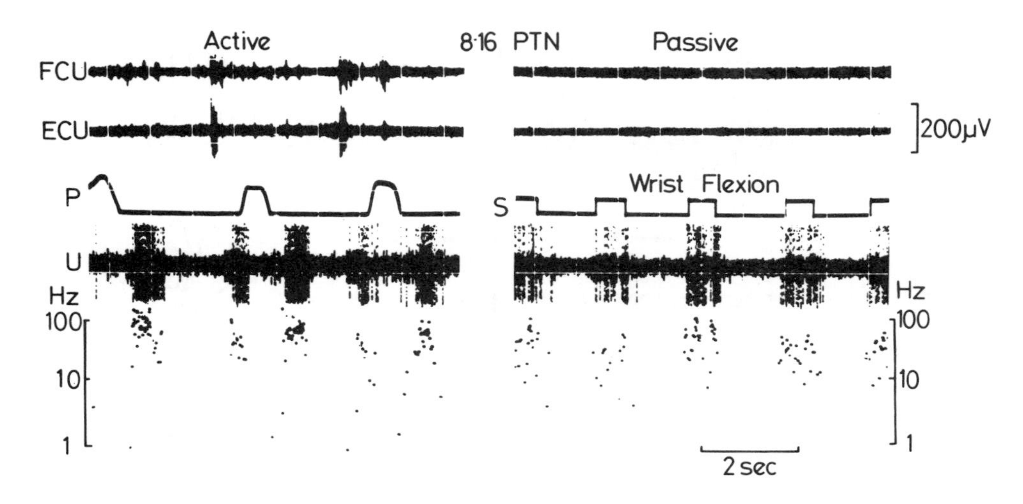

Fig. 1. Comparison of discharge frequencies for a PTN under active conditions (left) when the monkey was pulling the lever, and under passive conditions (right) when a stimulus was imposed upon the passive monkey. Records are, from top to bottom: EMG records from flexor carpi ulnaris (FCU), extensor carpi ulnaris (ECU), potentiometeric indication of lever movement (P), microelectrode recording (U) and the output of a logarithmic display module monitoring the frequency of the neurone, U. The signal, S on the right in this and subsequent figures gives an approximate indication of the duration of the imposed stimulus, in this case wrist flexion.

of these, 154 were only influenced by a single movement at that joint (e.g. wrist flexion, but not wrist supination). These results reflect the small size of the afferent input zones. Nearly all the neurones were directionally sensitive. Thus the neurone illustrated in Fig. 2 was excited by extension but not flexion of the wrist joint. The neurone in Fig. 1 provides a further example. Those neurones with a tonic discharge pattern were excited by joint movement in one direction and inhibited by movement in the opposite direction.

Of the 227 neurones influenced by joint movement, 196 were only affected by dynamic movement of the joint and not by any one static joint position. The remaining 31 neurones did show this latter feature, usually at one extreme of the range of joint motion. A typical example was a neurone with a high tonic discharge frequency with the elbow held at full flexion, a very low discharge frequency at full extension and inhibition and excitation during dynamic extension and flexion respectively.

It was sometimes found that the response of a neurone to movement of a joint could be modified by the position of adjacent joints, which were not themselves

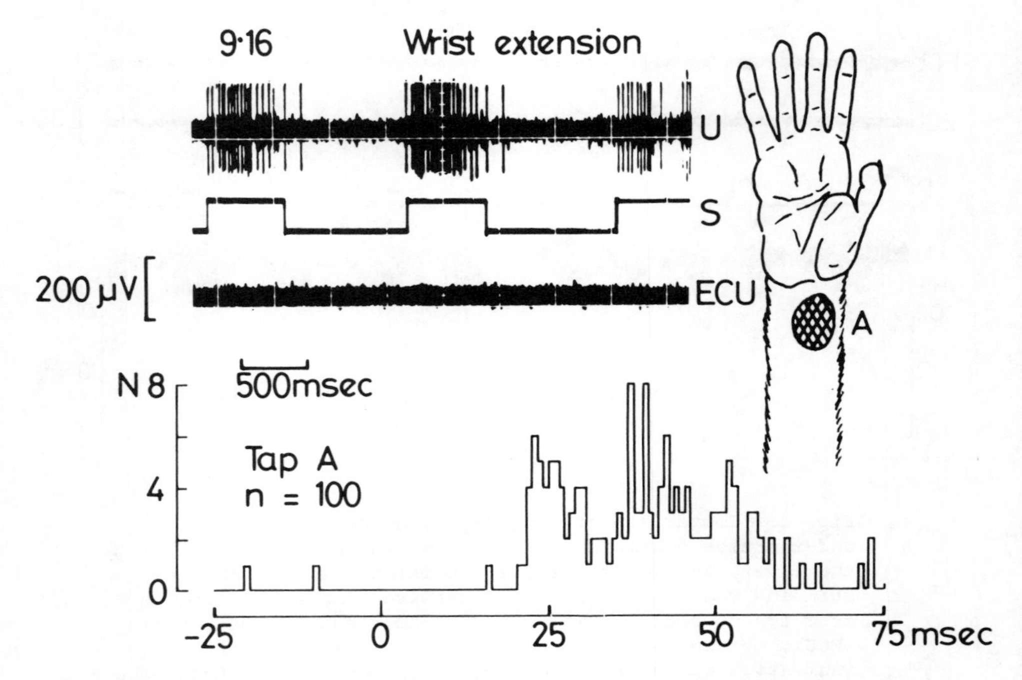

Fig. 2. Short-latency responses in a neurone with an input zone shown on the right (A) and influenced both by muscle taps and by extension of the wrist joint (top). The peri-stimulus histogram shows the number of responses (N) recorded 25 msec before and 75 msec after 100 brief taps were applied to the input zone at time zero.

included in the input zone. Thus the position of the wrist joint often influenced responsiveness to elbow movements in neurones unaffected by wrist movements alone. This could indicate a necessity to provide appropriate conditions of adjacent joints for movement of one of them to stretch particular muscles and activate particular muscle receptors. Alternatively, particular positions of adjacent joints could allow the movement to stretch sensitive parts of joint capsules, ligaments or fascial planes.

The discharge of all 227 joint-sensitive neurones was unaffected by palpation, prodding or squeezing the muscles acting at the joint. However a further 59 neurones did respond to a brief tap applied to a limited portion of the monkey's limb. Since light touch, sliding the skin over the underlying tissues and hair movement were ineffective stimuli for these neurones it was concluded that the stimulus was affecting muscle receptors. 41 of these neurones were also excited by the movement of the joint at which the muscle acted. In most cases the joint had to be moved in a direction that stretched the muscle containing the neurone's afferent input zone. An example is shown in Fig. 2. This neurone was excited by taps to the muscle bellies of the flexor aspect of the forearm and also by extension of the wrist joint which would have stretched these muscles. The remaining 18/59 neurones were not influenced by joint movement.

Cutaneous input zones were found for only 38/324 responsive neurones (12%). Cutaneous inputs thus represent only a small fraction of the afferent feedback to precentral neurones in the monkey. 28 of these neurones had input zones of less than 5 cm^2 in area, and 14 less than 2 cm^2. Most of these neurones had zones on the palmar surface of the hand and many were also influenced by movement of adjacent finger joints (for instance the neurone in Fig. 5). Stocking-type zones involving the whole arm surface were only found for 2 neurones. Only moving stimuli within a neurone's input zone were effective and once the probe came to rest, no clear response was obtained. There was no evidence for any directional sensitivity within the cutaneous input zones.

Response Latencies

Brief taps were applied with a touch-sensitive probe to the centre of the input zones of 34 neurones responding to a muscle tap and 14 neurones with cutaneous input zones. Each neurone characteristically gave a burst of several impulses following each stimulus. These stimuli were delivered at random intervals,

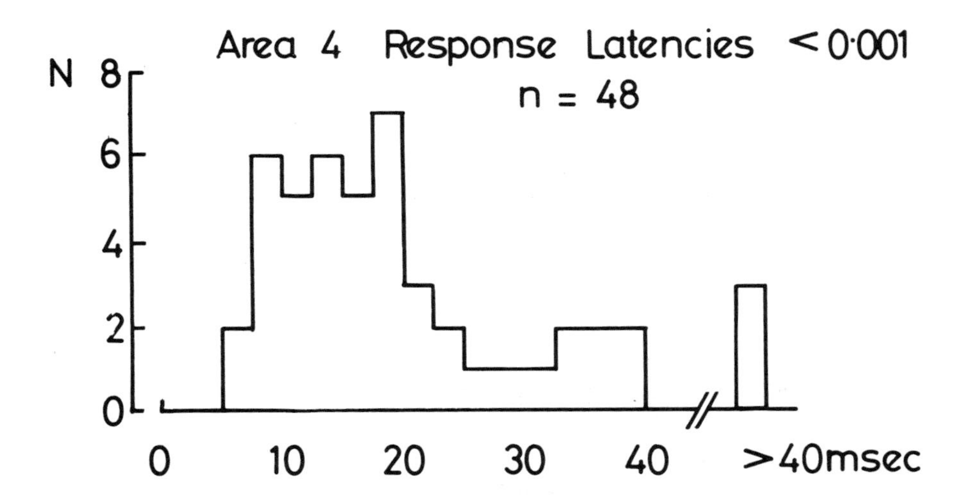

Fig. 3. Response latencies of 48 area 4 neurones to brief taps applied to the contralateral arm and hand. Latencies were taken at $P < 0.001$ level (see text).

with the shortest interval at 0.5 sec. Where possible, the limb was positioned so that the monkey could not see the application of the probe. The monkey remained relaxed throughout most stimulation periods; any responses recorded during struggling movements or periods of EMG activity were ignored.

Figure 2 shows a typical peri-stimulus histogram for a phasic neurone responding to a muscle tap. Note the low level of pre-stimulus activity. Figure 3 summarises the latencies obtained at the $P<0.001$ significance level for all 48 area 4 neurones. It can be seen that most of the sample had latencies of less than 25 msec, including some responding in 10 msec or less. In preliminary experiments 3 PTNs were found with very short latencies of 6, 8 and 9 msec (Lemon and Porter 1976a). More recent experiments have yielded very few PTNs with muscle or cutaneous inputs, and it has not been possible to confirm these short latencies. A further 4 PTNs have been found with latencies of 15-20 msec, and 1 PTN in the present sample had a latency of 53 msec. While no information is available on the latency for joint-sensitive neurones, the faithful and reproducible pattern of their responses to joint movement suggests a direct pathway which might also yield short-latency responses.

Effect Of Voluntary Movement On Neuronal Responsiveness

Since all the neurones in this sample showed clear changes in discharge

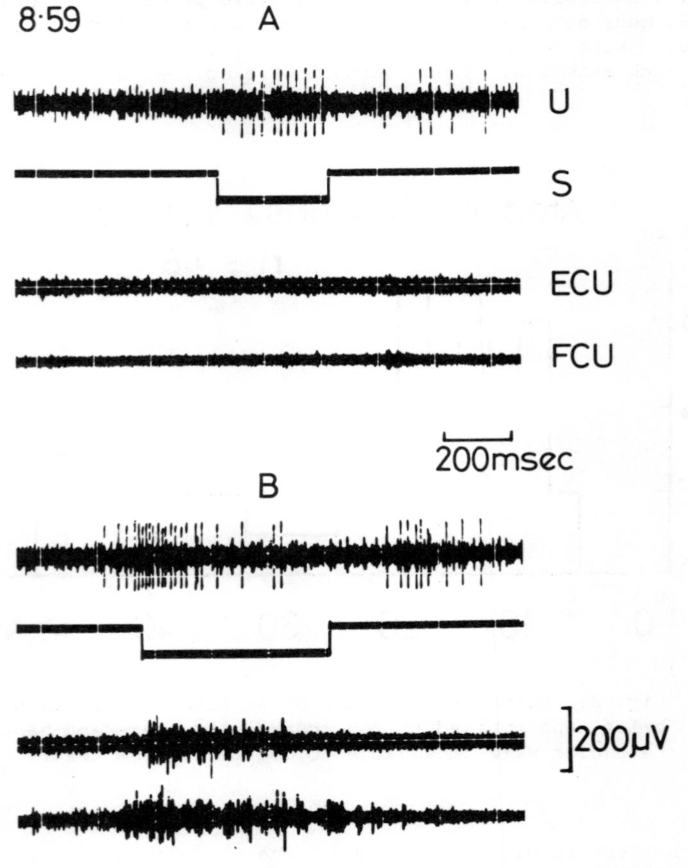

Fig. 4. Responses in an area 4 neurone to wrist flexion in the relaxed monkey (A) and a few seconds later when the monkey resisted the imposed movement (B).

frequency during voluntary movement, it would not be surprising to find a change in responsiveness to a stimulus applied by the experimenter during periods of active movement by the monkey. This point demonstrates the importance of working with a co-operative and fully relaxed animal in order to determine the true responsiveness to imposed stimuli. By comparison with this relaxed state, most neurones showed increased responsiveness to such stimuli during voluntary movement if the animal resisted the imposed movement. In Fig. 4 approximately similar flexion movements of wrist produced a much more intense response from the neurone illustrated when the monkey resisted the imposed movement by co-contraction of muscles acting at the wrist (Fig. 4B) than was found in the relaxed monkey, in which both EMGs were quiescent (Fig. 4A). A further complicating factor was that during active movements some neurones apparently showed a responsiveness to stimuli applied outside the input zone determined under passive, relaxed conditions. Thus the neurone illustrated in Fig. 4 also responded to elbow movements when the monkey was tense and resisting the experimenter. Such responses were not reproducible, which contrasts strongly with the findings in the relaxed monkey.

As stated above, the great majority of responsive neurones (over 90%) had stable input zones in the relaxed monkey. A small proportion behaved in a more labile fashion, and did not show reproducible responses. Many of these

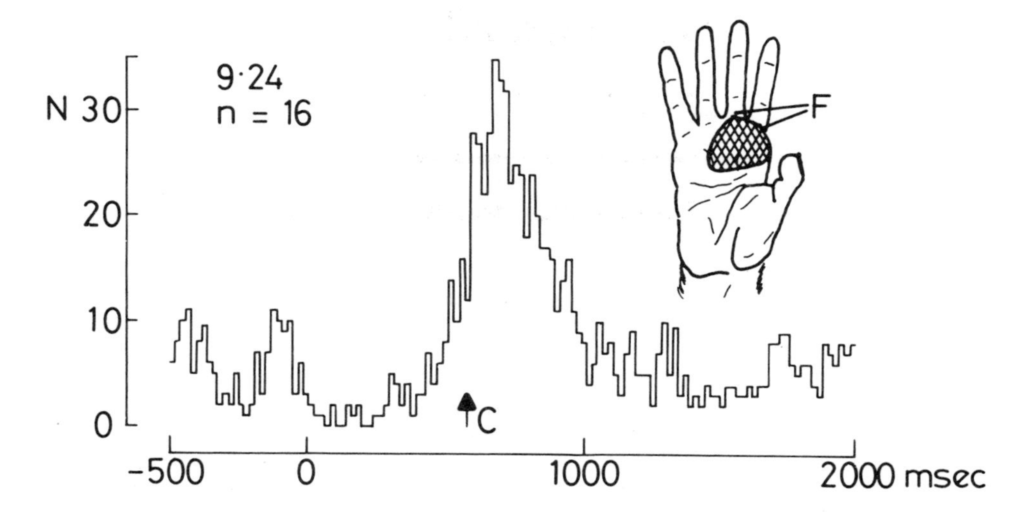

Fig. 5 Peri-response histogram for a neurone responding to light touch applied to the input zone illustrated, and to flexion (F) of the metacarpophalangeal joints of the index and middle digits. The neurone showed a burst of activity just prior to the beginning of lever movement (time zero). A much greater discharge was seen about 600 msec later during the collection and handling of the food reward. The arrow at C indicates the earliest time after lever movement that the monkey's hand made contact with the food reward.

neurones demonstrated attention-like behaviour. Thus they responded when the monkey was attending to the experimenter imposing the stimulus upon him (Fig. 6A), but the responses were often diminished when the monkey was distracted with a food reward to the opposite limb (Fig. 6B).

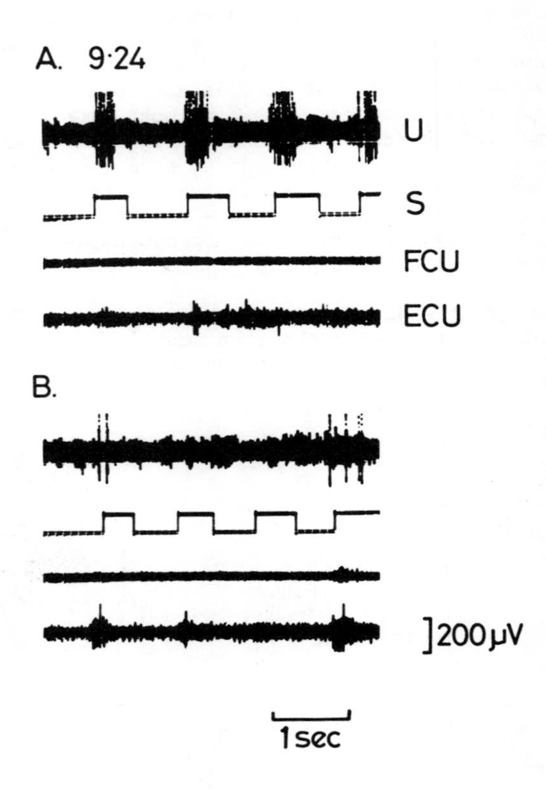

Fig. 6. Responses in the neurone illustrated in Fig. 5 to repeated flexion of the index finger (indicated by the signal S). Consistent responses were obtained when monkey was attending to the experimenter (A), but were diminished when the monkey was distracted by a food reward to the left limb (B).

In comparing responsiveness under 'active' and 'passive' conditions it is important to impose the same peripheral stimulus under both sets of conditions. For this reason electrode cuffs were implanted on the ulnar nerve (UN) at the wrist in 2 monkeys. In these experiments it was found that single, 50 µsec shocks at voltages about 2.5 times greater than that needed to excite a Gp I volley (range 3-5V) were sufficient to excite 9 area 4 neurones. All these neurones had input zones to natural stimulation near or at the wrist, or on the ulnar side of the hand. A further 6 neurones were found in these monkeys with zones on the radial side of the hand, and these were completely unresponsive to UN shocks, again confirming the small input zones in the conscious monkey. This contrasts with a more convergent pattern described by Wiesendanger (1973) in the anaesthetised primate.

Figure 7 and 8 are peri-stimulus histograms for two of the UN-responsive neurones. Figure 7 shows a neurone excited at short-latency (13 msec) by UN

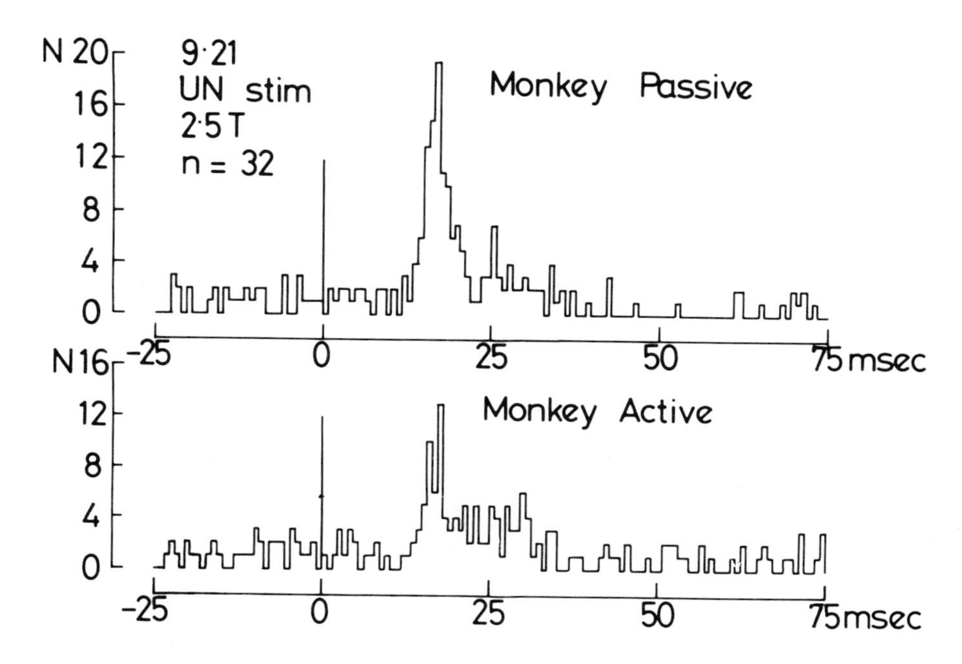

Fig. 7. Peristimulus histogram showing responses of an area 4 neurone to 32 successive shocks (duration 50 μsec, strength 2.5 x threshold (T) for eliciting a Gp. I volley) applied at random intervals to the right ulnar nerve (UN). UN shocks were delivered to the passive, relaxed monkey (top histogram) and immediately after when the monkey was actively pulling the lever etc. (bottom histogram).

stimulation. Stimuli were delivered randomly to the passive monkey, and then further responses were recorded using the same stimulus voltage during a period when the monkey was carrying out active movements and pulling the lever, collecting food rewards etc. Once again these stimuli were delivered randomly and were not related to any one phase of the task. All 9 neurones showed a diminished responsiveness to UN stimulation during such active movements. These effects have been seen in neurones with both phasic and tonic patterns of discharge. The former type showed only a small increase in the pre-stimulus activity found in active as compared with passive conditions (Fig. 7) while the latter type showed a marked increase in pre-stimulus activity (Fig. 8) reflecting the high discharge rates seen in these cells during voluntary movements.

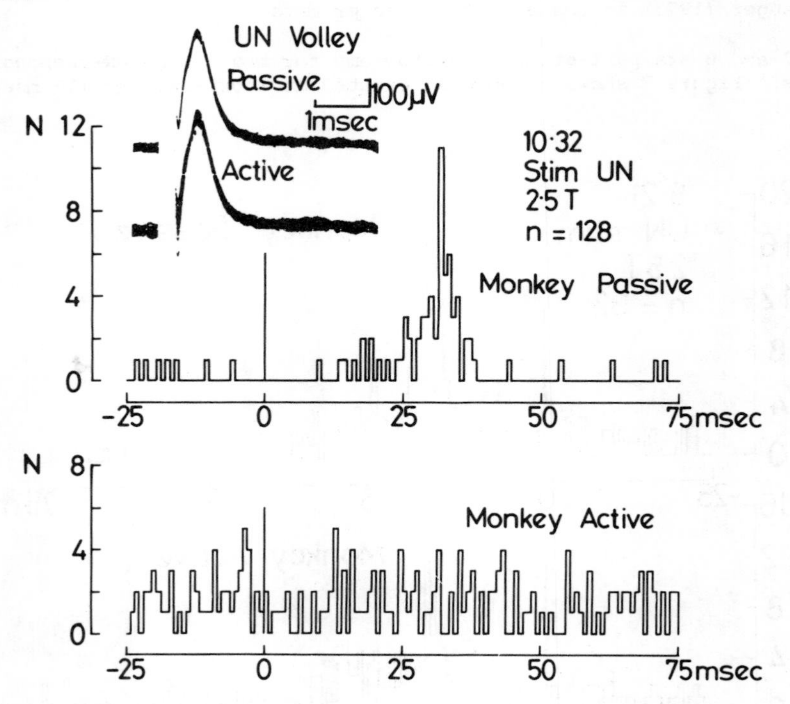

Fig. 8. Similar histograms (bottom) to those in Fig. 7. The inset shows the volley recorded from the UN (20 superimposed responses) that was set up by shocks of 2.5T applied to the nerve, under passive (top trace) and active (bottom trace) conditions.

In one monkey the UN volley set up by the stimulating electrodes was monitored by a pair of recording electrodes in a neighbouring cuff. This procedure showed that the UN volley was similar under both sets of conditions, although the responses of the precentral neurone differed greatly (Fig. 8).

DISCUSSION

The afferent input to neurones in area 4, including PTNs, comes from small specific input zones on the contralateral limb. The behaviour of these neurones is stable under passive conditions. The information reaching particular neurones concerns the direction of joint movement, muscle stretch and to a lesser extent, motion of objects over the tactile portions of the monkey's limb.

Many neurones receive this information at short latency, and this suggests a fast, direct pathway from the periphery to motor cortex. A trans-cerebellar route would seem to be an unlikely candidate for such a pathway, both because of the short-latencies and because of lack of evidence for involvement of n. ventralis lateralis (VL) of the thalamus. Neurones in VL appear to be unresponsive to peripheral stimuli in both the anaesthetised cat (Nyquist, 1975) and conscious monkey (Strick, 1976a). Since area 4 does not receive a direct projection from the 'lemniscal' portion of n. ventralis posterior lateralis (VPL_c), a cortico-cortical projection seems most likely, although there is other evidence that argues against such a route (Strick, 1976b). Further experiments on thalamic neurones especially in the transitional zone, VPL_o, in relaxed monkeys may help to solve this problem.

The latencies determined in this study seem to be considerably shorter than those described by Evarts and Tanji (1976). Their sample of 18 PTNs had latencies to a perturbation of movement introduced by a pulsed torque motor ranging from 20-48 msec, with a mean of 32 msec. Our sample of PTNs is too small for useful comparison but some responses were found earlier than 20 msec and the two methods of estimating the latencies are not strictly comparable. A sudden jerk to the limb during movement may produce a very complicated input quite unlike that provided exactly within the input zone of the passive animal's limb. Complicated interactions may occur in the cord and at higher levels between joint and muscle afferents caused to fire by the jerk. The effect may also be less synchronous for afferent fibres than for the tap stimuli used in these experiments.

The relationship between the afferent input zone of a particular area 4 neurone and its activity during voluntary movement has been explored by Lemon, Hanby and Porter (1976) and Lemon (1977). As far as tactile inputs are concerned the results confirm those of Rosén and Asanuma (1972) in that neurones with cutaneous zones on the palm of the hand were active during voluntary flexion of the digits, a movement that would have advanced the input zone towards the object. In the same way neurones with zones on the hand dorsum were active during voluntary opening of the hand. The "field of influence" of the peripheral receptors on precentral neurones was within or closely related to the "field of influence" of the precentral neurone on natural movement performance. Many neurones in area 4 have been shown by Evarts (1966, 1974) to be active before the onset of muscular activity, and also presumably before any afferent feedback is set up by this activity. For neurones with tactile inputs from the hand and digits, there is evidence of discharge before any contact with the input zone is achieved, although once this has occurred there is often a considerable increase in discharge frequency (Fig. 5). Such a system may be employed during exploratory behaviour for the reinforcement of movements yielding useful tactile information.

The nature of the afferent input during voluntary movements is difficult to assess. It seems clear that the powerful input to area 4 neurones (Fig. 1) is considerably modified during such movements. Thus neurones responsive to movement of a joint in one direction under passive conditions were often active during voluntary movement of the same joint in the opposite direction (Lemon, Hanby and Porter, 1976). There are also increases in responsiveness of some neurones during periods when the monkey resisted passive disturbances applied to his limb. For 9 neurones excited by stimulation of the contralateral ulnar nerve, there was a consistent suppression of responsiveness during movement, although the afferent volley was apparently unchanged. This finding underlines the differences that exist between the responses and response latencies of motor cortical neurones in the moving animal as compared to those found in the passive, relaxed animal.

These results further suggest that there is an important change in the functional role of afferent feedback to the motor cortex during movement. Once voluntary movement has begun, there is a greatly increased barrage of information reaching the CNS. Control over this input is exerted, in part, by the pathways descending from the motor cortex to different levels of the afferent system. One might expect these pathways to exert their greatest influence on incoming information just prior to and during movement. Further experiments are needed to assess the nature and direction of the changes in precentral responsiveness during voluntary acts.

ACKNOWLEDGEMENTS

We would like to thank Julie Cookson and Beverley Uttley for their expert assistance and to acknowledge a generous research grant to R. N. Lemon from the National Fund for Research into Crippling Diseases.

REFERENCES

Evarts, E. V. Pyramidal tract activity associated with a conditioned hand movement in the monkey. J. Neurophysiol. 29, 1011-1027 (1966).

Evarts, E. V. Precentral and postcentral cortical activity in association with visually triggered movement. J. Neurophysiol. 37, 373-381 (1974).

Evarts, E. V. & Tanji, J. Reflex and intended responses in motor cortex pyramidal tract neurons of monkey. J. Neurophysiol. 39, 1069-1080 (1976).

Fetz, E. E. & Baker, M. A. Response properties of precentral neurons in awake monkeys. Physiologist, 12, 223 (1969).

Lemon, R. N. Activity of neurones in the precentral gyrus during active and passive movements in the conscious monkey. 6th Congr. Internat. Primatol. Soc. (in press) (1977).

Lemon, R. N. & Porter, R. Afferent input to movement-related precentral neurones in conscious monkeys. Proc. R. Soc. Lond. B. 194, 313-339 (1976a).

Lemon, R. N. & Porter, R. A comparison of the responsiveness to peripheral stimuli of precentral cortical neurones in anaesthetised and conscious monkeys. J. Physiol. 260, 53-54P (1976b).

Lemon, R. N., Hanby, J. A. & Porter, R. Relationship between the activity of precentral neurones during active and passive movements in conscious monkeys. Proc. R. Soc. Lond. B. 194, 341-373 (1976).

Nyquist, J. K. Somatosensory properties of neurones of thalamic nucleus ventralis lateralis. Exptl. Neurol. 48, 123-135 (1975).

Phillips, C. G. The Ferrier Lecture. Motor apparatus of the baboon's hand. Proc. R. Soc. Lond. B. 173, 141-174 (1969).

Rosén, I. & Asanuma, H. Peripheral afferent inputs to the forelimb area of the monkey motor cortex: input-output relations. Exp. Brain Res. 14, 257-273 (1972).

Strick, P. L. Activity of ventrolateral thalamic neurons during arm movement. J. Neurophysiol. 39, 1032-1044 (1976a).

Strick, P. L. Anatomical analysis of ventrolateral thalamic input to primate motor cortex. J. Neurophysiol. 39, 1020-1031 (1976b).

Wiesendanger, M. Input from muscle and cutaneous nerves of the hand and forearm to neurones of the precentral gyrus of baboons and monkeys. J. Physiol. 228, 203-219 (1973).

MOTOR CORTEX RESPONSES TO KINESTHETIC INPUTS DURING POSTURAL STABILITY, PRECISE FINE MOVEMENT AND BALLISTIC MOVEMENT IN THE CONSCIOUS MONKEY

Christoph Fromm* and Edward V. Evarts**

*University of Düsseldorf, Düsseldorf, West Germany
**National Institute of Mental Health, Bethesda, Maryland 20014, USA

INTRODUCTION

Pyramidal tract neurons (PTNs) of monkey motor cortex exhibit "reflex" responses to peripheral somesthetic inputs with latencies as short as 20 ms. These reflex responses vary with the intensity and direction of the somesthetic input (1,2,3,4,5,6) and are thought to have a role in tactile exploration (7,8) and load compensation (9,10). In the present study we have sought to find out if the strength of cortical reflexes depends on the mode of operation of the motor system at the time the reflexes are elicited. The three modes in which we have tested reflex responsiveness are 1) postural stability, 2) precise fine movement, and 3) rapid ballistic movement.

It has already been shown that a variety of segmental reflexes (unloading, stretch, and tonic vibration) fail to modify the initial EMG pattern during a rapid movement, though such reflexes can be induced shortly before and during the latter part of rapid movement (11,12,13). The present investigation sought to detect an analogous change in responsiveness of precentral neurons. It was hypothesized that rapid ballistic movements might depend on central programs operating relatively "open loop," whereas small precisely controlled movements might utilize the continuous feedback provided by the "servo" properties of transcortical reflexes. Were this the case, motor cortex reflexes would be heightened during the closed-loop servo mode occurring for small precise movements, while these reflexes would be attenuated during the open-loop mode underlying ballistic movement. The results to be presented are consistent with this hypothesis.

METHODS

Experimental Paradigm

The procedures used have been described in two reports (14,15) on experiments designed to compare motor cortex activity with large and small movements. For these and for the present experiments monkeys (Macaca mulatta) were trained to hold a handle within a small zone and to rotate it by supination-pronation movements of the arm. They viewed two horizontal rows of lamps (track and target lamps), each lamp in the track row being directly above a corresponding lamp in the target row. At all times only one lamp in each row was on. Output from a position transducer coupled to the handle selected which of the track lamps was illuminated. While the track display informed the monkey of the actual handle orientation, the location of the target lamp was programmed by the experimenter and indicated how the handle should be positioned. The behavioral sequence established by operant conditioning methods involved:

1) A holding period of continuous correct alignment varying in duration from 1.5 to 2.5 s. The monkey held the handle vertically to align the central track with the central target lamp. In the figures which follow, a horizontal position (i.e., potentiometer) trace indicates steady holding, and the thickness of the horizontal line in superimposed position traces reflects the width of the hold zone (1.5°).

2) Small error-correcting supination-pronation movements of 1-2° occurred when a misalignment of track and target lamps was caused by the monkey's failure to keep the handle within the narrow hold zone. Figure 1F shows a series of pronation drifts corrected by subsequent supination movements carried out by the monkey in order to realign the track with the fixed target lamp. The same type of small corrective movements could also be triggered by a shift of the track display resulting from an error signal added to the output of the position transducer which controlled the track display. Addition or removal of this voltage step is seen as abrupt deflection in the position trace (for example, Fig. 1D), although there has been no actual movement of the handle. It resulted in a shift of the track display from the center lamp to the one on its right or left, thus requiring of the monkey a corrective movement in the opposite direction to realign track with the fixed target at the center. The position records (see Fig. 1D,E,F and Fig. 4, left column) demonstrate how precisely controlled these small movements were: overshooting of the small target zone rarely occurred.

3) Fast supination-pronation movements of large amplitude (20°) were triggered by a jump of the target to the extreme right or left of the panel. The monkey was rewarded with a drop of juice after the handle had been rapidly rotated in the direction of the target jump; movement time had to be less than 150 ms for reward to be delivered. The monkey's "ballistic" movements were not precisely controlled as to termination (cf. Fig. 1A,B), since target zones at both extreme positions were bounded by a physical barrier. These movements commonly reached velocities which were 20-30 times greater than the peak velocities occurring for the small controlled movements.

4) Immediately after a ballistic movement, the target jumped back to the center and the monkey returned the handle to the central hold zone to initiate a new cycle (step 1). Figure 1C shows that this type of large movement is slower than the preceding ballistic movement, becoming progressively slower as the target zone is approached.

5) The handle was coupled to the axle of a DC torque motor which produced a constant torque during steady holding. This torque required either steady supinator or pronator force and resulted in an effective force of 90 gr applied to the monkey's hand. Units were usually tested under both load conditions. In addition, 50-ms rectangular torque pulses could be delivered to the handle via the motor to study unit sensitivity to peripheral inputs. Randomly alternated pronating and supinating torque pulses were applied a) after 700 ms of steady holding, b) at a fixed delay of 120 ms after a target jump and thus immediately preceding the ballistic movement, c) at the start of a ballistic movement, and d) at the start of a small movement. The latter pulses (c and d) were triggered at the time the handle left the central hold zone.

For successive cycles of the paradigm, directions of target jumps and triggered movements of torque pulses and of perturbed and unperturbed movements were varied in a pseudo-random order.

Data Recording and Processing

Precentral single unit activity of the contralateral arm area was recorded in the standard manner (16). In two monkeys, PTNs were antidromically identified by stimulation of the medullary pyramid. Antidromic latencies (ADL) ranged from 0.6-5.8 ms. Original spikes, their discriminator pulses, handle position and velocity, and code signals corresponding to stimulus and response events of our paradigm were recorded on magnetic tape. A PDP-12 computer was used for data analysis and display. In rasters each row is a trial, with dots representing individual spikes. Average discharge frequency (Hz) is shown for the histograms, which usually have 10 ms bin widths. For each unit at least 20 trials were obtained for pronation as well as supination for each of the four kinds of movements described (see Fig. 1B,C,E,F), and for both directions

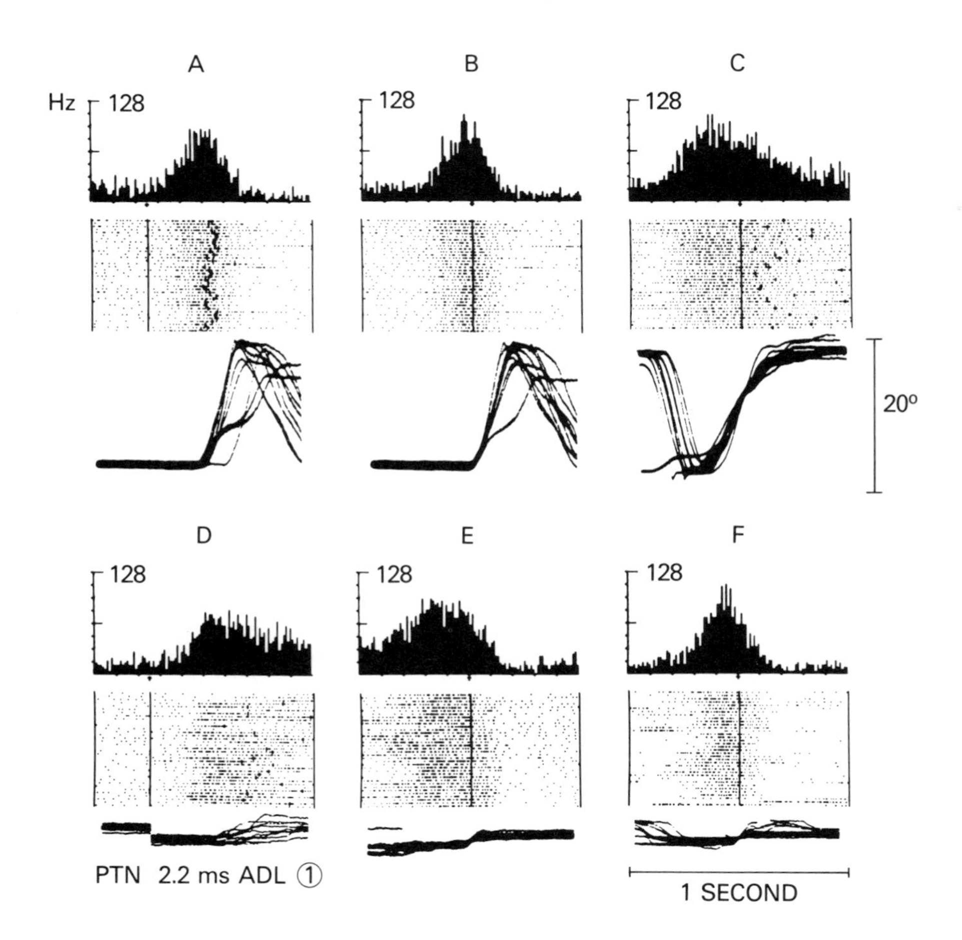

Fig. 1. Same peak frequency of a PTN for large rapid (A, B), large controlled (C), and small controlled supination movements (D-F), the latter being either triggered (D,E) or self-initiated corrections (F).

of torque pulse. A neural response in relation to movement or perturbation was defined as a deviation from control frequency values in the peri-response displays at the $p < 0.01$ level of statistical significance; the same criterion applied to the computation of post-stimulus response latencies (cf. ref. 17).

The center line in most of the rasters marks either onset of torque pulse (for example, Fig. 2, right column) or the time when the handle leaves the hold zone. For the small movements (Fig. 1E,F) this moment is identical with the handle entering the correct zone to realign track and target lamps, whereas for the large rapid movements (Fig. 1B) it corresponds to movement onset. This vertical line was taken as the reference point for the superimposed position traces. In all figures, upward deflection of the position record indicates supination, while downward deflection indicates pronation. In rasters A (same trials as B) and D (same trials as E) of Fig. 1 the vertical line indicates occurrence of either a target jump (A) or an error signal added to the position display control (D), both triggering the respective supination movements; the single heavy dot in each row marks the handle leaving the hold zone.

RESULTS

Relation of Pre-central Units to Large Ballistic and Small Accurately Controlled Movements

Over 700 neurons were analyzed for the different types of movement studied in this paradigm. In about 30% of these units significant responses were detected for both small and large movements, while another 30% changed exclusively in relation to the fast large movements. The former group was characterized by the following features (14):

1) Reciprocal relation to supination and pronation, i.e., decrease of activity for one direction and increase of discharge for the other; such a relation was invariably found in PTNs.

2) These units used their full dynamic frequency range even for the smallest ($< 1^{o}$) and slowest (1-5°/s) detectable supination-pronation movements. Thus, the peak change in frequency with small movements equalled the maximum changes associated with rapid uncontrolled movements (cf. Fig. 1).

3) Unit discharge with precisely controlled movements was more prolonged (compare B and E in Fig. 1), although detectable movement time for both kinds of movement was approximately the same. Unlike neuronal activity with small movement, the burst discharge associated with ballistic movement stopped prior to completion of the actual movement. This has also been shown for the single motor unit's burst preceding a ballistic movement in man (18).

4) Within the group of identified PTNs, slower conducting tonic PTNs (particularly those with ADLs around 2.0 ms) appeared to be preferentially engaged in control of precise fine movements.

Correlation between Torque Pulse Sensitivity and Discharge with Small Precisely Controlled Movements

The responsiveness of 149 PTNs and 440 non-PTNs to pronating and supinating torque pulses applied to the handle during steady holding was systematically tested in one monkey. These pulses produced a handle displacement of 5° (see Fig. 4) and were thus smaller than the disturbances used in comparable previous studies (1,2,3). Early reflex responses occurred at 20 ms. Of the total sample, 124 non-PTNs and 55 PTNs showed significant changes of discharge frequency with small controlled movements, and 96% of these PTNs (82% of the non-PTNs) were clearly influenced by the perturbation. Conversely, only 31% of 54 PTNs (35% of 150 non-PTNs) which were recruited solely with the large rapid movements exhibited significant responses to the kinesthetic input. Furthermore, it has been demonstrated (15) that the majority of those neurons discharging intensely with fine movements were reciprocally related to the active movement and also to the two oppositely directed torque pulses (cf. the two PTNs illustrated in the following figures), whereas in units solely related to the large ballistic movement, nondifferential sensory responses (i.e., same sign of change in frequency) to either direction of torque pulse prevailed. This double reciprocity to small passive and active supination-pronation found in a defined group of motor cortex neurons may reflect a close input-output relation organized in columnar subdivisions as stated by Rosén

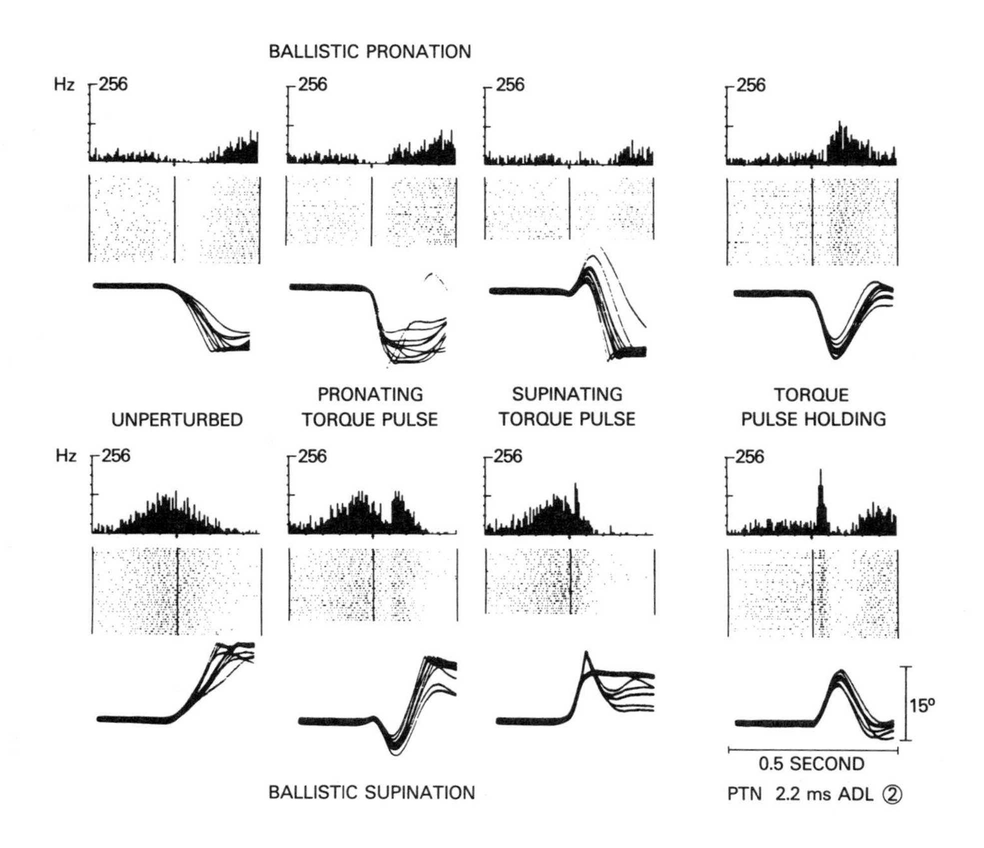

Fig. 2. Reduced reflex responses to unexpected disturbance of ballistic movements. Same unit as in Fig. 1. Note that time calibration (0.5 s) and bin width (5 ms) of histograms is different from the other illustrations.

and Asanuma (8), a view which is supported by the fact that these neurons were predominantly encountered within a restricted focus found at the same coordinates in all three monkeys (centered at A16, L20, with a diameter of 1.5 mm and situated 2 mm anterior to the central sulcus).

The point we would like to stress from these results is the finding of almost invariable reflex driving of those units (particularly tonically discharging PTNs) most involved in control of small precise movements. This appears analogous to the great sensitivity to segmental reflex inputs in those alpha motoneurons first to be recruited in the course of movement (19).

Torque Pulse Injection into Ballistic Movements

In two monkeys we compared the reflex responses of pre-central neurons to sudden perterbations delivered during steady holding with the reflex responses to the same torque pulses delivered unexpectedly at the start of a ballistic movement. Units were tested at two different torque pulse intensities, the greater one producing a handle displacement of 12-15° (Figs. 2 and 3), the smaller of 5° (Figs. 4 and 5).

The PTN of Fig. 2 exhibited a strong 20-ms excitation following the supinating displacement of the handle during accurate holding (4th column, bottom plot) and an early inhibitory response to the pronating torque (4th column, top plot). The later unit activity showed a "mirror image" response when the handle movement changed its direction at the end of the 50-ms torque pulse (see also ref. 2). The excitatory supinating torque pulse was less effective when inserted into the rapid ballistic movement which opposed pronation and assisted supination (3rd column), but the silence in discharge associated with pronation was not markedly altered, and the excitatory reflex response, although superimposed on persisting increased activity with active supination, was not algebraically summed and was not consistent from trial to trial. However, the pronating torque pulse which opposed ballistic supination inhibited unit discharge as effectively as during holding (2nd column, bottom plot). Thus, ballistic movement was associated with attenuation but not elimination of torque pulse responsiveness.

Discharge of the PTN shown in Fig. 3 increased with active pronation and decreased with supination (left column), and this PTN was strongly excited by a supinating perturbation during steady holding (right column). Nevertheless, unit activity during either of the ballistic movements was almost totally unaffected by the injected torque pulse, although the respective movements were severely disturbed (middle column). The inhibitory pronating torque pulse was also without effect.

We have analyzed 87 non-PTNs and 36 PTNs which showed significant ($p < 0.01$ in the post-stimulus histogram) responses to at least one direction of torque pulse applied during holding and for which complete raster plots for all four movement-perturbation combinations were obtained. In 28 PTNs and 60 non-PTNs (about 70%) there was no difference between discharge pattern with the unperturbed control and the perturbed ballistic movement. Sixteen non-PTNs and 6 PTNs were significantly affected by the kinesthetic input during the ballistic movement, but the response was less than the response elicited by the same torque pulse during steady holding. Only 11 non-PTNs and 2 PTNs showed equal reflex responses for both modes of torque pulse application. It was found that the attenuation of cortical reflexes during ballistic movement was relative, primarily depending on the intensity of perturbation (see also ref. 2).

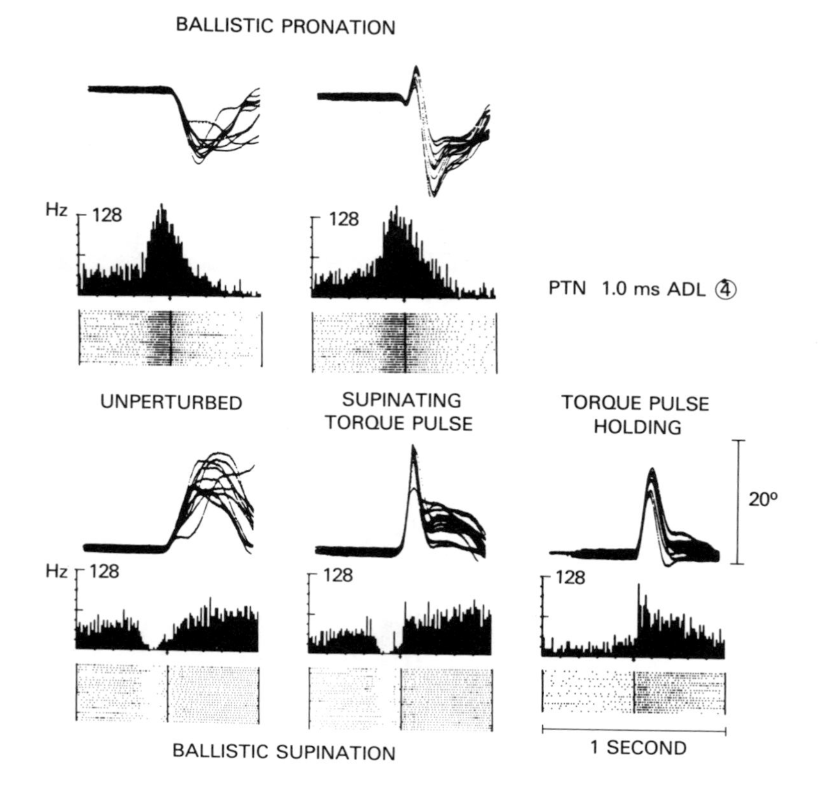

Fig. 3. Attenuation of response to torque pulse injected into ballistic movements.

For example, reducing the torque amplitude shown in Fig. 2 by half still yielded significant early responses for that unit during steady holding, but no longer modified unit activity when inserted into the ballistic movement.

Torque Pulses Superimposed on Precise Small Movements and Triggered before Ballistic Movements

Shortly after collection of data for the records shown in Fig. 3, torque pulse strength was lowered and the displays constituting Figs. 4 and 5 were obtained for the same PTN. The reduced supinating pulse was still suprathreshold for this neuron, and the pronating perturbation was followed by a weak early inhibition (bottom and top displays in the right column of Fig. 4). The same perturbations were superimposed on the small pronation-supination movements which were accompanied by the typically prolonged and reciprocal change in frequency (control, left column). Those torque pulses assisting the respective small movements are depicted in the middle column of Fig. 4. Heightened sensitivity to disturbance of fine movements--particularly obvious for the excitatory reflex--is reflected not only in the increased magnitude and duration of

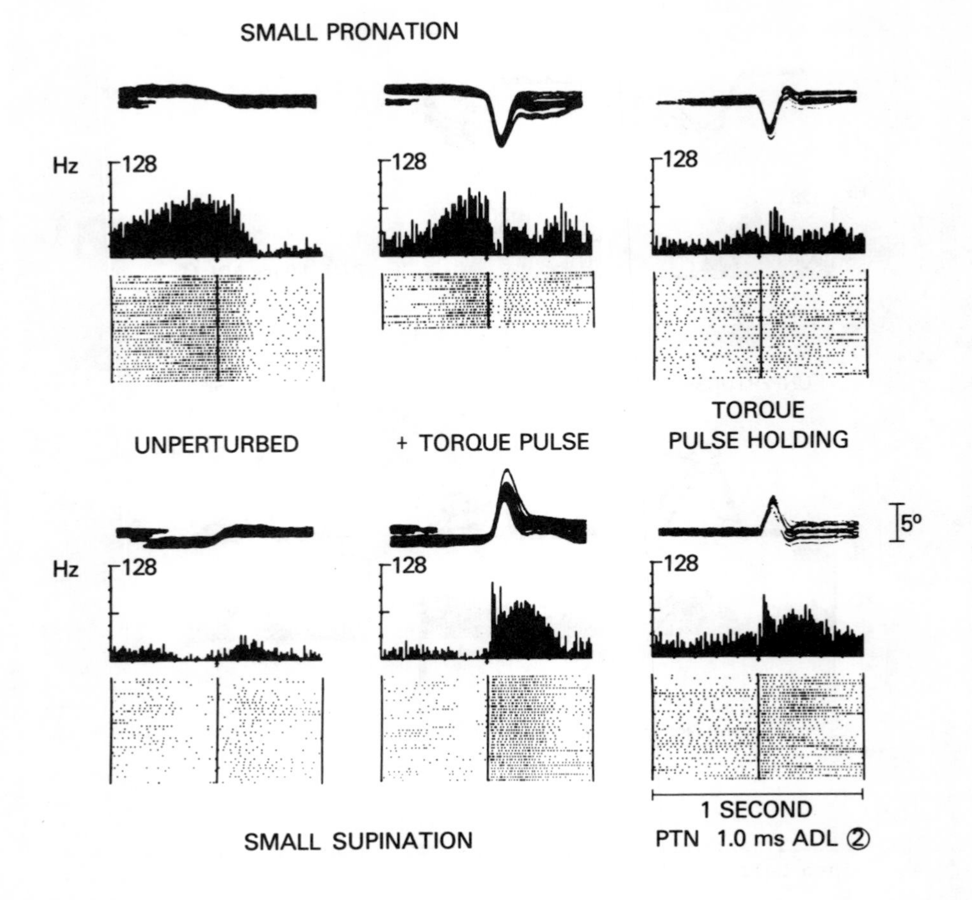

Fig. 4. Increased reflex responses to disturbance of small precisely performed movements. Same unit as Fig. 3.

discharge but also by a discharge pattern showing fluctuations which are tightly time-locked to the torque pulse. Moreover, it should be noted that both the inhibitory and excitatory sensory inputs during precise movement had to override the oppositely directed change of unit activity called for by the "movement program." This result indicates that different PTN excitability (membrane polarization) at the time the somesthetic input reaches motor cortex during a perturbed movement can hardly account for the differential responsiveness to disturbances during ballistic and fine movements or postural stability (cf. Figs. 3 and 4).

This conclusion is also reached from results in Fig. 5 where only unit activity with matching directions of active and passive supination is illustrated to allow for direct comparisons. Two sets of rasters and histograms have been added to the three bottom displays of Fig. 4, the two added sets showing the application of torque pulses prior to ballistic movement (2nd column) and the corresponding control ballistic movement (1st column).

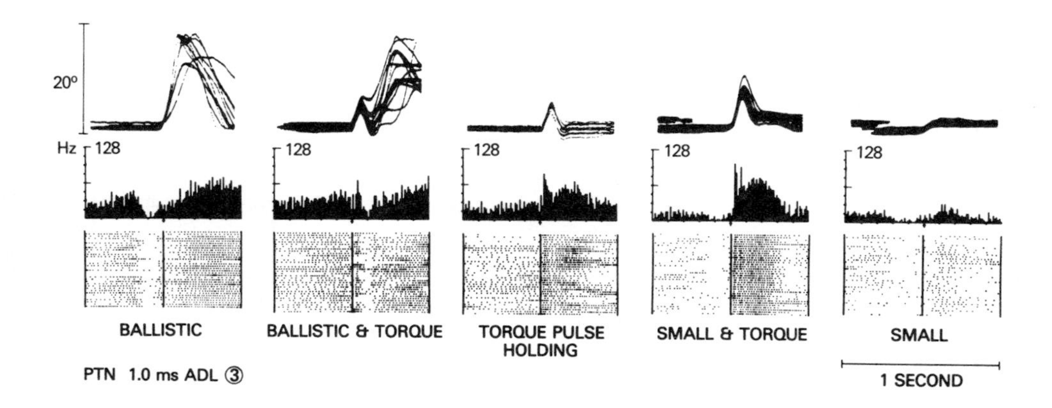

Fig. 5. Differential sensitivity to the same perturbations when applied immediately prior to ballistic movement, or during steady holding and accurate small movement. Same unit as in Figs. 3 and 4.

As is apparent from the position record, timing of the pre-ballistic torque pulse was such that it immediately preceded the start of the ballistic movement, coinciding with the occurrence of unit inhibition associated with the active supination movement. The response to the pre-ballistic torque pulse is attenuated and variable from trial to trial. Of 37 units (19 PTNs and 18 non-PTNs) studied as in Fig. 5, about 60% showed an enhanced reflex response to perturbation of fine movement.

DISCUSSION

The present investigation revealed relative enhancement of precentral neuron responses to kinesthetic stimuli during accurate positioning and controlled fine movements with depressed responsiveness just before and during a rapid ballistic movement. It would seem that different motor cortex neuron excitability at the time sensory input impinges on the motor cortex neurons cannot explain this finding, but peripheral mechanical factors cannot be ruled out, since torque pulses injected into an ongoing rapid movement might have caused less peripheral receptor activation than pulses injected during accurate positioning. However, such alterations in the peripheral effects of the perturbation cannot account for the suppressed reflex response to the pre-ballistic torque pulse. It would thus seem that variations of cortical reflexes seen with the two types of movements may be due in part at least to the nature of the different central programs underlying ballistic and fine movements.

There is mounting evidence for a different organization of motor control in ballistic as contrasted to continuously controlled movements and posture (14, 18,20,21). It has also been established that central motor programs modify somatosensory transmission. As formulated by Adkins et al. (22): "The pyramidal tract...constitutes one route by which the cerebral cortex can modify its own afferent input." PTN axons are known to terminate on various sensory relay nuclei; corticofugal control of the dorsal column-medial lemniscus system and of the fusimotor system is of particular interest in this regard (cf. for review ref. 23).

While the gamma-spindle loop is probably operating and highly sensitive to small changes in muscle length during precisely controlled movements and postural stability (24,25), this loop may be disabled for a short period of time (100 ms) when a preprogrammed rapid movement is being initiated (26): the initial part of such a movement is then under direct supraspinal control via the alpha efferent pathway (alpha-controlled, see also ref. 27,28) and is relatively speaking proprioceptively open loop and therefore less modifiable by sensory inputs (11,13). The switching of the motor system from the "gamma" mode of precise movement and posture to the "alpha" mode of rapid movement may correlate with the "pre-motion silence" in the agonist muscle prior to a fast movement (29). The magnitude of this inhibition seems to be related to the velocity of the subsequent voluntary movement (30).

Moreover, depression of lemniscal responses has been found to occur shortly before and during a voluntary movement (31). This inhibition of sensory transmission in the lemniscal pathway was found to be positively correlated to the amplitude of EMG activity (31) and velocity of a rapid movement (32). And in the present symposium, it has been reported that perception of tactile stimuli is diminished before and during a ballistic tracking movement (33).

The present findings demonstrate that the transcortical servo-loop is continuously active and powerful during precisely guided fine movements and posture. The intense and prolonged discharge of pre-central neurons during those fine movements must partly reflect the central motor program and partly reflect sensory feedback in a closed loop. The relative contribution of the two inputs to the motor cortex output remains to be fractionated. The relative lack of sensory feedback to motor cortex just before and during a ballistic movement appears to depend on the nature of the central preprogram (34) and on the specific parameters of the movement itself. Finally, it should be noted that for sufficiently intense inputs there is an early cortical reflex response even during ballistic movement (cf. Fig. 2). Even though attenuated, this reflex response to mismatch between the intended movement and the actual displacement might play a role in load compensation.

REFERENCES

(1) Evarts, E. V., Motor cortex reflexes associated with learned movement, Science 179, 501 (1973).

(2) Conrad, B., Meyer-Lohmann, J., Matsunami, K., Brooks, V. B., Precentral unit activity following torque pulse injections into elbow movements, Brain Res. 94, 219 (1975).

(3) Evarts, E. V., Tanji, J., Reflex and intended responses in motor cortex pyramidal tract neurons of monkey, J. Neurophysiol. 39, 1069 (1976).

(4) Porter, R., Influences of movement detectors on pyramidal tract neurons in primates, Ann. Rev. Physiol. 38, 121 (1976).

(5) Lemon, R. N., Porter, R., Afferent input to movement-related precentral neurones in conscious monkeys, Proc. R. Soc. Lond. B. 194, 313 (1976).

(6) Lemon, R. N., Hanby, J. A., Porter, R., Relationship between the activity of precentral neurones during active and passive movements in conscious monkeys, Proc. R. Soc. Lond. B. 194, 341 (1976).

(7) Clough, I. F. M., Kernell, D., Phillips, C. G., The distribution of monosynaptic excitation from the pyramidal tract and from primary spindle afferents to motoneurons of the baboon's hand and forearm, J. Physiol. 198, 145 (1968).

(8) Rosén, I., Asanuma, H., Peripheral afferent inputs to the forelimb area of the monkey motor cortex: input-output relations, Exp. Brain Res. 14, 257 (1972).

(9) Conrad, B., Matsunami, K., Meyer-Lohmann, J., Wiesendanger, M., Brooks, V. B., Cortical load compensation during voluntary elbow movements, Brain Res. 71, 507 (1974).

(10) Phillips, C. G., Motor apparatus of the baboon's hand, Proc. R. Soc. Lond. B. 173, 141 (1969).

(11) Hallett, M., Shahani, B. T., Young, R. R., EMG analysis of stereotyped voluntary movements in man, J. Neurol. Neurosurg. Psychiat. 38, 1154 (1975).

(12) Garland, H., Angel, R. W., Spinal and supraspinal factors in voluntary movements, Exp. Neurol. 33, 343 (1971).

(13) Hopf, H. C., Lowitzsch, K., Schlegel, H. J. (1973) Central versus proprioceptive influences in brisk voluntary movements, New Developments in Electromyography and Clinical Neurophysiology, vol. 3, Karger, Basel (p. 273).

(14) Fromm, C., Evarts, E. V., Relation of motor cortex neurons to precisely controlled and ballistic movements, Neurosci. Lett. (1977) (in press)

(15) Evarts, E. V., Fromm, C., Sensory responses in motor cortex neurons during precise motor control, Neurosci. Lett. (1977) (in press)

(16) Evarts, E. V., A technique for recording activity of subcortical neurons in moving animals, Electroenceph. clin. Neurophysiol. 24, 83 (1968).

(17) Tanji, J., Evarts, E. V., Anticipatory activity of motor cortex neurons in relation to direction of an intended movement, J. Neurophysiol. 39, 1062 (1976).

(18) Desmedt, J. E., Godaux, E., Ballistic contractions in man: characteristic recruitment pattern of single motor units of the tibialis anterior muscle, J. Physiol. 264, 673 (1977).

(19) Henneman, E. (1974) Principles Governing Distribution of Sensory Input to Motor Neurones, The Neurosciences, Third Study Program, MIT Press, Cambridge (p. 281).

(20) Stetson, R. H., McGill, J. A., Mechanisms of the different types of movement, Psychol. Mono. 32, 18 (1923).

(21) Kornhuber, H. H., Motor functions of cerebellum and basal ganglia: the cerebellocortical saccadic (ballistic) clock, the cerebellonuclear hold regulator, and the basal ganglia ramp (voluntary speed smooth movement) generator, Kybernetik 8, 157 (1971).

(22) Adkins, R. J., Morse, R. W., Towe, A. L., Control of somatosensory input by cerebral cortex, Science 153, 1020 (1966).

(23) Wiesendanger, M., The pyramidal tract. Recent investigations on its morphology and function, Ergebn. Physiol. 61, 72 (1969).

(24) Matthews, P. B. C. (1972) Mammalian Muscle Receptors and Their Central Actions, Arnold, London.

(25) Vallbo, A. B. (1973) The significance of intramuscular receptors in load compensation during voluntary contractions in man. Control of Posture and Locomotion. Advances in Behavioral Biology, vol. 7, R. B. Stein, K. G. Pearson, R. S. Smith, J. B. Redford (eds.), Plenum Press, New York (p. 211).

(26) Navas, F., Stark, L., Sampling or intermittency in hand control system dynamics, Biophys. J. 8, 252 (1968).

(27) Dijkstra, S. J., Denier van der Gon, J. J., An analog computer study of fast, isolated movements, Kybernetik, 12, 102 (1973).

(28) Stark, L. (1966) Neurological feedback control system, Advances in Bioengineering and Instrumentation, F. Alt, New York.

(29) Ikai, M., Inhibition as an accompaniment of rapid voluntary act, Nippon Seirigaku Zasshi 17, 292 (1955).

(30) Yabe, K., Premotion silent period in rapid voluntary movement, J. App. Physiol. 41, 470 (1976).

(31) Coulter, J. D., Sensory transmission through lemniscal pathway during voluntary movement in the cat, J. Neurophysiol. 37, 831 (1974).

(32) Ghez, C., Pisa, M., Inhibition of afferent transmission in cuneate nucleus during voluntary movement in the cat, Brain Res. 40, 145 (1972).

(33) Dyhre-Poulsen, P. (1977) Perception of tactile stimuli during movement, Symposium on "Active Touch," Pergamon Press, Oxford.

(34) Stetson, R. H., Bouman, H. D., The coordination of simple skilled movements, Arch. Néerl. Physiol. 20, 177 (1935).

MOVEMENT PERFORMANCE AND AFFERENT PROJECTIONS TO THE SENSORIMOTOR CORTEX IN MONKEYS WITH DORSAL COLUMN LESIONS

J. Brinkman and R. Porter

Department of Physiology, Monash University, Clayton, Australia

INTRODUCTION

Although Ferraro & Barrera (12) and Gilman & Denny-Brown (14) reported profound disturbances of spatially projected movements of the forelimbs in monkeys with dorsal column lesions, others (4, 5, 21) have noted little or no movement disability after similar lesions. Difficulties in tactile discrimination tasks have however been characteristic in monkeys tested for such abilities (4, 29).

In man, dorsal column lesions are seldom seen in the absence of other neurological disorders, but Marsden, Merton & Morton (20) found that a patient with a localized region of demyelination of the dorsal columns lacked the normal rapid, automatic correction for an imposed disturbance of a simple flexion movement of the fingers. The normal automatic correction could involve a "transcortical loop" through the sensorimotor cortex (1) because a patient with a post-central lesion also showed diminished automatic motor adjustments. Recent evidence from a number of laboratories has indicated short latency responses of pyramidal tract neurones in the motor cortex of conscious monkeys following brief perturbations of movement performance (10, 24). The timing of these effects could be consistent with their participation in automatic adjustments to perturbations via a transcortical loop. If they were so involved, they too might depend on intactness of dorsal column afferents.

Most neurones in both the principal "motor" and the principal "sensory" region of the "arm area" of sensorimotor cortex receive projections from peripheral receptors in the forelimb. The influences produced in precentral neurones by natural activation of peripheral receptors in the forelimb have been described in a number of studies (2, 13, 18, 19, 27, 32). Recognition of this input to the motor cortex, mainly from receptors activated by passive joint movement, has come much later than the classic descriptions of cutaneous projections to neurones in the sensory cortex (25).

Very few observations have been made of the effects of dorsal column lesions on the projections to neurones in the sensorimotor cortex of monkeys or on the receptive properties of these neurones. Phillips, Powell & Wiesendanger (22) reported that the responses of neurones in area 3a of baboons to stimulation of group 1 afferents in muscle nerves were abolished by dorsal column section. Dreyer, Schneider, Metz & Whitsel (9) examined the properties of neurones in areas 3, 1 and 2 of the leg representation of the primary sensory cortex in monkeys and found that many neurones continued to receive afferent projections after dorsal column lesions, but the quality of these projections had changed.

Our experiments have had as their objective the definition of the behavioural effects produced in monkeys by interruption of the cuneate fasciculi on both sides. In the same animals, we have attempted an evaluation of the receptive properties of single neurones in both the "motor" and "sensory" areas of the arm sensorimotor representation in an attempt to define neuronal correlates of the behavioural effects. We have concluded that complete interruption of the cuneate fasciculi does not impair skilled movement performance using the forelimbs. Even relatively independent movements of the fingers are well performed. In the absence of vision there is a minimal disturbance of contactual placing responses and of tactile discrimination if this has to be performed using the information available with only a limited surface area of skin (e.g. on the finger tip) in contact with the object. But the movements themselves are well executed.

Complete section of the cuneate fasciculi removes the short-latency feedback to neurones in the motor cortex of precise information about joint position, muscle length or skin contact. But the discharges of precentral neurones which accompany self-paced learned movement performance are indistinguishable from those recorded in normal animals (7). So the "long-loop" feedback to the motor cortex can not have a major role in determining the natural activities of these neurones in relation to natural movement performance.

The minimal deficit of discriminative ability is found to be associated with a changed responsiveness of post-central neurones to natural peripheral stimuli delivered to the forelimb of the conscious, cooperative, relaxed monkey.

METHODS

Four male cynomologus monkeys (*M. fascicularis*) were trained with food rewards to pull a small horizontal lever repetitively with the right arm. They were also trained to sit quietly and accept, without struggling, passive manipulation of all the joints of both limbs, tapping of muscles and natural stimulation of the skin (18). Noxious stimuli were not used. All training employed food rewards and was continued for a period of several weeks until the animal remained relaxed and cooperative through long periods of repeated manipulation of the limbs. This relaxed state and cooperation were essential for the testing of influences of natural stimuli, and the results to be described here could not have been obtained without this prolonged preparation. The animal's motor performance and acceptance of manipulation were recorded on movie film.

At this stage, an operation was performed under general anaesthesia and with full aseptic precautions. In three animals the uppermost levels of the spinal cord were exposed through the atlanto-occipital membrane. Through an incision in the dura and using a small guarded iridectomy knife the cuneate fasciculi on both sides were divided under microscopic visual control at about C1 - C2 level. Extreme care was taken to avoid damage to dorsal root filaments entering the spinal cord and to small local blood vessels. The knife-cut was extended medially and especially laterally in an effort to completely sever all dorsal column fibres in the cuneate fasciculus; in particular the fibres from the cervical enlargement (30). In the fourth animal after laminectomy, the cuneate fasciculi were sectioned at the lower part of C5 spinal level.

On the day following the operation the animals were replaced in the holding area where social contact with other monkeys was resumed and training in the movement task was continued. In contrast to the reports of others describing the effects of dorsal column lesions in monkeys (12, 14), the first three animals exhibited almost no ataxia of the forelimbs after recovery from the operation. They quickly demonstrated an apparently normal repertoire of movements in their free-range behaviour. Within a few days they were performing the learned movement task in a manner indistinguishable from that which had existed pre-operatively. These movements were recorded on movie film for comparison with the pre-operative performance.

The animals were retrained in the movement task and in the acceptance of manipulation for a period of about 2 weeks. Monkey 4 with a lesion at C5/6 initially had signs of ataxia (see below) and needed a longer retraining period.

At a second operation under general anaesthesia and with full aseptic ritual, a specially designed headpiece was implanted on the skull giving access, through a skull defect, to the arm area of the motor cortex of three of the monkeys and the sensory cortex of one (23). Electromyographic leads were brought subcutaneously from representative flexor and extensor muscles in the right forearm to a multi-pin socket within the headpiece. Stimulating electrodes were directed stereotaxically towards the left pyramidal tract to allow subsequent identification of pyramidal tract neurones according to their responses to stimulation of axons at this level.

During two recording sessions on each day for a period of 6 to 8 weeks, the activities of individual precentral or postcentral neurones in the left cerebral cortex were studied using tungsten microelectrodes while the animal carried out repetitive performances of the movement task, as well as a number of other movements outside the formal test situation. This enabled an association to be established between a neurone's firing and an aspect of movement performance (17). The responses of each of these cells were then tested with a variety of natural stimuli delivered to the skin, joints and muscles of the right arm and hand in order to attempt to define the effective natural stimulus for causing responses in the neurone and the peripheral zone from which these responses could be obtained. In this way, for each neurone, its response to natural peripheral stimulation could be classified and compared with the response of the same cell during an aspect of active movement.

At the completion of the recording period, the animal was again anaesthetized, marker electrodes were inserted stereotaxically into the brain, and the brain was perfused with fixative. Serial sections were cut and stained for identification of electrode tracks and for examination of the region of the lesion in the cuneate fasciculi.

RESULTS

Behaviour of the Monkeys Following Section of the Cuneate Fasciculi

Motor behaviour. It has been stated by Gilman & Denny-Brown (14) that high cervical dorsal column section disturbs all spatially projected movements of the forelimbs and causes particularly severe defects in contactual orienting and placing reactions. But the monkeys which were the subjects for our experiments, especially those with lesions at the C1 level, were not seriously handicapped in any motor activities. Repeated study of the animals and careful examination of their free-range motor behaviour revealed that, even on the first few days after the operation to divide the cuneate fasciculi, the forelimbs could be projected into space successfully, were accurately directed towards a food target or the handle of the lever, formed an appropriate grasp on the object once the contact was made and performed subsequent flexion movements in a well coordinated manner. The lever pulling movement task was performed in a way indistinguishable from the normal action observed preoperatively in all three animals.

In the first two animals, histological examination of the lesions showed that some cuneate fibres had escaped damage. Sparing was estimated to be 10% and 15% respectively. In the third animal, however, the lesion was complete and the motor performance in this monkey was not in any way different from that of the first two. In monkey 4, the lesion was at C5/6 level and was found to extend beyond the dorsal columns to involve most of the dorsal horn on both sides. As a consequence, it probably damaged some of the dorsal root fibres which travel in the dorsal horn at that level (28). Immediately postoperatively, this animal showed ataxia of the forelimbs and trunk which disappeared within three weeks. It also had some difficulty in fine manipulations. With retraining, movement performance was restored to normal levels for the lever pulling task.

Even when vision was occluded each animal projected his arm to the lever, located the handle (even when the lever had purposely been placed in a different position), oriented and closed his hand on this appropriately and carried out the flexion task accurately to bring the handle into the target zone. Trajectories of the hand plotted using frame-by-frame analysis of the movie films and rates of angular movement at the shoulder, elbow and wrist were not disturbed by the section of cuneate afferents, even in the two animals in which the lesions were found to be complete.

Special attention was paid also to the capacity of the monkeys to execute discrete movements of the distal extremeties, in particular of the relatively independent movements of the digits (16), since according to Gilman & Denny-Brown (14) no such movements of the forelimbs for use in extrapersonal space are observed in monkeys with dorsal column lesions. None of the first three monkeys, including the animal with complete section of the cuneate fasciculi (monkey 3) showed any deficits in the execution of such movements and again frame-by-frame analysis of films taken pre- and postoperatively failed to reveal any changes. Figure 1 shows drawings of representative frames which show the finger movements of monkey 3 before (Fig. 1A) and after (Fig. 1B) the lesion (inset). In both instances, the animal extends the individual index finger and opposes the thumb, keeping the other fingers flexed and out of the way, in order to retrieve a raisin from a small food well.

<u>"Sensory" behaviour</u>. With vision occluded by taping the eyes shut, contactual placing could still be performed even by the animals subsequently demonstrated to have complete cuneate lesions. Small pieces of food, held in long handled forceps, were touched against the hairs on some part of the body surface and then moved slightly away but still kept in the vicinity of the point contacted. The animal immediately brought his hand to the region of the food reward, located the food or the forceps, manipulated the food from the forceps and rapidly placed it in his mouth (Fig. 2A, B). The accuracy of projection of the hand to the vicinity of the surface point contacted was slightly

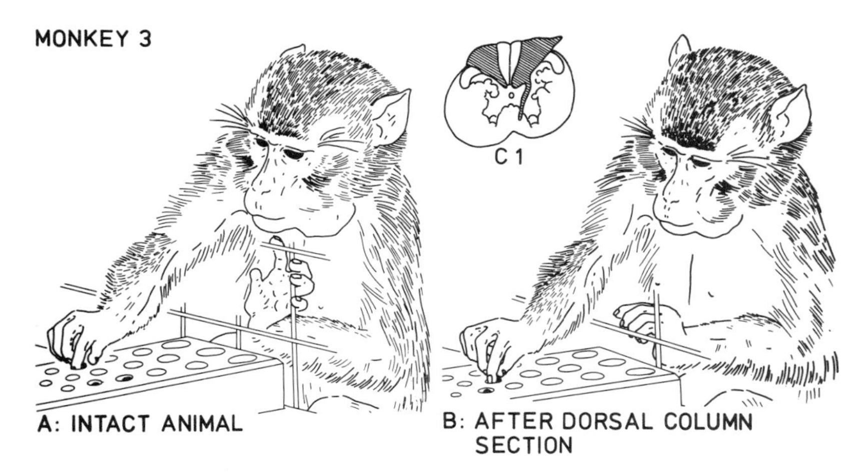

Fig. 1. Drawings from a film showing the execution of relatively independent finger movements in order to retrieve a raisin from a small well in monkey 3 before (A) and two weeks after (B) complete section of the cuneate fasciculi at C1 (inset). In both instances, the animal extends the index finger and opposes the thumb, keeping the other fingers flexed and out of the way.

impaired in relation to the performance of a normal animal and contact with the hand or fingers was sometimes neglected and elicited no movement responses. The reaching and grasping were most accurate when the point contacted by the food was on the same side of the body surface as the hand which was moved to it. "Crossed" contactual reaching and grasping directed to a point on the opposite side of the body to the moving limb, was less accurately performed (Fig. 2C and D). But the impairment was minimal; the animal responded to contact and the movement performance it produced was well directed and correctly executed, even when the limb was placed in a different starting position by the experimenter.

Exploration of a surface to locate food rewards hidden in small "gutters" of such a size that they admitted only the index finger was poorly performed by these monkeys after vision was occluded, but normal monkeys also found this discrimination difficult in the absence of vision. The operated animals would insert the finger tip into gutters on the board but, unlike the normal animals, on most occasions they apparently failed to recognise and retrieve raisins which were contacted by the finger. Food rewards, even small items such as raisins, were readily manipulated and picked up from a flat surface, and for this motor performance, independent finger movements were executed with near normal precision. The animal's ability to recognize the texture of the food within a gutter when only a small surface area of contact was provided was impaired.

Hence the study of the behavioural performance of the monkeys after dorsal column lesions and in the absence of vision indicated an impairment of discriminative ability - detection with the finger tip of the difference between an empty gutter and a gutter containing a raisin. Related to this may have been a minimal impairment of contactually-directed movements, particularly when the hand had to be directed to a distally located point on the wrist, hand or fingers of the opposite side of the body. These results, showing a defect of discriminative ability seem to be consistent with those of Azulay et al (4). But natural and learned movements of the forelimb projected into space, contact-directed manipulation of the hand, grasping, and flexion movements of the whole arm were performed in a manner indistinguishable from normal.

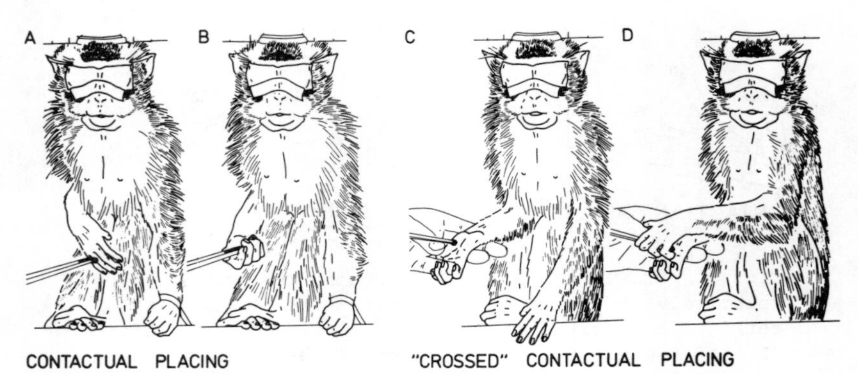

Fig. 2. Drawings from a film showing accuracy of contactual placing (A and B) and "crossed" contactual placing (C and D) in the same animal as Fig. 1. In A, the hairs on the dorsum of monkey's right hand were lightly touched with a piece of food held by forceps. This was then moved away from the hand for about 2 cm. Almost immediately, the animal made a slight withdrawal movement, supinated the hand and brought it forward. Upon making contact with the forceps, the fingers closed accurately around it to grasp the bait (B). Touching other parts of the right forelimb also resulted in excellent localization. Stimulation of proximal sites often elicited a response more quickly than that of distal ones. In C, the animal's left hand is touched with a piece of food. Although the reaching of the contralateral limb to the site stimulated is fairly good, the hand is misplaced with respect to the bait (D). In general, "crossed" contactual placing was performed less accurately compared to the ipsilateral performance, especially when the point stimulated was located on the hand or the wrist.

Responses of Cortical Neurones to Natural Stimulation of Peripheral Receptors

Responses of precentral neurones. The activities of 321 precentral neurones have been examined in detail and are discussed in the paper by Brinkman, Bush & Porter (7). Each precentral neurone produced a characteristic increase in discharge in association with a particular aspect of the learned movement task whenever this element of movement performance was executed (see ref. 17). Two important observations have been made in relation to the recorded activity of these precentral neurones, all of which were located within area 4 of the precentral gyrus. In the first place, the discharges of individual neurones were representative of all the patterns of activity seen in the normal animal and were indistinguishable from the patterns seen in the normal animal. Increases in the discharges of individual precentral neurones accompanied one of the separable aspects of the movement - arm protraction, elbow extension, finger flexion, elbow flexion, finger extension, etc., whenever this occurred (Fig. 3). In the second place, and in marked contrast to the findings in the normal animal, division of the cuneate fasciculi had removed the clearly reproducible short latency responses of these neurones to natural stimulation of localized, peripheral receptors by manipulation of joints, palpation of muscles or activation of cutaneous receptors in circumscribed regions of the forelimb (cf. ref. 18). All those influences which were detected after division of the cuneate fasciculi could have arisen in intact dorsal column afferents which escaped damage at the initial operation (monkeys 1 and 2) or could have entered the spinal cord above the site of the dorsal column lesion, in monkey 4. In this last animal, only 9 out of 171 units investigated were affected by a peripheral stimulus and all of these effects were from passive elbow or shoulder movements. Figure 3 shows correlograms of the activity of three precentral units with the performance of the learned motor task found in monkey 4. None of these units could be driven by stimulation of peripheral receptors when the animal was passive and relaxed.

Hence the normal movement performance in these experimental animals was correlated with apparently normal precentral neuronal activity in association with natural, spontaneous and self-paced movements. This was in spite of the deficient feedback of particular, specific, short-latency information from receptors in the moving limb to precentral neurones. The conclusion must be that natural, spontaneous, self-paced movement performance and the functions of the precentral motor cortex in relation to this do not depend directly on "long-loop" reflexes utilizing dorsal-column afferent projections. This does not, however, deny a role for these long loop influences in rapid, automatic adjustments of the musculature to perturbations of movement performance.

Responses of postcentral neurones. Electrode penetrations were made in areas 3a, 3b, 1 and 2 of the postcentral gyrus in monkey 3 with a complete cuneate lesion. The activities of 92 neurones were studied in this monkey, but only 68 of these could be studied for long enough to allow for complete sensory testing. The behaviour of these postcentral neurones requires comment first in relation to the learned movement task and secondly in their relation to their responsiveness to natural activation of peripheral receptors.

Like precentral neurones, postcentral cells demonstrated modulation of their natural discharge in relation to self-paced movements of the contralateral forelimb but the number of such units recorded was different for the different cytoarchitectonic subdivisions of the postcentral gyrus. No neurones were recorded in area 3a, but they were present in all other subdivisions. Only a few responsive neurones were found in area 3b, and hence most of the cells studied were in areas 2 and 1.* In all cytoarchitectonic areas, activity of most cells was found to be related to movements of the distal rather than the proximal limb.

---

*A large number of tracks were made in areas 3a and 3b of monkey 4, also with a complete cuneate lesion, but in this animal no responsive units were found in these zones.

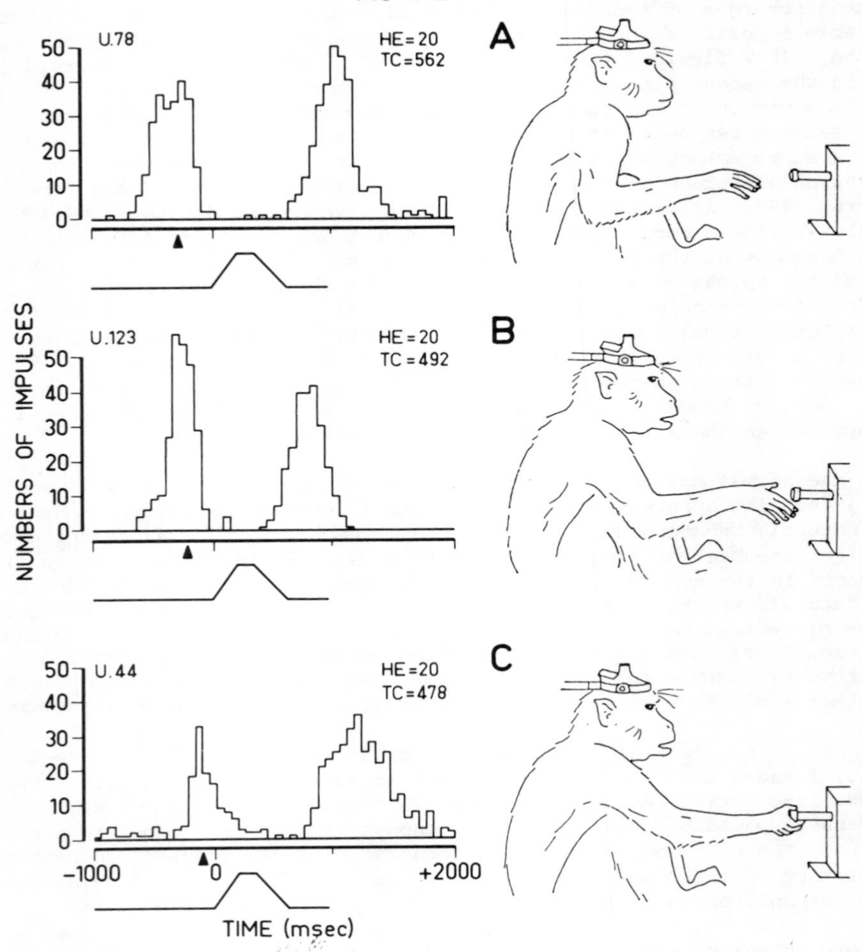

Fig. 3. Examples of patterns of natural activity occurring with movement performance in sample pre-central neurones studied in a monkey with complete section of the cuneate fasciculi at C5/6 (Monkey 4). Each of the histograms on the left side is the peri-response-time histogram of the discharges of a single cortical neurone during 20 repetitions of the lever pulling and food collecting movements (HE). The average duration of the lever displacement is indicated by the line below each histogram. O indicates the time at which the lever movement started. The total number of nerve impulses recorded during the repetitions of the movement is shown as total counts (TC). Drawings on the right are taken from single frames of a

movie film and show the animal at different times during the execution of the motor task. The times of the representative frames are indicated by the triangles below the respective histograms. A, Histogram of a neurone whose discharges were always associated with finger extension, and not with any other movement occurring during the execution of the pulling task and subsequent food collection. The picture on the right shows the extension of the fingers when the animal reached for the handle of the lever, and was taken from the frame on which this movement was maximal. Extension could first be detected about 400 msecs before the start of the pull. Discharge level dropped to zero when the hand reached the lever. A second peak of activity of this cell occurred in association with finger extension to release the lever and when reaching for the food reward; activity dropped when the reward was retrieved. B shows the histogram for a neurone whose increase in discharge was always associated with extension movements of the thumb. This occurred prior to grasping the lever, as shown in the drawing on the right, and after the pull when releasing the lever. The neurone illustrated in C discharged always before and to some extent during finger flexion movements which occurred when the animal grasped the lever and again when it retrieved its reward after the pull. Note that the maximum discharge for this cell associated with finger flexion followed that for a neurone associated with finger extension (A) or thumb extension (B). None of these units could be driven by natural stimulation of peripheral receptors when the limb was manipulated passively while the animal sat quietly and relaxed. 95% of the 171 precentral units found in this monkey with dorsal column section had no peripheral input.

Of the 68 cells which could be studied for a period long enough to allow for complete testing of afferent input from peripheral receptors, 42 (60%) showed responses to natural stimulation but 26 (40%) did not. Nevertheless the discharge of the latter group was clearly modulated during movement performance. Comparison of the timing of the discharges of the units in this second group with that of appropriate EMG activity recorded simultaneously, suggested that at least for a number of these cells the discharge could precede the onset of EMG activity - i.e. the behaviour of such units was similar to that of precentral neurones. For others, the onset of cellular activity and that of the appropriate EMG coincided (cf. ref. 11). Figure 4 shows the correlograms for three units of this group. Virtually all of the units of this group were found in area 2 and to a lesser extent in area 1.

Neurones which received afferent input were found in areas 3b, 1 and 2 but the units in the different areas differed with respect to the stimulus which was effective in causing them to discharge. Cells in area 3b responded only to cutaneous stimulation, whereas neurones in area 2 were most often influenced by moving joints and prodding muscles while the animal was relaxed. In area 1 both cells which were driven by cutaneous stimulation and cells responding to activation of deep receptors were found. For almost all the 42 cells, localization of the effective stimulus was restricted: with one exception, all

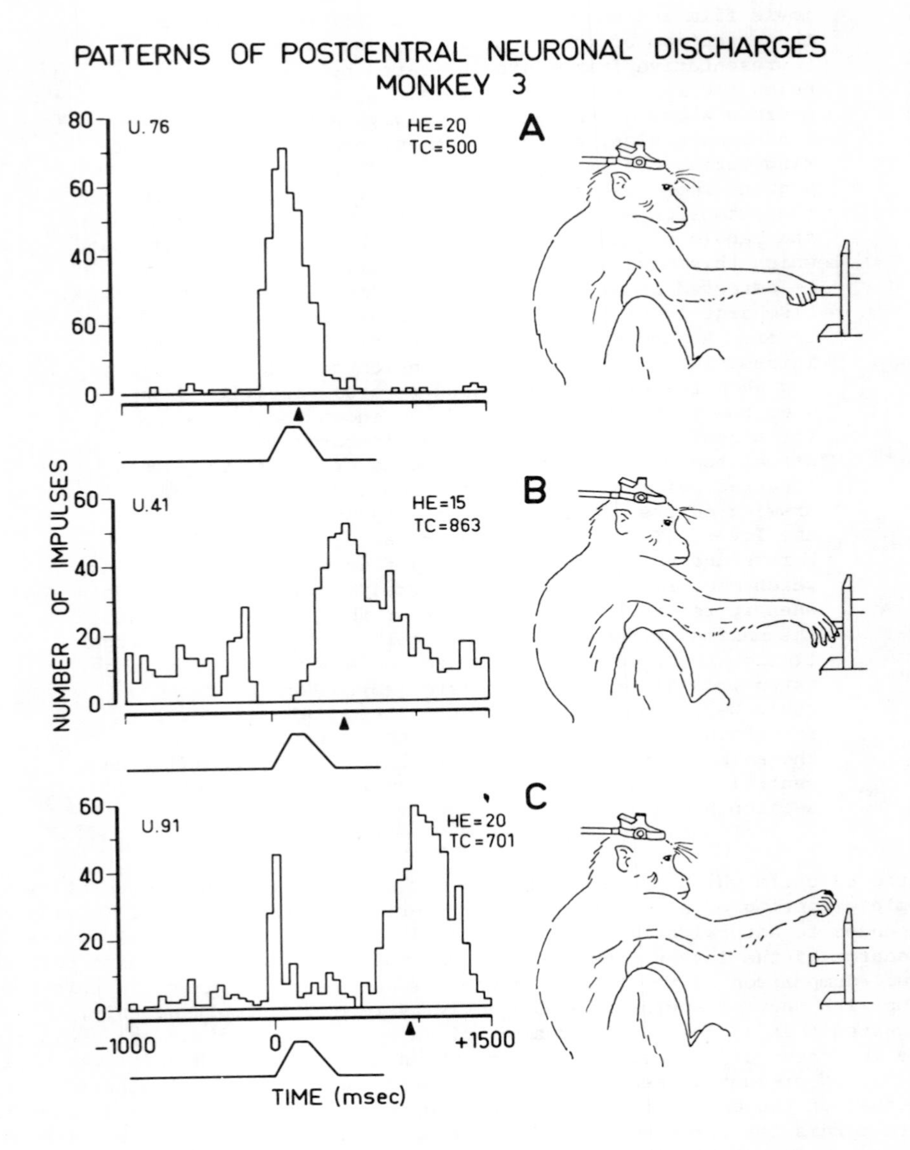

Fig. 4. Patterns of discharges of three neurones in the postcentral areas 2(A) and 1 (B and C) in monkey 3 with complete section of the cuneate fasciculi at C1, during performance of the lever-pulling task and during food retrieval afterwards. The neurone shown in A was associated with flexion of the elbow only during the pull, but not with the flexion of the elbow which occurred after the pull to bring the food to the mouth. The discharge dropped off sharply after the plateau phase of lever displacement

but some impulses were still occurring while the lever returned to its original position. The drawing on the right shows the maximal elbow flexion which occurs during the plateau phase and the timing of this frame is indicated by the triangle (▲). In B, the discharge of the neurone was associated with extension of the fingers reaching for and releasing the handle, while the cell illustrated in C increased its discharge for opposition of the thumb during grasp of the lever and during food retrieval. None of these neurones could be driven by natural stimulation of peripheral receptors when the animal was relaxed, even though their activity was clearly modulated during the active movement. About 40% of the neurones recorded in the postcentral gyrus of this monkey behaved in this fashion. Analysis of the onset of EMG activity in some representative muscles during performance of the task suggested that, for some of these cells, onset or increase of discharge could precede EMG activity (cf. ref. 11).

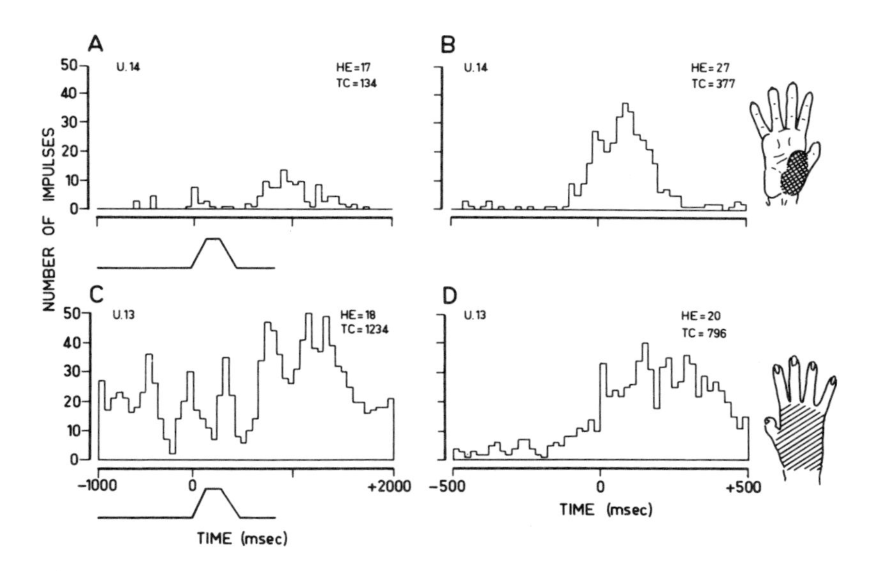

Fig. 5. Peri event-histograms of two cells recorded in area 3b in monkey 3. Responses during the active movement are shown on the left (A and C). Upon examination for peripheral input in the passive animal, both cells were found to be influenced from the hand and the post-stimulus time histograms for each neurone as well as a figurine delineating its effective afferent input zone are shown on the right. Unit 14 could be discharged at very short latency by tapping the skin of the thenar eminence (B) while unit 13 responded after a short latency to movement of hairs on the dorsum of the hand (D). Comparison of the "active" and "passive" histograms for each neurone suggests that discharges occurring during the active movement performance can be

explained as results of afferent input to the cell produced by the movement. Note the different time-scale of the post stimulus histograms, and the restricted effective afferent input zone.

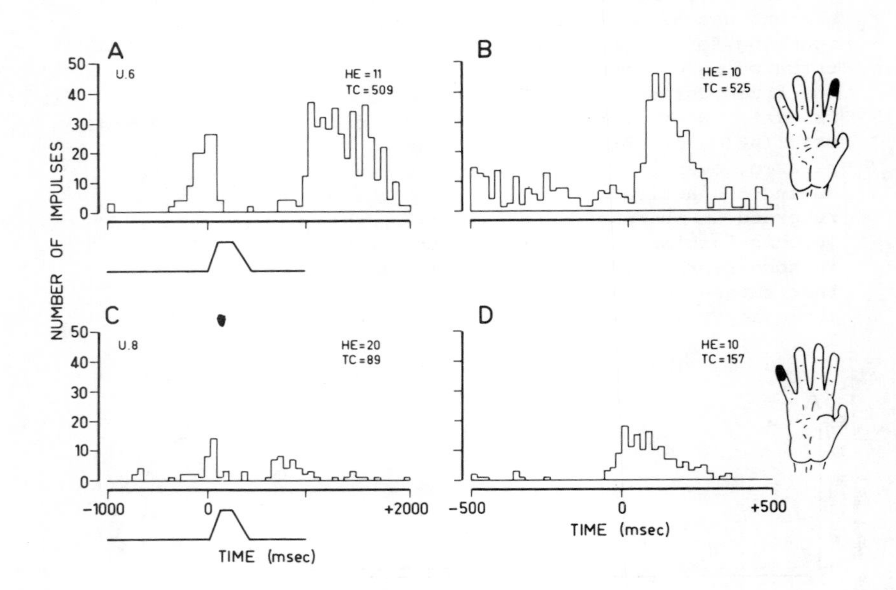

Fig. 6. Examples of two neurones recorded in area 1 with cutaneous afferent inputs. Comparison of the histograms of the discharges during the execution of the movement (A and C) and the post-stimulus time histograms (B and D) indicate that the "active" histograms are most likely the result of afferent inputs caused by peripheral stimuli occurring with the movement. Note small size of cutaneous afferent input zones (receptive fields).

cutaneous receptive fields were less than 5 cm^2 and most were less than 2 cm^2. For neurones driven by joint movements the stimulus was mostly restricted to movement at one joint in one direction only. Figures 5, 6, 7 and 8 show correlograms for units in the three cytoarchitectonic subdivisions of the post central gyrus both during the active movement of the monkey in performance of the lever-pulling task and during natural stimulation of peripheral receptors in the passive, relaxed animal. Inspection of the shape and timing of these and other correlograms suggested that the neuronal discharges seen during the performance of the task could be explained as resulting from stimulation of peripheral receptors during the execution of the movement. It is clear, however, that a very much smaller population of responsive neurones was encountered in area 3b than would have been expected in the normal conscious and cooperative animal without a lesion of the cuneate fasciculi.

The conclusion must be that the individual responses of post-central neurones, which can still be demonstrated after complete cuneate interruption, and which are most numerous in areas 1 and 2, are not sufficient to enable the monkey to demonstrate full discriminative ability. Afferent projections remain but responses are difficult to find in area 3. Individual post-central neurones still receive inputs from particular classes of receptors localized within circumscribed regions of the contralateral forelimb. But in tests which require the animal to detect brief, weak stimuli applied to the distal part of the limb, or which require discriminations to be made by utilizing only a small surface area of contact with the test object (e.g. that provided by the tip of the index finger) the afferent projections to the sensory cortex are inadequate to maintain full capacity for the sensory system.

DISCUSSION

Motor Behaviour

Monkeys with almost complete or complete lesions of the cuneate fasciculi at C1 (monkeys 1-3) showed no disturbances of their free-range motor behaviour and tests which specifically required the use of relatively independent movements of the fingers (16) failed to reveal any deficits. These findings

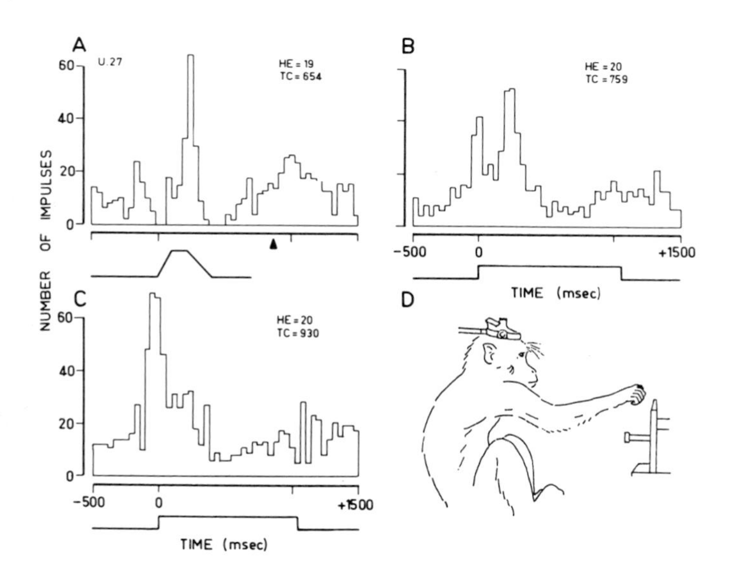

Fig. 7. Example of a neurone recorded in area 1 which could be driven by passive rotation of joints. A, Histogram of discharges occurring with active movement; B, Histogram obtained during 20 repetitions of a passive pronation movement. Average duration of each passive movement is indicated by the step in the trace below. This unit could also be driven by passive flexion of the metacarpophalangeal joint of the thumb (C). Peaks in activity in histogram A could be explained as a result of the afferent input from these two types of movement which may occur together as is shown in D when the monkey pronated and then opposed the thumb during food retrieval. (Triangle (▲) under A shows the

timing of this frame of the movie film).

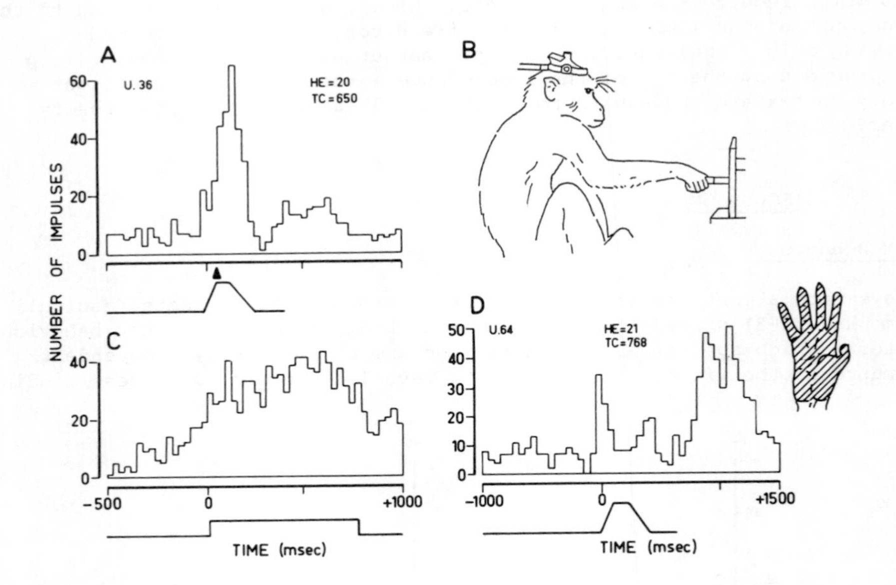

Fig. 8. Examples of two neurones with sensory inputs recorded in area 2 of monkey 3. Unit 36 responded vigorously to passive flexion of the elbow as shown in C; average duration of the flexion is indicated by the step in the trace below. This passive response pattern is mirrored by that observed during active elbow flexion during the pull (A) especially after the start of the plateau phase of lever movement (▲). The angle of the elbow at this stage is shown in B. D shows the activity of a neurone in area 2 which had a cutaneous afferent input zone involving all glabrous skin on the ventral surface of the hand. When brief taps were applied to the skin, the discharge of the unit showed a clear on-off response. This is reflected in the histogram obtained during active movements as a paired burst for grasping and releasing the lever, and also for grasping and releasing the food.

are in marked contrast to those of Ferraro and Barrera (12) and of Gilman and Denny-Brown (14) who reported severe loss of motor control after such lesions. However, they are in agreement with the work of Beck (5) in the squirrel monkey and of Mettler and Liss (21) in the rhesus monkey. Lack of motor disturbance is also mentioned in passing in a study of stumptail monkeys with similar lesions (4). Since in many of the monkeys used in the Gilman and Denny-Brown study the lesions were made at the C3 and C4 level, in one of the animals of the present study (monkey 4) the cuneate fasciculi were purposely interrupted at C5/6 but most of the dorsal horn at the level of the lesion was found to have been damaged as well. Although this animal showed transient signs of ataxia, it did not display permanent severe motor disturbances and the ataxia probably resulted from localized damage to dorsal root fibres since ataxia is a prominent symptom in monkeys with deafferented limbs (6). It is difficult to reconcile the data of the present study and those of others (4, 5, 21) with the findings of Gilman and Denny-Brown but it is possible that in their long term experiments fibre systems other than the dorsal columns had been damaged e.g. by interference with the cord's vascular supply, and that the disappearance of these fibres went undetected in their histological examination of the lesions. Different methods of testing in their experiments and the ones reported on here could be another important factor. The monkeys for this study were housed individually but had social contacts with other monkeys and they were encouraged to use their forelimbs actively and to move around.

Sensory Behaviour

The deficit in the monkeys, when blindfolded, for distinguishing the "texture" of a raisin placed in a gutter so that only a limited surface was available for palpation with the tip of the index finger contrasts with the proficiency shown in recognizing these same raisins when presented on a flat surface, and the ability to recognize shapes such as the handle of the lever when grasping, or when "fingering" down a forceps to retrieve a food reward at its tip. A similar deficit was found in the contactual placing tests when the site stimulated was located on the hand especially on its glabrous skin. These deficits seem to be of the same kind as that found in another study in which monkeys were not permanently impaired on shape discriminations after dorsal column lesions but consistently failed to recognize shapes cut out in small disks (4).

Responses of Precentral Neurones

After complete lesions of the cuneate fasciculi the responses of the neurones in the precentral gyrus to natural stimulation of receptors in the forelimb of the passive, relaxed animal were abolished. In intact animals, such responses are found in 80% of the neurones (18). Nevertheless, in the operated animals the patterns of modulation of the activity of these neurones during a variety of movements was indistinguishable from those found in intact monkeys during the performance of a learned motor task, and so was the motor behaviour of the operated animals in general. This raises the question of the role which 'long-loop' reflexes through the cuneate nuclei and precentral motor cortex play in the coordination of motor behaviour. Preliminary analyses of EMG activity of normal and operated monkeys performing the lever-pulling task while a brief perturbation is imposed upon the lever movement show that, in both groups, the EMG activity patterns obtained for normal lever pulling performance and for pulls with disturbances superimposed do not differ significantly. This is in contrast to some findings in humans (1).

Responses of Postcentral Neurones

The sample of postcentral neurones which showed modulation of their activity in association with the learned motor task of monkey 3 was found to consist of two groups, one without and one with afferent input. Evarts (11) studied the behaviour of postcentral neurones in the conscious normal monkey and related this behaviour to the onset of EMG activity. Unfortunately in his experiments no attempt was made to assess afferent input to the neurones. In the present study some of his results have been confirmed in animals with lesions of the cuneate fasciculi: postcentral cells tended to discharge later than precentral cells during the same type of movement. In contrast to his findings, some cells for which no afferent input could be detected discharged before the onset of EMG activity while for other neurones in this group increase in discharge coincided with EMG onset. It would be of interest to reinvestigate the behaviour of postcentral neurones in the normal awake, behaving monkey since, in the operated animals, these cells formed a sizeable proportion (40%) of the sample, and their behaviour and the timing of their changes in discharge would suggest that they could be part of a precentral-postcentral "corollary discharge" channel.

Organization of Afferent Inputs to the Postcentral Gyrus

No responsive neurones were found in area 3a after lesions of the cuneate fasciculi, in agreement with earlier findings (9) and only a few in area 3b. Sensory inputs to the latter area were exclusively cutaneous. Those to area 1 were either cutaneous or derived from "deep" receptors especially those activated by joint movement. Deep receptor input from joint movement dominated in area 2 where most of the responsive neurones were found. This distribution of different kinds of afferent input to different cytoarchitectonic subdivisions found in monkeys with lesions of the dorsal column system parallels that found in normal monkeys (25) except that cutaneous input seems to be greatly diminished. This lack of cutaneous input may be reflected in the deficits seen behaviourally in the contactual placing tests. Conversely, the input from deep receptors directed especially to area 2 may be responsible for sparing of the capacity for shape discriminations in monkeys with dorsal column lesions (4, 29). Lesions of this area, which apparently receives input from deep receptors both via lemniscal and extralemniscal pathways, results in deficiencies of shape discrimination but not of discrimination of texture or roughness (26).

Extent of Afferent Input Zones

All postcentral neurones which responded to natural peripheral stimulation, did so only for stimuli within limited afferent input zones. Cutaneous receptive fields were very small, and driving from deep receptors usually occurred during movement of a single joint in one direction only. Although no data are available for the forelimb, spinothalamic afferents from the monkey hindlimb in acute preparations have been shown often to possess very restricted afferent input zones, both for cutaneous and deep receptors (3, 33). Monkey spinocervical neurones also often possess very localized small receptive fields (8). If the same applies to the monkey forelimb these two pathways could account for some of the results obtained in the present study.

Somatotopic Localization within the Postcentral Gyrus

Modulation of the discharge of most of the postcentral neurones was associated with movements of the distal limb and they were encountered, be it in variable numbers, in a zone which extended in an anterior-posterior direction through all cytoarchitectonic subdivisions of the postcentral gyrus. Neurones associated with proximal movements were found both medial and lateral to this band. Again these findings parallel those obtained in acute experiments in intact normal monkeys (31).

ACKNOWLEDGEMENTS

This work was supported by grants from the National Health and Medical Research Council and from the Australian Research Grants Committee.

REFERENCES

(1) J. Adam, C.D. Marsden, P.A. Merton and H.B. Morton, The effects of lesions in the internal capsule and the sensorimotor cortex on servo action in the human thumb, J. Physiol. 254, 27-28P (1976).

(2) D. Albe-Fessard and J. Liebeskind, Origine des messages somato-sensitifs activant les cellules dur cortex motor chez le singe, Exp. Brain Res. 1, 127-146 (1965).

(3) A.E. Applebaum, J.E. Beall, R.D. Foreman and W.D. Willis, Organization and receptive fields of primate spinothalamic tract neurons, J. Neurophysiol. 38, 572-586 (1975).

(4) A. Azulay and A.S. Schwartz, The role of the dorsal funiculus of the primate in tactile discrimination, Exp. Neurol. 46, 315-332 (1975).

(5) C. Beck, Forelimb performance by squirrel monkeys (Saimiri sciureus) before and after dorsal column lesions, J. Comp. Physiol. Psychol. 90, 353-362 (1976).

(6) J. Bossom, Movement without proprioception, Brain Research 71, 285-296 (1974).

(7) J. Brinkman, B.M. Bush and R. Porter, Deficient influences of peripheral stimuli on precentral neurones in monkeys with dorsal column lesions, Manuscript submitted for publication (1977).

(8) R.N. Bryan, J.D. Coulter and W.D. Willis, Cells of origin of the spino-cervical tract in the monkey, Exp. Neurol. 42, 574-586 (1974).

(9) D.A. Dreyer, R.J. Schneider, C.B. Metz and B.L. Whitsel, Differential contributions of spinal pathways to body representation in post-central gyrus of Macaca mulatta, J. Neurophysiol. 37, 119-145 (1974).

(10) E.V. Evarts, Motor cortex reflexes associated with learned movements, Science 179, 501-503 (1973).

(11) E.V. Evarts, Precentral and postcentral cortical activity in association with visually triggered movement, J. Neurophysiol. 37, 373-381(1974).

(12) A. Ferraro and S.A. Barrera, Effects of experimental lesions of the posterior columns in Macacus rhesus monkeys, Brain 57, 307-332(1934).

(13) E.E. Fetz and M.A. Baker, Response properties of precentral neurons in awake monkeys, Physiologist, Washington 12, 223P (1969).

(14) S. Gilman and D. Denny-Brown, Disorders of movement and behaviour following dorsal column lesions, Brain 89, 397-418 (1966).

(15) S. Goldring and R. Ratcheson, Human motor cortex : Sensory input data from single neuron recordings, Science 175, 1493-1495 (1972).

(16) D.G. Lawrence and H.G.J.M. Kuypers, The functional organization of the motor system in the monkey. I. The effects of bilateral pyramidal lesions, Brain 91, 1-14 (1968).

(17) R.N. Lemon, J.A. Hanby and R. Porter, Relationship between the activity of precentral neurones during active and passive movements in conscious monkeys, Proc. Roy. Soc. B. 194, 341-373 (1976).

(18) R.N. Lemon and R. Porter, Afferent input to movement-related precentral neurones in conscious monkeys, Proc. Roy. Soc. B. 194, 313-339(1976).

(19) C.L. Li and T.M. Tew Jr., Reciprocal activation and inhibition of cortical neurones and voluntary movements in man: Cortical cell activity and muscle movement, Nature 203, 264-265 (1964).

(20) W. Marsden, P.A. Merton and H.B. Morton, Is the human stretch reflex cortical rather than spinal?, The Lancet i, 759-761 (1973).

(21) F.A. Mettler and H. Liss, Functional recovery in primates after large subtotal spinal cord lesions, J. Neuropath. exp. Neurol. 107, 509-516 (1959).

(22) C.G. Phillips, T.P.S. Powell and M. Wiesendanger, Projection from low threshold muscle afferents of hand and forearm to area 3a of baboon's cortex, J. Physiol. 217, 419-446 (1971).

(23) R. Porter, M. McD. Lewis and G.F. Linklater, A headpiece for recording discharges of neurones in unrestrained monkeys, Electroenceph. clin. Neurophysiol. 30, 91-93 (1971).

(24) R. Porter and P.M.H. Rack, Timing of the responses in the motor cortex of monkeys to an unexpected disturbance of finger position, Brain Research 103, 201-213 (1976).

(25) T.P.S. Powell and V.B. Mountcastle, Some aspects of the functional organization of the postcentral gyrus of the monkey: A correlation of findings obtained in a single unit analysis with cytoarchitecture, Bull. Johns Hopk. Hosp. 105, 133-162 (1959).

(26) M. Randolph and J. Semmes, Behavioural consequences of selective subtotal ablations in the postcentral gyrus of Macaca mulatta, Brain Research 70, 55-70 (1974).

(27) I. Rosén and H. Asanuma, Peripheral afferent inputs to the forelimb area of the monkey motor cortex: input-output relations, Exp. Brain Res. 14, 257-273 (1972).

(28) J.E. Shriver, B.M. Stein and M.B. Carpenter, Central projections of spinal dorsal roots in the monkey. I. Cervical and upper thoracic dorsal roots, Am. J. Anat. 123, 27-74 (1968).

(29) C.J. Vierck, Alterations of spatiotactile discrimination after lesions of primate spinal cord, Brain Research 58, 69-79 (1973).

(30) A.E. Walker and T.A. Weaver Jr., The topical organization and termination of the fibers of the posterior columns in Macaca mullata, J. Comp. Neurol. 76, 145-158 (1942).

(31) B.L. Whitsel, D.A. Dreyer and J.R. Roppolo, Determinants of body representation in postcentral gyrus of macaques, J. Neurophysiol. 34, 1018-1034 (1971).

(32) M. Wiesendanger, Input from muscle and cutaneous nerves of the hand and forearm to neurones of the precentral gyrus of baboons and monkeys, J. Physiol. 228, 203-219 (1973).

(33) W.D. Willis, D.L. Trevino, J.D. Coulter and R.A. Maunz, Responses of primate spinothalamic tract neurons to natural stimulation of hindlimb, J. Neurophysiol. 37, 358-372 (1974).

INTERPRETATIONS OF THE SENSORY AND MOTOR CONSEQUENCES OF DORSAL COLUMN LESIONS

Charles J. Vierck, Jr.

Department of Neuroscience, University of Florida College of Medicine, Gainesville, Florida 32610, U.S.A.

The intent of this article is to consider the functions served by the dorsal column - medial lemniscus pathway in the context of the substantial projection of this system to the cerebral cortex. For some time, it was thought that the primary somatosensory cortex (SI) received nearly all of its thalamocortical input from the dorsal column (DC) - lemniscal system. However, the earliest physiological studies of single cortical units in animals with DC lesions did not find global deafferentation of the SI or SII cortical areas, either in terms of space (i.e., areas with cells that were not driven by peripheral stimulation) or submodality composition (e.g., loss of small-field light touch units; Refs.16, 21, 41). These results have led to several suppositions: (1) that DC afferents converge extensively onto thalamic and cortical units, such that the effects of DC lesions on the receptivities of these cells are more subtle than had been expected; extensive convergence of the DC-lemniscal system with other ascending pathways appears to be a likely possibility, in view of their overlapping projections to the thalamus (Ref. 46); and (2) that there is a good deal of redundancy of information transmitted by the separate spinal pathways; this proposition has been amply confirmed by physiological studies of single cells in the ascending somesthetic pathways (see Ref. 8).

These revised interpretations of the dependency of somatosensory cortex on DC input, coupled with numerous behavioral investigations that have failed to reveal expected deficits in DC-lesioned animals (Refs. 10, 13, 14, 15, 34, 43, 54, 61, 62, 64, 69), have diminished an exalted reputation of the dorsal columns. For a long time, the dorsal columns were thought to be the only pathway responsible for the discriminative aspects of touch and proprioception, when the discriminations required resolution either in space or time (Ref. 52). With the discovery of other ipsilateral pathways that join the medial lemniscus anatomically and functionally (Refs. 8, 12, 51), and with investigations depicting some "epicritic" capacities for the crossed pathways (Refs. 43, 69, 70, 74), it has become clear that the dorsal columns are not preponderant across the board for discriminative somesthetic sensations. In the wake of this realization, we have clung to a saving notion - that our premier pathway is the channel for sensorimotor rather than purely sensory functions (Refs. 63, 76). But now that we have had time to closely observe the performance of lesioned animals on a variety of tasks requiring coordination, precision and speed of motor movements as well as sensory evaluation of the consequences of these activities (Refs. 5, 61, 73), it is apparent that many aspects of somatomotor integration are not disrupted by DC lesions. Thus, it is time to begin construction of a new story with the pieces left from experiments guided by grandiose illusions for our favored spinal channel. It is the author's impression that recent evidence confirms some of the past expectations, in that the DCs do contribute critical information for discriminative somatosensory and sophisticated somatomotor capacities. The major difference from the classical story is that these DC functions are highly specialized, and they neither represent all discriminative somesthetic attributes nor fit easily into a singular category of coding operations.

Investigators who have used a fine-grained single unit analysis of somatosensory cortex have now demonstrated that, although DC lesions do not deafferent all of SI, they do significantly alter a portion of the SI cortex in primates (Refs. 17, 42). After a DC lesion, there is a dearth of touch sensitive units in cytoarchitectural areas 1 and 3b, in the portions of SI given over to the distal extremities, but the surrounding areas 3a and 2 are little affected. The input to areas 1 and 3b in normal animals is predominantly from skin afferents, while units in areas 3a and 2 respond primarily to stimulation of muscle or joint receptors. This finding of a precisely delimited gap in SI is of considerable interest for those who inquire into the functions of somatosensory cortex as well as those who seek to reveal the contributions made by the dorsal columns. The boundaries of the DCs are clearly defined for surgery and histology, while the limits of the somatosensory cortical areas (and sub-areas) are not. In addition to this problem of confining a lesion within the surface boundaries of a physiologically defined cortical area, direct cortical lesions can unintentionally have remote effects by involving the white matter, disturbing the blood supply or inducing edema. It is also the case that spinal lesions present some difficulties of interpretation, since the DCs do not project exclusively to SI cortex, and thus it is important to compare the results of spinal lesions with those following ablation of restricted portions of SI. To the extent that the central processing of DC input depends on the SI cortex, the physiological studies predict that DC lesioned animals should have touch but not proprioceptive deficits; the impairments should be restricted to the distal extremities; and the specific characteristics of the spinal deficits should be reproduced following lesions restricted to areas 3b and 1 of the distal extremity portions of the SI map.

One example of a DC function that fits the above predictions has been revealed by studies of motor capacity, which we would expect to uncover proprioceptive contributions of the dorsal columns. Much attention has been directed toward characterization of the importance of DC input for motor coordination or for sensory discriminations that are made during active tactile exploration. This approach was engendered by early reports that expected deficits in discrimination of passively received tactile stimuli are not observed with surgical DC lesions, but gross motor impairments have been described. For a period as long as several months following DC section, monkeys are ataxic, are reluctant to use the affected limb(s), display catatonic fixity of posture, and appear deficient in the initiation and guidance of projected movements in space (Refs. 26, 28). These findings would be expected to follow an interruption of proprioceptive information that specifies the positions of the limbs in space. However, observation of DC lesioned monkeys for months following surgery reveals a gradual return of gross motor function in a free-ranging situation to the point that the limbs ipsilateral to a lesion do not exhibit impairment during climbing, running and other complex motor activities (Refs. 73, 75).

As an aside, it should be useful at this point to categorize the major classes of behavioral effect that can be expected to result from interruption of a CNS pathway: (1) No deficit can be observed when the lesioned tract does not provide relevant information or when equally effective signals are offered by intact pathways. The latter explanation for no effect doesn't prove that there is redundancy of inputs from the pathways in question, because the organism may be able to make a certain discrimination by utilizing different sets of cues. (2) In the latter case, when several pathways provide cues that can be utilized for a common function, it should often be the case that a temporary loss of the function will result from interruption of either tract. This can result from a variety of causes and has been observed in several instances following DC section (Refs. 70, 74). Disruption of one set of cues can alter the relevant sensations to the extent that the animal must learn to use the "new" cues, or the lesion might produce a condition of "shock" or imbalance of inputs that temporarily disrupts interpretative neural functions, in spite of the availability of sufficient information to support the behavior. (3) In order to show that a given function depends critically upon unique contributions of a pathway, it is necessary to observe the animals over a long period of time and to motivate them strongly to perform the task(s) that are presumed to measure that function.

Evidence that the dorsal columns do not provide critical inputs for fast and accurate projection of the limbs toward a target has been provided by studies in which animals have been trained to execute a variety of motor acts. In these situations, only transient deficits of limb projection are seen, regardless of whether the movements are of the ballistic or pursuit varieties, and even when the task requires accurate direction to different points in space, without visual control (Refs. 5, 22, 73, 75; see Fig. 1). Beyond indicating that the precise control of limb movements does not depend upon the somesthetic feedback provided by the dorsal columns, the investigations of trained motor capacities suggest that the detailed topographic organization of the dorsal column system is not crucial for maintenance of a body image. That is, in order to program a movement toward an object without visual guidance, it must be necessary for the organism to conceptualize the spatial relationships of the body parts and the external object.

While the evidence reviewed to this point indicates that chronic DC lesioned animals exhibit normal sequencing of individual muscles to direct a limb toward a target and normal coordination of the different limbs in running and climbing, some complex motor tasks have revealed DC deficits (Refs. 18, 49, 50). This presents difficulties analytically, because complex tasks are more likely than simple tasks to contain some component feature that is sensitive to the lesion, and it is difficult to unveil the relevant component as distinct from a generalized inability to handle complexity. The test situations that are sensitive to DC lesions involve operant responses to moving targets, so that the animal must project a limb (and sometimes the entire body) to meet the moving object at some point in space. Deficits on these tasks could indicate an impaired ability to anticipate. This in turn can be imagined to result from an interruption of the somatosensory cortical relay to the prefrontal cortex, where predictions of future contingencies may be processed. Or the difficulty could result from defective correlation of somatosensory and visual inputs (e.g., by the posterior parietal cortex). Also, it has been suggested that complex <u>sequences</u> of motor activity (e.g., directed jumping followed by directed reaching and then landing) are not programmed appropriately in the absence of the dorsal columns. While it is difficult to rule out

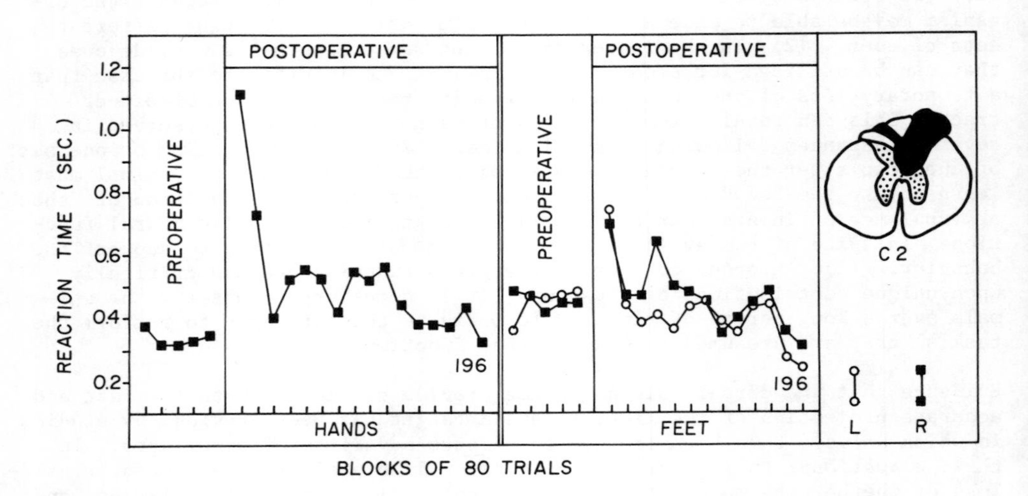

Fig. 1 Ballistic direction of either hand or foot to different points in space, without visual control. After a recent DC lesion at C_2, this monkey demonstrated complete recovery of the speed of the movements, over 196 calendar days.

these interpretations entirely, there is evidence that primary motor control is disrupted at several specific instances in the complex sequences.

When the tested behavior involves leaping, DC lesioned animals do not generate the normal profile of force on takeoff, as measured on a platform with pressure sensors (Ref. 48). While this could result from impaired anticipation and a resultant disruption of the motor program, it seems more reasonable to propose that generation of the proper takeoff force depends on tactile feedback from the foot pads (Ref. 48). This would permit the animal to compensate for movement of the surface leaped from, and it would appear to be necessary to account for the ability of animals to jump appropriately either from a standing start or when running. This explanation is certainly consistent with introspective analysis of the exquisite tactile sensitivity that is required to accurately project an environmental object a certain distance or speed. At the point of release, when throwing a ball, there appears to be a critical comparison of the force being exerted and the resistance that is offered by the mass of the ball. Such a feedback would have to be provided by a system with characteristics similar to those possessed by the dorsal columns. The input should be rapidly and securely conducted and precisely timed, and topographic organization would be required to specify the direction of compensatory motor responses. For example, if a supporting surface tilted as an animal jumped, the compensatory muscular actions would differ greatly according to the direction of tilt, which could be specified by the topography of pressure on the foot. In addition, the requisite afferent channel should project onto a motor system that regulates the force of muscular contractions, particularly of the distal extremity. This requirement is met by cortico-spinal projection neurons of the motor cortex, which are reciprocally connected with SI and discharge in proportion to the force of contraction generated in the distal muscles they innervate (Ref. 24). An apparent contradiction to this analysis is offered by the finding that dogs are able to maintain a standing posture despite perturbations of a supporting platform (Ref. 59). However, it has been demonstrated recently that a component of the reflex response to perturbations affecting distal extremity flexors is lost following interruption of long ascending spinal pathways (including the dorsal columns; Ref. 45). Thus, it appears that simple recovery of a stable posture following small perturbations is redundantly directed by several sources of input and is not critically dependent upon the spinal-cortical-spinal component of the reflex adjustment (although the waveforms of compensatory force applied to the platform are altered by the DC lesions; Ref. 59).

Several studies have shown that the terminal phase of a motor response is disrupted by DC lesions, when it is required that the animal grasp an object. This deficit is obvious when the target is moving (Refs. 5, 18), but it is also evident when the object is stationary and small (or otherwise demands a precise response; Refs. 73, 75). Observation of the animals as they perform on a variety of tasks suggests that there may be two kinds of grasp deficits. One type of deficiency clearly is present for the forelimbs but is difficult to demonstrate for the hindlimbs. The greater involvement of the forelimbs could result from interruption of cuneocerebellar fibers or from a broader representation of slowly adapting touch and proprioceptive afferents in the cervical fasciculus cuneatus (most of the slowly adapting afferents do not project to cervical levels of the fasciculus gracilis; Refs. 56, 59). Whatever the explanation, DC lesioned monkeys have great difficulty in executing a facile grasp response (Ref. 73). With visual control permitted, the most enduring deficiency is in opposition of the tips of the thumb and forefinger. After months of postoperative practice, the animals can use the hands reasonably well, but they employ scraping movements to dig food out of depressions in a surface, and they oppose the sides of the thumb and forefinger. Also, without visual control, the monkeys are unable to rapidly execute the terminal movements required to place the involved hand in a small compartment. These impairments are consistent with a well documented involvement of the cortico-spinal system in manual dexterity and particularly in the fractionation of digital motor activities (Refs. 6, 39).

Another apparent deficit is difficult to separate unequivocally from primary difficulties in execution of a grasp response. It is observed in monkeys that are trained to grasp objects with their feet, with visual control permitted (Ref. 75). On many trials, the initial response is quick and accurate, and the target is acquired with no readily discernible abnormality of movement.

However, over months of testing on this simple task, the animals consistently make more errors with the involved limb. The errors consist of trials on which the first pass does not acquire the pellet, knocking it through a trap door and out of reach (Fig. 2). The existence of this impairment for the hindlimbs implicates the DC-lemniscal system (as distinct from spinocerebellar projections), as did the altered force profile that was detected for cats when they jumped (Ref. 48). If the errors of grasp do not reflect some insufficiency in the initial programmed set of movements, then they may be attributable to interruption of a dorsal column - pyramidal tract linkage that has been implicated as the circuitry responsible for adjustive motor responses to small perturbations during performance of motor tasks (Ref. 25). For example, SI cortical lesions eliminate an early EMG response to distal extremity stimulation in a situation where that stimulation serves as the signal for a discrete adjustive response (Ref. 65). Again, it is the fast conduction and the topographic resolution of the DC - SI - motor cortex - pyramidal tract linkage that would seem to be required to appropriately modify an existing motor program for the extremities on the basis of unpredictable events at the site of motor action. Thus, when a pellet is insecurely grasped by the DC-lesioned animals, they appear unable to make a quick adjustive response and prevent the object from slipping away. In order to compensate effectively, the animal would have to appreciate the direction in which the object moved, relative to the hand.

If the DC-lemniscal - pyramidal system is critical for adjustive motor responses to tactile stimulation, then it seems reasonable to expect deficits of tactile placing with lesions of this long reflex circuit. Several investigations of tactile placing confirm that expectation and claim an absence of placing following dorsal column or SI cortical lesions (Refs. 4, 28), but there is dissent, particularly with regard to the DC contribution (Refs. 9, 44). It is likely that the disagreement stems from the fact that tactile placing is a complex response, and consequently only certain components may depend upon DC input. For example, an analysis of the placing reaction has shown that EMG activity of the biceps muscle consists of clearly dissociable early and late components during placing, and it is the late component that is more highly correlated with the latency of the withdrawal portion of the response (Ref. 1). Lesions of the ventrobasal thalamus (VB) or the somatosensory cortex ordinarily obliterate tactile placing, but the instances of sparing are perhaps the most instructive for the present discussion. The animals that retain placing reactions postoperatively do not exhibit the early EMG response normally present, and their movements are poorly directed (Ref. 1). These are precisely the components of the placing response that should be most affected by DC lesions. The dorsal column relay to motor cortex occurs quickly and appears to specify the direction of fine distal movements rather than the reaction of the entire limb.

The preceding account of motor-related dorsal column functions has stressed the direct communication of the DC-lemniscal system with the corticospinal system and has reviewed evidence for precise communal regulation by these pathways of the force and direction of distal extremity movements in relation to environmental objects. Several additional factors that must be taken into account in this formulation of long cortical reflex functions have to do with specifics of the anatomical and physiological interactions between the afferent and efferent pathways. First of all, while a localized deafferentation of the tactile "cores" of the distal extremity portions of the SI map fits with the DC lesion deficits, several investigators have shown that afferent input

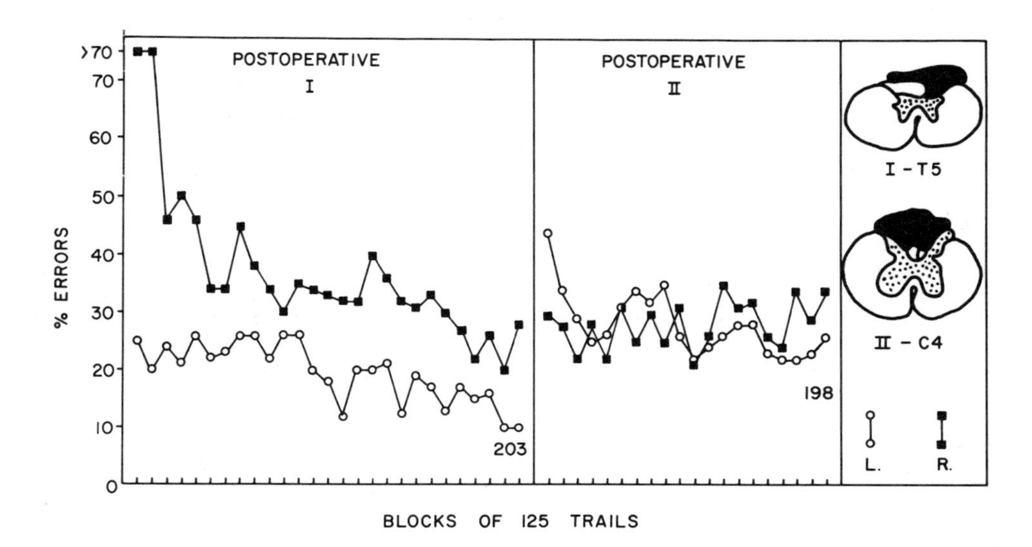

Fig. 2 Percent errors of a Macaca speciosa monkey attempting to grasp a food pellet (resting on a trap door) with either foot. Although preoperative data were not available, the performance of the involved sides (R. foot postop. I, R. and L. feet postop. II) was clearly inferior to that of the uninvolved left foot (during the postop. I period).

to the motor cortex is not exclusively relayed through the SI cortex (Refs. 53, 60). However, a report presented at this symposium shows that precentral neurons do not respond to somesthetic stimulation following DC section (Ref. 7). This finding underscores the importance of the DC system for feedback regulation of pyramidal functions, and it suggests that the systems interact via collaterals to the motor cortex from thalamic cells that project to the SI cortex and receive input from the dorsal columns.

A second, more difficult question of DC-pyramidal interactions is presented by efferents from the motor cortex to the DC - lemniscal system at medullary, thalamic and cortical levels (see Ref. 66). Since both inhibitory and facilitatory influences of pyramidal stimulation on DC - lemniscal conduction have been described, it is to be expected that certain qualitative features of the ascending conduction are accentuated during motor activity, while other presumably irrelevant features are blanked out. This could increase the signal to noise ratio in the afferent channel as a whole, and it could reduce the probability that competing response tendencies would be triggered by peripheral stimulation incidental to the intended response. In terms of the evidence presented so far it would be predicted that tactile input from the distal extremities that projects to VB (and then SI) would be immune from inhibition, relative to proprioceptive and proximal inputs.

The anatomical arrangement of inputs to the dorsal column nuclei (DCN) of cats appears to be consistent with the notion of a selective passing of distal extremity touch through the DC - lemniscal system (e.g., Ref. 31). The mid-dorsal regions of the DC nuclei contain clusters of cells that receive dense projections from cutaneous receptors located distally on the limbs, send axons almost exclusively to the ventrobasal thalamus, exhibit a high synaptic security and receive few corticofugal terminations. In contrast, the rostral and ventral portions of the DCN receive less dense dorsal root terminations and more non-primary spinal afferent input, project more diffusely to extra-lemniscal regions of the brain stem, the thalamus and cerebellum and receive corticofugal terminations. However, if some of the rostral and ventral DCN cells serve as interneurons for corticofugal inhibition of the mid-dorsal, cell-nest region (Ref. 2), then it would be the touch input to VB that is gated out during pyramidal activity. This has been confirmed at the thalamic level in cats, where VB cells responding to movements of joints are excited by stimulation of the pyramidal tract, while cutaneous (hair) units are inhibited by pyramidal stimulation (Ref. 67). There are indications that cats and primates differ markedly in the physiological patterns of inhibition and facilitation onto cells in the DC - lemniscal system (Ref. 66), suggesting that the prevalent motor activities of these animals are sufficiently distinct to require qualitative differences in the sensory information gated during corticospinal activation. Thus, it is dangerous to relate cat and primate data on the question of corticofugal modulation of somesthetic input. Nevertheless, the behavioral data supporting the importance of touch for distal extremity motor control come from cats as well as primates. Furthermore, perception of tactile stimulation is diminished in human subjects during performance of a motor task (Ref. 20), so the dilemma cannot be erased simply by invoking species differences.

The possibility that inhibition serves to gate out touch and accentuate proprioception could have some validity in the present context, since the majority of afferent influences on motor cortex cells are proprioceptive (Ref. 40). If this is the case, then it is important to determine whether the fast feedback regulation of distal movements depends upon joint and muscle receptors rather than the tactile input that has been emphasized. Another possibility is that inhibition of a given submodality serves to make that information more apparent under certain circumstances (Ref. 11). For example, if tonic activity of a group of cells is suppressed except when conditions permit peripheral stimulation to break through, then the resultant burst of activity would be salient by contrast. This saliency could be expressed spatially if the inhibition limits the responsivity of central cells to those with receptive fields centered at the locus of stimulation. Or the inhibition could be selective temporally. For example, it is reasonable to expect that DC input would be sampled only at the beginning and end points of discrete motor responses; but the time course of the inhibitory effects that have been seen argues against this (Refs. 11, 20).

The most plausible resolution of the puzzle of corticofugal control of dorsal column input is offered by an experiment presented at this symposium (Ref. 27). Motor cortex units that respond to peripheral input exhibit an enhanced excitability during controlled small movements of a response manipulandum, but these units are inhibited before and during ballistic movements. That is, the specificity of the corticofugal influences may be related more to the nature of the motor act than to some portion of the ascending channel or some aspect of the afferent signal. While feedback regulation is required to precisely control direct interactions with the environment, reflex adjustments during ballistic motor acts could be quite disruptive. For example, when an animal runs through the woods or swings through the trees, unavoidable contacts of the animal and vegetable limbs should not redirect the trajectories of the component movements, once they are initiated. Thus, it is likely that pyramidal activity predominates during fine movements of the distal extremities that operate directly on environmental objects or surfaces, and the corticospinal system must enhance conduction, at these times, over the DC - lemniscal projection to VB, SI and area 4. If this is the case, then inhibition of DC - lemniscal conduction during ballistic acts should be mediated by "extrapyramidal" systems. In this regard, it will be important to determine how these influences interact when animals emit rapid sequences of ballistic and finely controlled movements. It seems as if the corticospinal effects would be tightly locked in time to pyramidal motor activity and would override the influences seen during ballistic phases of movement. This combination of effects would increase the saliency (signal to noise ratio) of feedback appropriate for fine adjustive movements of the distal extremities. Also, it explains the inhibition of touch that has been observed at the initiation and termination of motor acts; the passing of DC input should ordinarily be at these times, but only when the movement requires precise interaction with the environment.

Given a fairly coherent body of evidence favoring the dorsal columns as the critical channel of somesthetic input to the corticospinal system, it is compelling to inquire as to the types of somesthetic cue which are utilized for pyramidal motor feedback. This is a difficult question to answer directly, but a useful approach to understanding the DC-lemniscal system might be to cross-predict between studies of motor and sensory function. It is now clear that the dorsal columns provide input to the pericruciate cortex that is essential for certain sensory discriminations as well as for specific motor reflexes, and although the sensory and motor functions are not <u>necessarily</u> related, it is reasonable to expect that they will be similar by virtue of arising from common evolutionary pressures. For example, if the motor feedback provided by the dorsal columns guides movements of the distal extremities when interacting with the environment, then the sensory cues provided by the dorsal columns may be those which would normally occur during active movement of an extremity when in direct contact with an object or surface (i.e., during <u>active touch</u>). Some of these sensory cues could depend upon active participation (e.g., as a result of corticofugal influences), while others might be demonstrated by passive stimulation that mimics conditions produced by active movement.

If the DC - lemniscal system has evolved in pace with capacities to manipulate environmental objects, then it seems likely that tactile and proprioceptive identification of these objects would require DC input (Refs. 63, 76). However, studies of monkeys with DC lesions have shown that active discriminations of size, shape and texture are not severely disrupted (Ref. 61). In terms of the present discussion, the principal lesson to be learned from these and related studies is that stereognosis can be accomplished by utilization of a variety of cues, and not all of the relevant information is carried by the DC - lemniscal system. Thus, it becomes important to analyze the component sensory attributes that permit stereognostic recognition.

When monkeys are required to tactually discriminate shapes cut in relief, so that the cues are acquired by moving the surface of the skin along edge contours, this form of stereognostic recognition is severely impaired by DC lesions (Ref. 3). There are, of course, a number of ways in which this task differs from the more usual method of testing stereognosis, where the animals are permitted to grasp a 3-dimensional object. The cut-out shapes cannot be appreciated on the basis of proprioceptive information that specifies the positions of the fingers during grasp or from holistic tactile impressions of the relative positions of edges or other salient features of the stimulus. That is, the animals cannot take a series of somesthetic "pictures" in which certain relationships are simultaneously apparent. The forms must be built up over time from a sequence of limited tactile cues that follow a sequence of chained movements. As an animal learns, an important strategy might be to seek confirmation of a series of expectancies that are specified in part by motor outputs (with associated corollary discharges).

Some of the cues provided by movement of the skin over an object can be isolated and delivered passively, to determine whether a motor component must be present for DC deficits to be observed. Although one of these, simple perception of movement across the skin, is not disrupted by DC lesions, discrimination of the direction of stimulus motion is permanently obliterated (Ref. 71). This finding confirms the expectation that DC sensory functions fit the "needs" of the corticospinal system for feedback relative to movement, without the requirement that the movement be actively produced by the animal. More generally stated, it may be valid to characterize dorsal column functions as spatiotemporally integrated. Fast conduction and precise spatial and temporal resolution can be seen as important attributes for corticospinal feedback and for determination of the direction of stimulus movement across the skin. In primates, only fast adapting afferents conduct the entire length of the dorsal columns (Ref. 59), and touch sensitive neurons in the motor cortex only respond to moving stimuli (Ref. 40). Not only are the DCs topographically ordered and comprised of units with small receptive fields, but the afferent inhibitory influences on DC cells operate precisely in space _and_ time (Refs. 30, 38). Thus, DC input to SI and the motor cortex from a moving point stimulus should have the following characteristics: stimulation of each point on the skin surface should produce a brief discharge in a small set of spatially coded neurons; and the onsets of discharge in successive sets should be precisely related to the rate of stimulus movement.

If we have learned anything from the dorsal columns, it is that no one principle will explain all of its functions, but it is likely that lack of attention to the temporal characteristics of stimuli utilized in behavioral tasks can explain some of the negative findings with DC lesions. For example, a variety of psychophysical and physiological experiments have combined to give the impression that the dorsal columns are not critical for proprioception. Dorsal column lesions do not impair position sense (Refs. 62, 69); slowly adapting joint receptors supplying the hindlimb exit the dorsal columns to synapse on cells that project into the lateral column (Refs. 56, 59); and the importance of joint receptors for position sense has been seriously questioned (Refs. 29, 47). Nevertheless, fast adapting joint receptors do project through the DC - lemniscal system to the SI and motor cortices (Refs. 40, 81). The properties of these afferents do not suggest that they would serve the perception of static limb position but rather that they would signal velocity or acceleration. Also, it is likely that feedback concerning rate of limb movement would be useful to the corticospinal system that can generate these rates by determining the force of muscular contractions. Thus, it will be important to determine whether DC lesions affect discriminations of the rate of movement at a given joint.

Another form of proprioceptive input that could provide useful feedback for pyramidal motor activities is provided by muscle receptors that project via fasciculus cuneatus and the medial lemniscus to area 3a of the pericruciate cortex (e.g., Ref. 37). Although there has been little behavioral work that would directly test the importance of muscle input for pyramidal regulation, it is of considerable interest that weight discrimination is not impaired by DC lesions (Ref. 13). While it seems that weight discrimination might be accomplished by a comparison of peripheral resistances to certain amounts of motor output, it appears instead that the animals monitor the amount of effort (motor output as sensed through corollary motor circuits) required to hold, lift or move the weights (Ref. 47).

As discussed earlier, a tactile feedback signal that might be required to precisely regulate the force of distal extremity motor activity is an accurate sense of pressure, and if this is the case, then DC-lesioned animals would be expected to show impairments on discriminations of tactile pressure. This expectation has been confirmed in part. Although absolute thresholds for touch are unaffected by DC lesions, discrimination of the amount of pressure applied to the skin (with von Frey hairs) is severely disrupted for several months after surgery in monkeys (Ref. 74). However, with continued postoperative training, different thresholds (DLs) for tactile pressure return to normal values. It is easy to explain the recovery of DLs on the basis of the projection of slowly adapting touch afferents onto ascending pathways other than the DCs, but it is more difficult to understand why the animals did not rely on the slowly adapting input all along. Only the type I and type II slowly adapting afferents code indentation over a wide range of values (Refs. 32, 78), and force has been assumed to be closely related to indentation. In fact, this assumption is questionable (Ref. 57), and the DC lesion study indicates that judgments concerning the intensity of a tactile stimulus are normally made during indentation, when fast adapting DC afferents are active and before the SA units have reached their steady state level of discharge. That is, the animals may naturally attend to force as it develops in time rather than static force. After the DC lesions, the animals could have relearned the DC task on the basis of static pressure cues that were available. This argument is strongly reinforced by a recent demonstration that DC lesioned monkeys are deficient at tracking an oscillating manipulandum, when they must rely on tactile feedback to generate constant force on the manipulandum (Ref. 19). Oscillation of the manipulandum prevents the development of static force cues. It is almost certainly pertinent to this discussion that the discharge of corticospinal projection cells of the motor cortex is related not simply to the force produced (and experienced) in displacing a manipulandum, but to a combination of absolute force and changes in force with time (Ref. 24).

In view of the above emphasis on spatiotemporal coding by the dorsal columns, one of the more puzzling examples of no deficit following DC lesions is presented by investigations of roughness discrimination in cats and monkeys (Refs. 15, 34, 61). It seems as if roughness would be specified by an interaction of spatial texture and the frequency of stimulation at a given spot, as the skin sweeps over a surface (Ref. 59). In addition, there is the motor

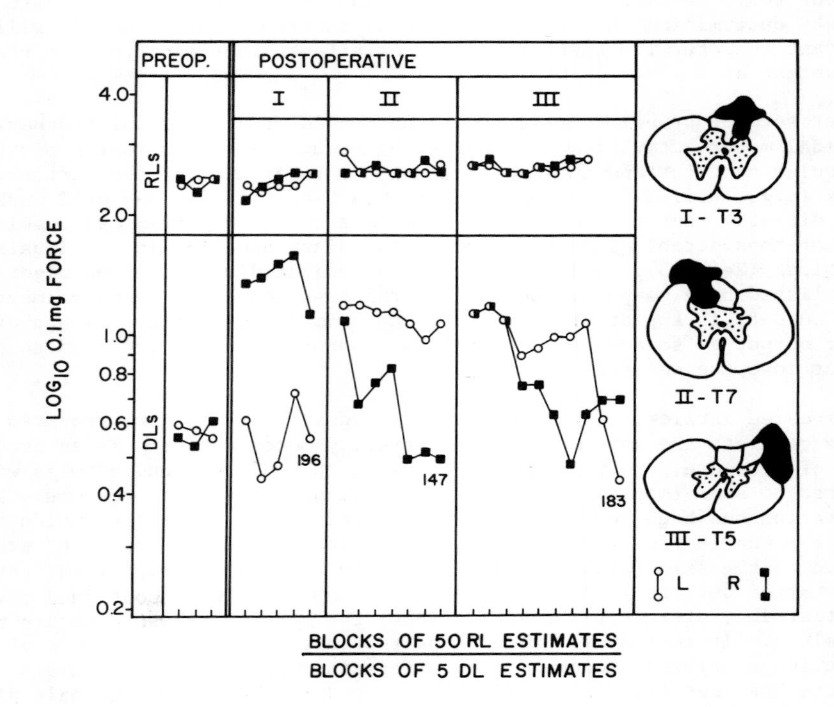

Fig. 3 Absolute thresholds (RLs) and difference thresholds (DLs) for touch pressure on the feet following sequential lesions to the R. dorsal column, then the L. dorsal column and then the R. dorsolateral column of a monkey. The DLs were elevated by DC lesions but recovered to normal values after extensive post-operative testing

search pattern that has been permitted in studies of roughness and has been thought to be a determinant of DC sensory functions. In order to analyze this problem, it will be necessary to deliver the stimuli passively and to separate out identifiable components to see if the presence of certain cues has masked deficits that would have been apparent with isolation of other cues. For example, it would be useful to test for discrimination between spatial textures, without contamination by variables such as sharpness, which differs among the sandpapers that have been used for roughness stimuli. It is quite possible that the primary cues provided by movement of textured stimuli across the skin could involve Pacinian afferents that are not conducted exclusively via the dorsal columns (Refs. 8, 62).

It is difficult to say, at this time, whether DC - lesioned animals are impaired in any discriminations that test pure somesthetic resolutions of either space or time. In the temporal domain, lesioned animals are able to detect the presence of vibration (Ref. 62), but it is not known whether they can discriminate different frequencies or intensities of oscillation. The question of purely spatial functions is more complicated on several grounds. First of all it is misleading to say that the spatial characteristics of a somesthetic stimulus can be varied independent of temporal cues, because the animal can choose to make judgments during the application of the stimulus, when tensions at different skin points will vary as a function of time. Also, when any stimulus with a significant spatial extent is applied to the skin, it will be important for the animal to determine whether the area stimulated represents the confines of the object or whether a smaller object has moved across the skin. This requires a temporal analysis of whether contact of different skin regions is simultaneous or successive. In view of these considerations, the fine-grained topographic resolution of the DC - lemniscal system, and the importance of accurately localized feedback to the pyramidal system, it is expected that most spatial somesthetic discriminations will be affected by DC lesions. Thus, a surprising negative finding in the DC literature has been the demonstration that 2-point thresholds are not significantly elevated in humans or monkeys following DC section (Refs. 10, 43).

Even though there are distortions of the body proportions as represented in the SI topographic map, spatiotactile patterns of stimulation have been thought to be faithfully represented in many respects by the configuration of unit activities in SI. Thus, it has seemed reasonable to assume that many spatial discriminations could be read off the map by a scanning process that detects the relative positions of active cells. However, recent physiological studies have indicated that given points on the body surface are multiply represented in SI (Ref. 17), and there may be a segregation of fast and slowly adapting units in areas 1 and 3b, respectively (Ref. 55). These data suggest that the cortex as a whole does not serve to represent tactile space in the true sense of a map, but that sub-areas of SI provide orderly sources of input to other groups of cells for specialized purposes. For example, cytoarchitectural area 1 appears to supply the motor cortex for fine motor feedback. Also, cells in lamina IV of the somatosensory cortex supply other SI and SII cells that respond selectively to certain directions of movement across the skin surface (Ref. 80). Results of this kind lead to the prediction that partial lesions or deafferentation of SI could affect certain spatiotactile functions without generally affecting spatial resolution. On the other hand, if perception of tactile space depends upon a holistic sensing of activity within SI, then the gap created in SI by DC lesions should impair all spatiotactile discriminations on the distal extremities.

The two-point tactile test appears to provide an optimal assessment of spatial resolution that would be served by the small-field DC afferents from the distal extremities. This expectation follows from a correlation that has been observed between 2-point thresholds from different body parts and the average receptive field sizes (and innervation densities and areas of cortical representation) for those regions (Refs. 68, 77). However, small receptive fields are not necessarily a major determinant of spatial resolution (Ref. 23), and the small fields of the distal extremities are not supplied exclusively by DC afferents (Ref. 8). Furthermore, powerful afferent inhibitory influences (as observed in the DC - lemniscal system) may interfere with 2-point discrimination (Ref. 33).

In discriminating between one and two points of tactile stimulation, it is assumed that an animal learns to detect a gap between the two points so that they are felt as separate. It is possible that the animals can detect the presence of a gap between two contours but be deficient in recognizing the distance between the two stimuli. This would be somewhat comparable to other findings: that absolute thresholds for touch intensity are not affected by DC lesions, while difference thresholds are elevated, or that discrimination of the direction of movement across the skin is impaired, while detection of movement (across the skin) per se is not disrupted (Ref. 71, 74). This kind of effect is reminiscent of a frequent result of cortical lesion studies - that it is only the most "difficult" discriminations that are affected, implying that the cortex is involved with "higher order" processing of sensory input. To the extent that DC lesion effects occur on discriminations within certain somesthetic categories or continua, without effects on distinctions between stimuli of these categories, then we will not only define the functions of the dorsal columns but will also appreciate one aspect of what defines "difficulty" and what is meant by "higher order processing" for the SI cortex. Possibly the "submodality purity" of the DC - lemniscal system is maintained not to permit qualitative distinctions between these categories (submodalities) but to provide uncontaminated inputs for quantitative distinctions within continua. If this is the case, then it may be possible to define categories of spatial discrimination that are served by the topography of the DC - lemniscal system. This is confirmed by the finding that discrimination of the sizes of discs (i.e., distances between edge contours) impressed on the skin is severely disrupted following DC lesions (Ref. 70). As is the case with the discrimination of touch intensity (Ref. 74), thresholds for tactile size return to normal levels after several months of postoperative training, and the impurities in these tasks remain to be defined. Nevertheless, the contrast between the early postoperative performance on 2-point and size discriminations is striking.

The clearest prediction concerning spatial coding that comes from studies of motor performance following DC lesions is that localization will be impaired (at least for the distal extremities), because it is important that adjustive motor acts be appropriately directed in relation to the eliciting stimuli. Although it seems as if normal 2-point discrimination might depend on an appreciation of the absolute or at least the relative positions of the two stimuli, this is not necessarily the case. If the animals can sense that there are two stimuli but be unable to specify the distance between edge contours, then they may also be unaware of the precise locations of the spots touched. Ideally, it would be valuable to test for absolute localization of singular skin contacts, but this is difficult to achieve with laboratory animal subjects, and the closest approximation has been a modification of the point localization paradigm. In this situation monkeys are stimulated twice on each trial at different locations, and the animals must identify the order in which the proximal and distal spots are touched. That is, they must indicate the relative locations of the two sequential contacts. The point localization thresholds that are obtained by systematically varying the distance between the spots touched are markedly elevated by dorsal column lesions (Ref. 72).

In summary, it is now apparent that the dorsal columns are "responsible" for many of the sophisticated somesthetic functions that appeared for a while to have been falsely credited to them. In terms of sensory attributes, deficits following DC section have been demonstrated with passive reception of stimuli varying along several continua. Lesioned animals have difficulty identifying the relative positions of and distances between stimulations with point or

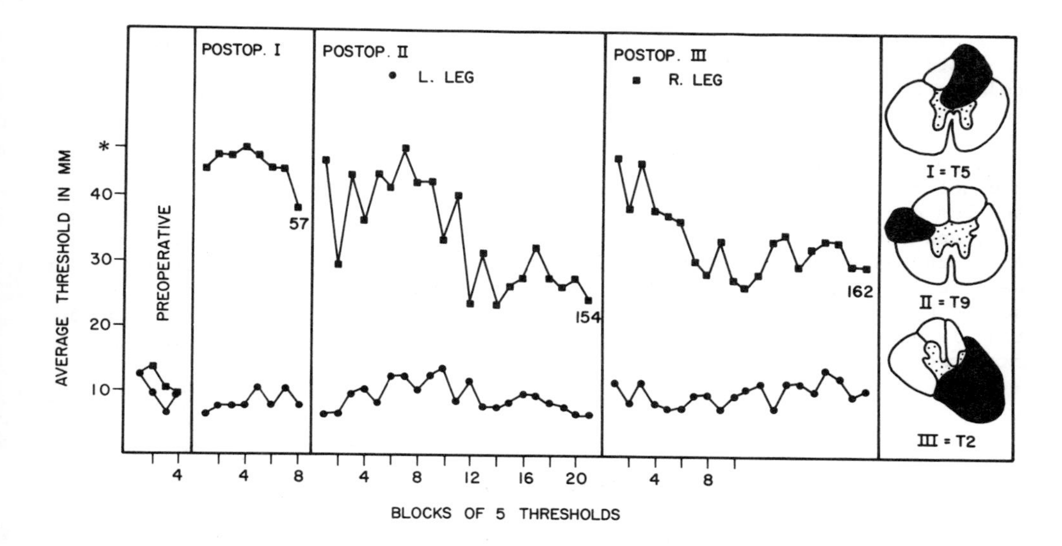

Fig. 4 Point localization thresholds on the legs before and after lesions to the R. dorsal column, the L. dorsolateral column and then the R. lateral and ventral columns of a monkey.

edge contours, even though they can discriminate between one and two contacts at the normal threshold of separation. Similarly, lesioned monkeys are able to discriminate between indentations of the skin with or without a minute component of movement across the skin surface, but they demonstrate an impaired appreciation of different forces of indentation or different directions of movement across the skin. In view of the pattern of negative and positive results of DC lesion studies over the years, it is highly unlikely that the sensory impairments which have been seen can be ascribed to any "non-sensory" deficit of movement, attention, motivation or learning. It is difficult to come up with a general description that neatly encompasses the principle features of all deficits observed to date, but several characteristics can be mentioned: (1) DC - lesioned animals have been shown to be temporarily deficient in several quantitative estimations - e.g., the amount of force applied to the skin or the amount of skin interposed between edge contours; and (2) several of the behavioral tests that have revealed enduring deficits suggest a particular involvement of the DC input in spatiotemporal integrations.

Although the DC input should not be considered to be strictly motor-related, in the sense that active movement is required for utilization of the ascending signals, all of the sensory deficits that have been described are consonant with apparent "needs" of the corticospinal system for peripheral feedback. Studies of motor behavior of animals with DC or SI lesions have revealed long term deficits in the abilities of animals to grasp (i.e., to fractionate finger movements), to precisely regulate the force of distal extremity movements and to adjust a motor act quickly in response to significant stimuli pertinent to the intention of the act. These impairments can be related in part to an absence of tactile feedback for specification of the location and force of stimulations impinging on the distal extremities. However compelling this correspondence of DC and pyramidal functions may be, it is important to note in closing that the sensory deficits following DC lesions are not restricted to stimulation of the distal extremities (Refs. 71, 72); the DCs contain a large number of afferents supplying hairy skin (Refs. 56, 80); and the DC - lemniscal system is well developed in cetaceans (Ref. 35). Also, the ascending projections of the dorsal columns are not restricted to the pericruciate cortex. Thus, it is likely that the full list of functions served by the dorsal columns extends well beyond the rudimentary description that is now available from experiments that have focused on a confined set of expectations.

REFERENCES

1. V.E. Amassian, R. Ross, C. Wertenbaker and H. Weiner, Cerebellothalamo-cortical interrelations in contact placing and other movements in cats, In: T. Frigyesi, E. Rinvik and M.D. Yahr (Eds.) (1972) Corticothalamic Projections and Sensorimotor Activities, p. 395, Raven Press, New York.

2. P. Andersen, J.C. Eccles, R.F. Schmidt and T. Yokota, Identification of relay cells and interneurons in the cuneate nucleus, J. Neurophysiol. 27, 1080 (1964).

3. A. Azulay and A.S. Schwartz, The role of the dorsal funiculus of the primate in tactual discriminations, Exper. Neurol. 46, 315 (1975).

4. P. Bard, Studies on the cortical representation of somatic sensibility, Harvey Lect. 33, 143 (1938).

5. C. Beck, Forelimb performance by squirrel monkeys (Saimiri Sciurius) before and after dorsal column lesions, J. Comp. Physiol. Psychol. 90, 353 (1976).

6. C.H. Beck and W.W. Chambers, Speed, accuracy, and strength of forelimb movement after unilateral pyramidotomy in Rhesus monkeys, J. Comp. Physiol. Psychol. 70, Pt. 2 (Monograph), 1 (1970).

7. J. Brinkman and R. Porter, Responses of cortical neurones to peripheral stimuli in monkeys with dorsal column lesions (this volume).

8. A.G. Brown, Ascending and long spinal pathways: dorsal columns, spinocervical tract and spinothalamic tract, In: A. Iggo (Ed.) Handbook of Sensory Physiology 2, 315 (1973).

9. J. Christiansen, Neurological observations of Macaques with spinal cord lesions, Anat. Rec. 154, 330 (1966).

10. A.W. Cook and E.J. Browder, Function of posterior columns in man, Arch. Neurol. (Chic.) 12, 72 (1965).

11. J.M. Coquery, Role of active movement in control of afferent input from skin in cat and man (this volume).

12. A.M. Dart and G. Gordon, Some properties of spinal connections of the cat's dorsal column nuclei which do not involve the dorsal columns, Brain Res. 58, 61 (1973).

13. J.L. DeVito, T.C. Ruch and H.D. Patton, Analysis of residual weight discriminatory ability and evoked potentials following section of dorsal columns in monkeys. Indian J. Physiol. Pharmacol. 8, 117 (1964).

14. I.T. Diamond, W. Randall and L. Springer, Tactual localization in cats deprived of cortical areas SI and SII and the dorsal columns, Psychon. Sci. 1, 261 (1964).

15. P.J.K. Dobry and K.L. Casey, Roughness discrimination in cats with dorsal column lesions, Brain Res. 44, 385 (1972).

16. P.J.K. Dobry and K.L. Casey, Coronal somatosensory unit responses in cats with dorsal column lesions. Brain Res. 44, 399 (1972).

17. D.A. Dreyer, R.J. Schneider, C.B. Metz and B.L. Whitsel, Differential contributions of spinal pathways to body representation in postcentral gyrus of Macaca mulatta, J. Neurophysiol. 36, 690 (1975).

18. B. Dubrovsky, F. Davelaar and E. Garcia-Rill, Role of dorsal columns in serial order acts, Exp. Neurol. 33, 93 (1971).

19. P. Dyhre - Poulsen, Impairment of tactile tracking after dorsal column lesions in monkeys, Exp. Brain Res. 23 (Suppl.), 64 (1965).

20. P. Dyhre - Poulsen, Perception of tactile stimuli during movement (this volume).

21. E. Eidelberg and C.M. Woodbury, Apparent redundancy in the somatosensory system in monkeys. Exper. Neurol. 37, 573 (1972).

22. E. Eidelberg, B. Woolf, C.J. Kreinick and F. Davis, Role of the dorsal funiculi in movement control, Brain Res. 114, 427 (1976).

23. R.P. Erickson, Stimulus coding in topographic and non-topographic modalities: on the significance of the activity of individual sensory neurons, Psychol. Rev. 75, 447 (1968).

24. E.V. Evarts, Relation of pyramidal tract activity to force exerted during voluntary movement, J. Neurophysiol. 31, 14 (1968).

25. E.V. Evarts and J. Tanji, Reflex and intended responses in motor cortex pyramidal tract neurons of monkey, J. Neurophysiol. 39, 1069 (1976).

26. A. Ferraro and S.E. Barrera, Effects of experimental lesions of the posterior columns in Macacus rhesus monkeys, Brain 57, 307 (1934).

27. C. Fromm and E.V. Evarts, Motor cortex responses to kinesthetic inputs during postural stability, precise fine movement and ballistic movement in the conscious monkey (this volume).

28. S. Gilman and D. Denny-Brown, Disorders of movement and behavior following dorsal column lesions, Brain, 89, 397 (1966).

29. G.M. Goodwin, D.I. McCloskey and P.B.C. Mathews, The contribution of muscle afferents to kinesthesia shown by vibration induced illusions of movement and by the effects of paralyzing joint afferents, Brain 95, 705 (1972).

30. G. Gordon and M.G.M. Jukes, Dual organization of the exteroceptive components of the cat's gracile nucleus, J. Physiol. 173, 263 (1964).

31. P.J. Hand and T. Van Winkle, The efferent connections of the feline nucleus cuneatus, J. Comp. Neurol. 171, 83 (1977).

32. T. Harrington and M.M. Merzenich, Neural coding of the sense of touch: human sensations of skin indentation compared with the responses of slowly adapting mechanoreceptive afferents innervating the hairy skin of monkeys, Exp. Brain Res. 10, 251 (1970).

33. M.B. Jones, C.J. Vierck, Jr. and R.B. Graham, Line-gap discrimination on the skin, Percept. Motor Skills 36, 563 (1973).

34. S.T. Kitai and J. Weinberg, Tactile discrimination study of the dorsal column-medial lemniscal system and spino-cervico-thalamic tract in cat, Exp. Br. Res. 6, 234 (1968).

35. L. Kruger, Specialized features of the cetacean brain, In: K.S. Norris (Ed.) (1966), p. 232, Whales, Dolphines and Porpoises, Univ. Calif. Press.

36. L. Kruger and J.A. Mosso, An evaluation of duality in the trigeminal afferent system, Adv. Neurol. 4, 73 (1974).

37. S. Landgren, H. Silfvenius and D. Wolsk, Somato-sensory paths to the second cortical projection area of the group I muscle afferents, J. Physiol. 191, 543 (1967).

38. S.E. Laskin, Cortical correlates of cutaneous masking: a comparison of psychophysical and electrophysiological inhibitory functions, Doctoral dissertation, New York University, 1975.

39. D.G. Lawrence and H.G.J.M. Kuypers, the Functional organization of the motor system in the monkey: 1. the effects of bilateral pyramidal lesions, Brain 91, 1 (1968).

40. R.N. Lemon and R. Porter, Short-latency peripheral afferent inputs to pyramidal and other neurons in the precentral cortex of conscious monkeys (this volume).

41. M. Levitt and J. Levitt, Sensory hind limb representation of SmI cortex in the cat after spinal tractotomies. Exp. Neurol. 22, 276 (1968).

42. M. Levitt and J. Levitt, Post central somatic mechanoreception: I. dorsal column and anterolateral lesions, Proc. Soc. Neurosc. 306 (1974).

43. M. Levitt and R.J. Schwartzman, Spinal sensory tracts and two-point tactile sensitivity. Anat. Rec. 154, 377 (1966).

44. A. Lundberg and U. Norrsell, Spinal afferent pathway of the tactile placing reaction, Experientia 16, 123 (1960).

45. C.D. Marsden, P.A. Merton, H.B. Morton and J. Adam, The effect of posterior column lesions on servo responses from the long thumb flexor, Brain 100, 185 (1977).

46. D.C. Mash, G. Worden and K.J. Berkley, Projections of the lateral cervical nucleus and the dorsal column nuclei to regions adjacent to and within the n. ventralis posterolateralis of the cat thalamus. Neurosc. Abs. 2, 943 (1976).

47. D.I. McCloskey and S.C. Gandevia, Role of inputs from skin, joints and muscles, and of corollary discharges, in human discriminatory tasks (this volume).

48. M. McCormack and B. Dubrovsky, Biomechanical and cinematographical analysis of a serial motor act. Effects of dorsal column section, Neurosc. Abst. 2, 525 (1976).

49. R. Melzack and J.A. Bridges, Dorsal column contributions to motor behavior, Exp. Neurol. 33, 53 (1971).

50. R. Melzack and S.E. Southmayd, Dorsal Column contributions to anticipatory motor behavior, Exper. Neurol. 42, 274 (1974).

51. F. Morin, A new spinal pathway for cutaneous impulses, Amer. J. Physiol. 183, 245 (1955).

52. V.B. Mountcastle, Some functional properties of the somatic afferent system, In: W.A. Rosenblith (Ed.) (1961) Sensory Communication, p. 403, John Wiley, New York.

53. J.T. Murphy, Y.C. Wong and H.D. Kwan, Afferent-efferent linkages in motor cortex for single forelimb mucles, J. Neurophysiol. 38, 990 (1975).

54. U. Norrsell, The spinal afferent pathways of conditioned reflexes to cutaneous stimuli in the dog. Exp. Br. Res. 2, 269 (1966).

55. R.L. Paul, M. Merzenich and H. Goodman, Representation of slowly and rapidly adapting cutaneous mechanoreceptors of the hand in Brodmann's area 3 and 1 of Macaca mulatta, Brain Res. 36, 299 (1972).

56. D. Petit and P.R. Burgess, Dorsal column projection of receptors in cat hairy skin supplied by myelinated fibers. J. Neurophysiol. 25, 849 (1968).

57. J. Petit and Y. Galifret, Sensory coupling function and the mechanical properties of the skin (this volume).

58. T.A. Quilliam, Structure of finger print skin (this volume).

59. P.J. Reynolds, R.E. Talbott and J.M. Brookhart, Control of postural reactions in the dog: the role of the dorsal column feedback pathway, Brain Res. 40, 159 (1972).

60. I. Rosén and H. Asanuma, Peripheral afferent inputs to the forelimb area of the monkey motor cortex: input-output relations, Exp. Brain Res. 14, 257 (1972).

61. A.S. Schwartz, E. Eidelberg, P. Marchok and A. Azulay, Tactile discrimination in the monkey after section of the dorsal funiculus and lateral lemniscus, Exp. Neurol. 37, 582 (1972).

62. R.J. Schwartzman and M.D. Bogdonoff, Proprioception and vibration sensibility discrimination in the absence of the posterior columns, Arch. Neurol. (Chic.) 20, 349 (1969).

63. J. Semmes, Protopathic and epicritic sensation: a reappraisal, In: A.L. Benton (Ed.) (1969) Contributions to Clinical Neuropsychology, p. 142, Aldine, Chicago, Ill.

64. D.M. Tapper, Behavioral evaluation of the tactile pad receptor system in hairy skin of the cat, Exp. Neurol. 26, 447 (1970).

65. W.G. Tatton, S.D. Forner, G.L. Gernstein, W.W. Chambers and C.N. Liu, The effect of post central cortical lesions on motor responses to sudden limb displacements in monkeys, Brain Res. 96, 108 (1975).

66. A.L. Towe, Somatosensory cortex: descending influences on ascending systems, In: A Iggo (Ed.) Handbook of Sensory Physiology 2, 701 (1973).

67. T. Tsumoto, S. Nakamura and I. Iwama, Pyramidal tract control over cutaneous and kinesthetic sensory transmission in the cat thalamus, Exp. Brain Res. 22, 281 (1975).

68. Å.B. Vallbo and R.S. Johansson, The tactile sensory innervation of the human hand (This volume).

69. C.J. Vierck, Jr., Spinal pathways mediating limb position sense, Anat. Rec. 254, 437 (1966).

70. C.J. Vierck, Jr., Alterations of spatio-tactile discrimination after lesions of primate spinal cord. Brain Res. 58, 69 (1973).

71. C.J. Vierck, Jr., Tactile movement detection and discrimination following dorsal column lesions in monkeys, Exp. Brain Res. 20, 331 (1974).

72. C.J. Vierck, Jr., Deficits in tactile direction sensitivity after dorsal column lesions in monkeys, Proc. Soc. Neurosc. 406 (1974).

73. C.J. Vierck, Jr., Proprioceptive deficits after dorsal column lesions in monkeys, In: H.H. Kornhuber (Ed.) (1975) The Somatosensory System, p. 311, Georg Thieme, Stuttgart, Germany.

74. C.J. Vierck, Jr., Absolute and differential sensitivities of touch stimuli after spinal cord lesions in monkeys, Brain Res. (In press).

75. C.J. Vierck, Jr., Comparison of forelimb and hindlimb motor deficits following dorsal column section in monkeys, Brain Res. (In press).

76. P.D. Wall, The sensory and motor role of impulses traveling in the dorsal columns towards cerebral cortex, Brain 93, 505 (1970).

77. S. Weinstein, Intensive and extensive aspects of tactile sensitivity as a function of body part, sex and laterality, In: D. Kenshalo (Ed.) (1968) The Skin Senses, p. 195, C.C. Thomas, Springfield, Ill.

78. G. Werner and V.B. Mountcastle, Neural activity in mechanoreceptive cutaneous afferents: stimulus-response relations, Weber functions, and information transmission, J. Neurophysiol. 28, 359 (1965).

79. B.L. Whitsel, L.M. Petrucelli and G. Werner, Modality representation in the lumbar and cervical fasciculus gracilis of squirrel monkey, Brain Res. 15, 67 (1969).

80. B.L. Whitsel, J.R. Roppolo and G. Werner, Cortical information processing of stimulus motion on the skin, J. Neurophysiol. 35, 691 (1973).

81. W.J. Williams, S.L. BeMent, T.C.T. Yim and W.D. McCall, Jr., Nucleus gracilis responses to knee joint motion: a frequency response study, Brain Res. 64, 123 (1973).

ROLE OF ACTIVE MOVEMENT IN CONTROL OF AFFERENT INPUT FROM SKIN IN CAT AND MAN

Jean-Marie Coquery

Laboratoire de Psychophysiologie, Université de Lille I, 59650 Villeneuve d'Ascq, France

The experiments to be reported here stem from this very simple observation : when a brief electrical stimulus, a train of 3 pulses with a frequency of 300/sec is applied to a fingertip at an intensity well above the threshold for perception as measured at rest, a flexion of the fingers abolishes the perception of this stimulus. Moreover, perception starts decreasing some time before the movement. Figure 1 gives an example of the results obtained in 6 subjects. It shows the time course of perception for a stimulus 1.5 times the threshold, before and during a flexion of the fingers. The subject had to flex his fingers every 1.5 second, the pace being set by a click over a loudspeaker ; the stimulus was delivered at different intervals before and after the movement. The subject had to tell whether he felt the stimulus or not by answering "yes", "no" or "weak". Time zero indicates the onset of EMG activity in the flexors of the fingers. The solid line gives the percentage of the "yes" and "weak" reports ; the hatched area indicates the percentage of the "weak" answers only. It is clear that perception is abolished during movement and starts decreasing more than 100 msec before the onset of EMG activity.

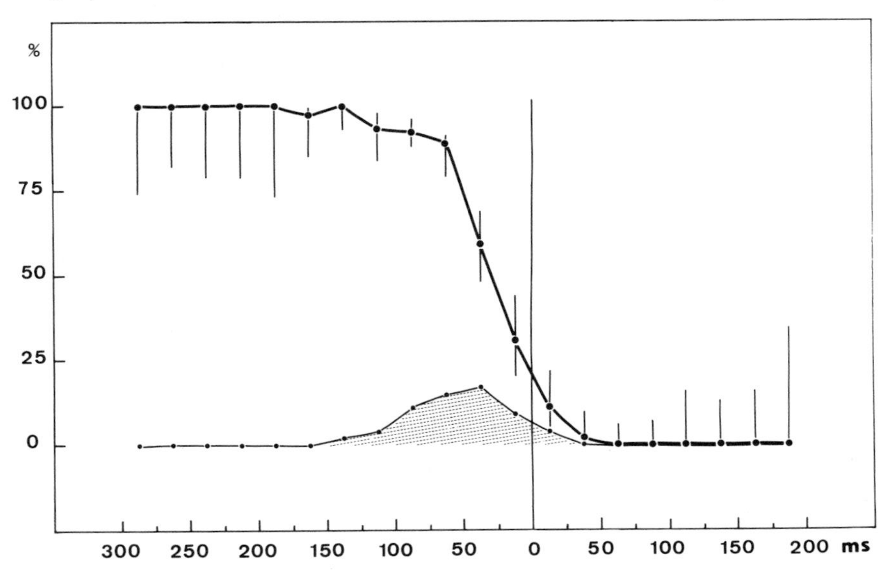

Fig. 1. Variations in the perception of an electrical stimulus before and during an active flexion of the fingers.

Such a perceptual suppression or depression may be accounted for by a number of factors, in particular by the afferents elicited by displacement of the stimulated area or by descending motor commands.

I - ROLE OF PERIPHERAL AFFERENTS IN THE PERCEPTUAL DEPRESSION DURING MOVEMENT.

Cutaneous and proprioceptive afferents set up by the displacement are likely to induce surround inhibition and this especially when their origin is close to the stimulated area. We have shown (ref. 7) that movement of a body segment decreases perception as a function of its proximity to the stimulus. Twelve subjects were stimulated on the ventral and dorsal surfaces of various parts of the upper limb : fingers, hand, forearm, arm. For each position of the stimulating electrodes, they had to extend or flex these four segments. Stimulation, using the same parameters as above, was triggered by the EMG of the active muscle with a 50 msec delay. Subjects had to subjectively rate the intensity of the stimulus on a four point scale. When the fingers were stimulated, the most effective movements for attenuating perception were, in decreasing order : movements of the fingers, of the hand, of the forearm, of the arm. When the hand was stimulated, this order was : hand, fingers, forearm, arm ; for the forearm : hand, forearm, arm, fingers ; for the arm : forearm, arm, hand, fingers.

These results are consistent with the possibility of lateral inhibition of the test volley during movement. However it is difficult to account for the diminution of perception which takes place before the occurence of the displacement and of the reafferences it triggers. We should involve in this case an

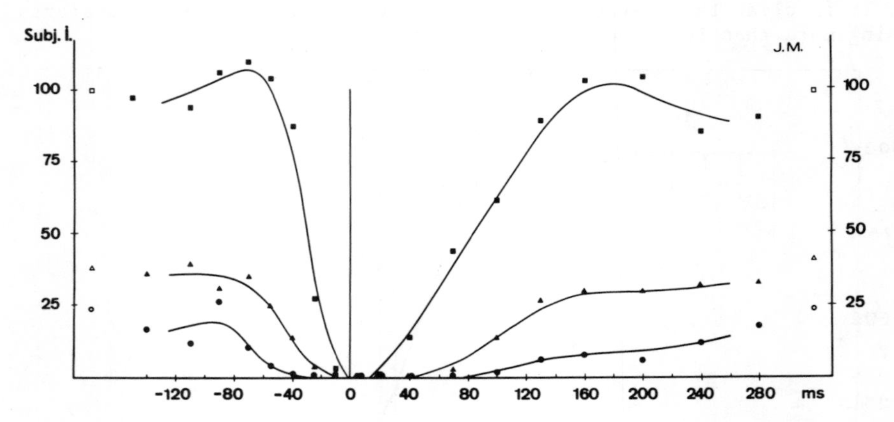

Fig. 2. Backward and forward masking of the perception of an electrical stimulus by a mechanical tap.

Abscissa : time interval in milliseconds between electrical test stimulus and mechanical masking tap ; onset of masking tap at time zero.
Ordinates : subjective intensities of the stimuli in percent of the control value of the strongest stimulus. Open symbols : control values of the three reference stimuli ; filled symbols : test values.

inhibition by the motor commands, which are present severals tens of milliseconds before any EMG activity (ref. 10, 4), unless we can demonstrate the same phenomenon before and during a passive movement.

As a passive movement would anyhow induce skin and joint stimulation, we decided to focus our investigation on the modifications of the perception of an electrical stimulus when preceded or followed by a light mechanical tap on the stimulated area (ref. 3). Six subjects were run, each during 16 sessions. Electrical stimuli were applied to the tip of the middle finger ; the mechanical taps, 5 msec in duration, were delivered by an electromagnetic hammer on the dorsal surface of the same finger. Three different intensities of the electrical stimulus were used, the highest one never being felt unpleasant by the subjects. After a warning signal, a first electrical stimulus, used as a reference, was given. Three seconds later, the test stimulus was given, preceded or followed at variable intervals by the mechanical slap. The nine possible pairs of reference and test stimuli were presented at random. Subjects had to report the subjective intensity of the test and reference stimulus as a ratio (ref. 9), the strongest member of the pair always being rated 10. Afterwards, the subjective values of the 3 stimuli, at the different intervals before and after the mechanical slap, were expressed as a percent of the mean value of the strongest stimulus as measured in the absence of the tap.

In the 6 subjects we found that perception was masked both before and after the application of the tap. For the weakest stimuli, backward masking may extend over 100 msec, forward masking over 250 msec (Fig. 2).

Taken together with the previous results, these data suggest that afferents elicited by active or passive displacement of a stimulated area may inhibit transmission of a test volley and/or interfere, at some unknown level, with the process of perceptual integration.

II - EFFECTS OF ACTIVE AND PASSIVE MOVEMENTS ON CORTICAL EVOKED POTENTIALS IN MAN.

Variations in the perception of an electrical stimulus during movement are sometimes correlated with alterations of the cortical potentials evoked in the somatic area by the test stimulus. Figure 3 gives an example of the evoked potentials elicited by stimulation of the middle finger at an intensity 2.8 times the threshold, given between 0 and 100 msec after the onset of EMG activity in the flexors of the fingers. Each trace is the average of 19 potentials. In A the subjects feels the stimulus ("yes" answers) ; in B perception is attenuated ("weak" answers) ; in C perception is obliterated ("no" answers). In this last condition the N1 wave of the evoked potential (peaking at 100 msec after the stimulus) is markedly reduced. However such a correlation between perception and the evoked potentials is not always evident. We simultaneously investigated in 13 subjects the changes in perception and the evoked potentials before and after a flexion of the stimulated hand (ref. 6). With respect to their amplitude at rest, evoked potentials (specially the N1 wave) are enhanced when elicited in the 200 msec before EMG onset. They start to decrease in the 100 msec after movement ; they are abolished in the 100 to 200 msec after EMG onset (Fig. 4, left). It should be noted that, although evoked potentials are increased before movement, perception is, at that time, attenuated as the subjects give some "weak" and "no" answers.

When movement is made by the hand controlateral to the stimulated one, evoked potentials are much less affected (Fig. 4, right) ; in 2 out of 5 subjects

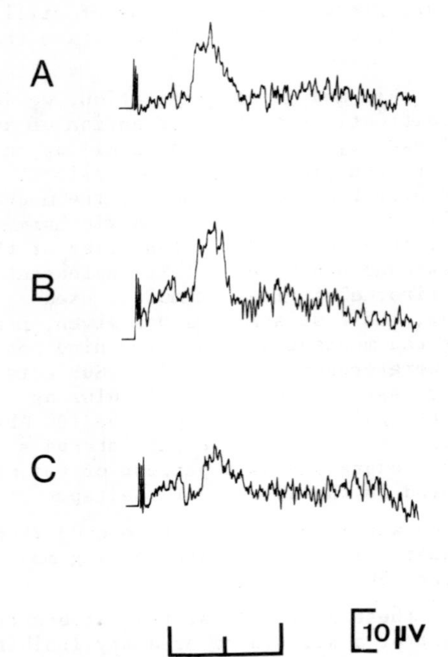

Fig. 3. Cortical evoked response as a function of stimulus perception.

In A, perception is normal ; in B, perception is attenuated ; in C, perception is abolished.

tested in this condition they even show a moderate increase (N1 and P2 waves) in the 200 msec before and the 100 msec after the movement. We did not observe any change in the perception of the stimulus in this situation.

When movement is performed by the foot ipsilateral to the stimulated hand, perception is not modified but evoked potentials elicited 110 msec after EMG onset are reduced or suppressed. Passive displacement of the foot gives the same results. These effects were consistently observed in 12 subjects.

The observations that evoked potentials are diminished during active and passive movements are in agreement with those of Broughton *et al.* (1) and of Giblin (13). As passive movements, or even only touching the stimulated area, attenuate evoked potentials, we may conclude that afferents elicited by a displacement are able to induce this decrease as they are able to reduce perception.

However, evoked potentials increase before movement. At that time there are no reafferences to interfere with the ascending test volley and the increased size of the evoked potentials may then indicate either a facilitation of the transmission of this volley or a greater excitability of the neurons of the sensorimotor cortex. According to Lee and White (14) this increase of the evoked potentials may even continue during the movement. These authors report that evoked potentials are consistently enhanced during movement, that they are not affected by passive movements and that subjects do not report any subjective change in the stimulus strength during movement. The discrepancy with our own results may be due to differences in the experimental conditions. We stimulated one fingertip only with three 1 msec pulses ; Lee and White stimulated three fingers simultaneously with one 0.3 msec pulse. In our experiments stimulus intensity was between 2 and 3 times the threshold for perception, that is usually no more than 2 mA ; although the data by Lee and White do not allow a direct comparison, it seems that they used a much higher stimulus strength (50 to 70 V, with the output from the stimulator being delivered through a 1 : 4 stepup transformer).

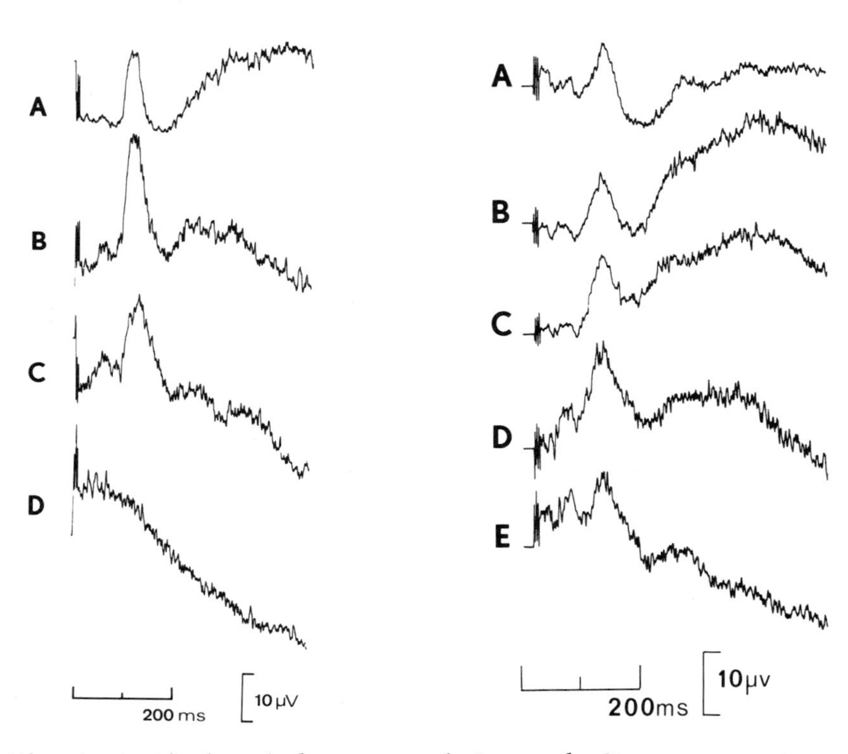

Fig. 4. Cortical evoked responses before and after a movement.

Left : Stimulation of the right middle finger ; flexion of the right hand. Stimulus intensity : 1.7 mA ; threshold for perception : 0.6 mA. Stimulation is given in A 200 to 100 msec before EMG onset, in B 100 to 0 msec before EMG, in C 0 to 100 msec after EMG, in D 100 to 200 msec after EMG. Each trace is the average of 65 responses.

Right : Same subject ; stimulation of the right middle finger, flexion of the left hand. Stimulus intensity : 1.8 mA ; threshold for perception : 0.6 mA. Stimulation is given in A in the absence of movement, in B 250 to 100 msec before EMG, in C 100 to 0 msec before EMG, in D 0 to 100 msec after EMG, in E 100 to 250 msec after EMG. Each trace is the average of 130 responses.

I would then suggest that active movement increases the reactivity of some of the somatosensory relays. But when the test volley is weak and originates from a restricted area this facilitation can be overcome by a surround inhibition brought up by cutaneous or proprioceptive afferents elicited by an active or a passive displacement of the stimulated segment or of a more distant area. Perception is not always reflected in the size of the evoked potentials. Backward masking may lead to a perceptual decrease at a time when evoked potentials are enhanced ; conversely depression of the evoked potentials during the displacement of a distant area is not associated with perceptual attenuation.

III - EFFECTS OF ACTIVE MOVEMENT ON TRANSMISSION IN THE LEMNISCAL PATHWAY OF THE CAT.

We tried to get some information about the effects of active movement upon the somatosensory relays by recording in cats the variations of a volley elicited

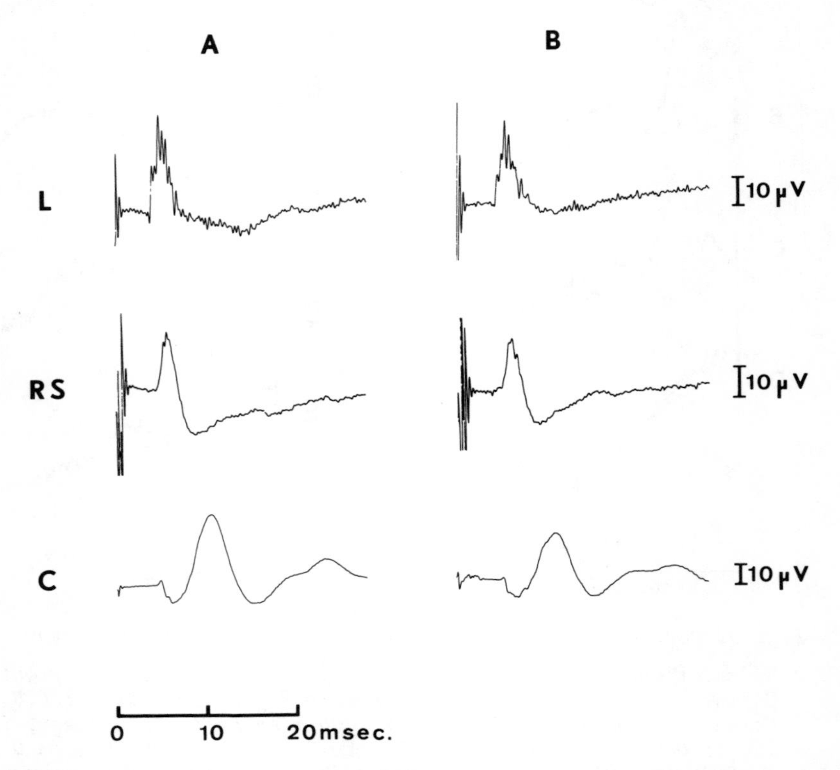

Fig. 5. Effects of a conditioned movement on the evoked responses recorded in the lemniscal pathway of the cat.

Each trace is the average of 100 responses to stimulation of the forepaw. Records are taken in the medial lemniscus (L), in the thalamocortical radiations (RS) and over the SI cortex. In A the animal is at rest ; in B he presses a lever with the stimulated paw. Notice the depression of the responses during movement.

by stimulation of the forepaw during a conditioned movement of the stimulated limb (ref. 2). Seven cats were implanted with macroelectrodes in the medial lemniscus, in the thalamocortical radiations and over the primary somatic cortex. They were trained to press a lever in response to a conditional sound and to stay still when presented with a stimulus of a different frequency. The forepaw was stimulated by means of two disc electrodes taped on the shaved skin. Stimulation was given when the cat was sitting still or at the time he touched the lever. Cats were tested during 10 sessions and got in each session about 100 stimulations at rest and during movement. Potentials recorded on magnetic tape were measured by numerical integration using a PDP-12 computer. For each session the median value of the potentials recorded at each site was computed for movement and no movement and for each animal the medians were averaged over the sessions.

The results show that, at the three recorded sites, the cutaneous volley is reduced during movement : across animals cortical potential decrease by 17.4 % on the average (range : + 1.5 % and - 28.6 %), lemniscal volley by 11.2 % (range : - 1.7 % and - 20.6 %), thalamocortical volley by 6.9 % (range : + 0.8 % and - 21.6 %), (Fig. 5).

Reduction during movement of the potentials recorded along the lemniscal pathway has been reported by several authors. Trouche *et al.* (16) in the cat,

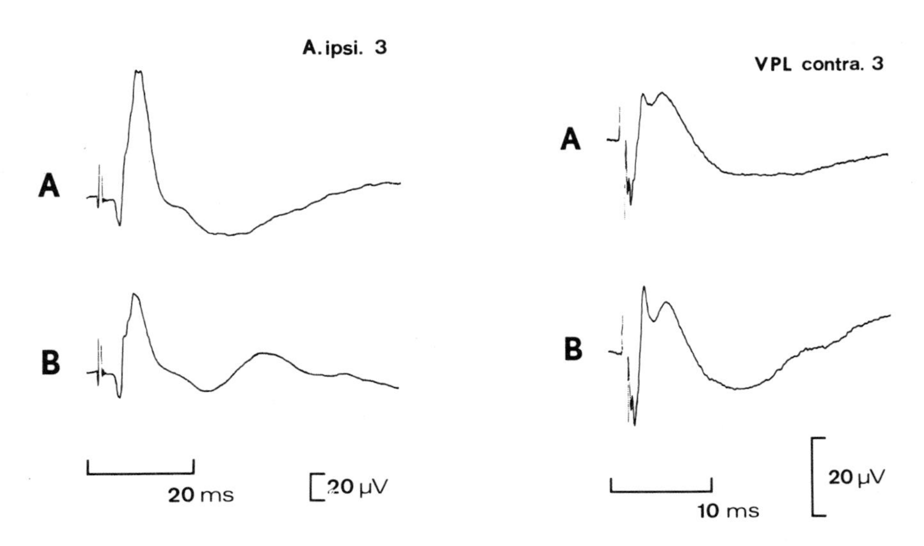

Fig. 6. Cortical evoked potentials elicited by stimulation of the paw or of the thalamus in the cat.

Left : Stimulation of the right forepaw with one 0.2 msec pulse. Stimulus intensity, 3.5 mA. A : at rest. B : during movement of stimulated paw.

Right : Stimulation of the left ventroporterolateral nucleus of the thalamus with one 0.2 msec pulse at 2.5 V. A : at rest. B : during movement of the right forepaw.

During movement evoked responses are decreased when elicited by a peripheral stimulation ; they are moderately increased when elicited by a thalamic stimulation.

Ricci and Valassi (15) in the monkey observed a reduction of the somaesthetic cortical potentials during a conditioned movement. Ghez and Lenzi (11), Ghez and Pisa (12), Coulter (8) indicate that in the cat movement of the stimulated limb reduced the size of a test volley recorded in the medial lemniscus and that this reduction is apparent several tens of milliseconds before the movement ; as a consequence motor commands must be involved in this depression, in addition to the peripheral afferents elicited by the displacement.

Our own data also indicate that the reduction of the test volley is not of the same magnitude at the three recorded sites. In particular, it is less pronounced after the thalamus than after the first relay. Although in our sample the percentage of change between rest and movement is not significantly different after the first and after the second relay, the smaller decrease of the thalamocortical volley fits other observations. In 5 cats, we compared the cortical potentials elicited by stimulating either the paw or the ventroposterolateral nucleus of the thalamus (ref. 5). During movement, cortical potentials were depressed to a lesser degree when they were induced by thalamic stimulation than when the paw was stimulated (Fig. 6) ; in 3 cats they even showed an increase. Ricci and Valassi (15) report similar findings in the monkey.

It is then reasonable to suspect that during an active movement the reduction of a cutaneous volley at the dorsal column nuclei is compensated by a facilitation at the thalamic relay. Such a mechanism should be ascertained during and before movement ; if real, it could play a role in the facilitation of the cortical evoked potentials observed in human subjects before an active movement.

To conclude, psychophysical and electrophysiological tests show that active movement has complex effects on the transmission and the integration of a cutaneous volley : motor commands appear to inhibit the first relay of the lemniscal pathway and possibly to facilitate the thalamic relay. Afferents elicited by the displacement also inhibit transmission of the test volley and may induce perceptual masking.

Reduction of transmission by descending or ascending influences has been interpreted as an improvement of the signal to noise ratio ; a reduction of transmittive capacity would indeed benefit the most concentrated and intense afferences like an electrically elicited volley. However what is noise and what is signal ? Should we consider the ascending volley as a message in a situation where the subjects, humans or animals, perform a ballistic movement not related to the stimulus being delivered ? We could also look at the motor program as the signal and at any afference occuring while it is carried out as noise. In view of the data reported by Fromm and Evarts (this volume) it is possible that skilled movements or movements of palpation in which motor programs have to adjust continuously to the consequences of the movement itself could yield different results.

REFERENCES

(1) BROUGHTON, R., REGIS, H., GASTAUT, H. Modifications du potentiel somesthésique évoqué pendant une bouffée de rythme mu et/ou la contraction des poings. *Rev. Neurol.* 111, 331 (1964)

(2) COQUERY, J.M. Selective attention as a motor program. In J. Requin (ed.) *Attention and Performance VII*, Lawrence Erlbaum, Hillsdale (in press).

(3) COQUERY, J.M., AMBLARD, B. Backward and forward masking in the perception of cutaneous stimuli. *Perception & Psychophysics* 13, 161 (1973).

(4) COQUERY, J.M., COULMANCE, M. Variations d'amplitude des réflexes monosynaptiques avant un mouvement volontaire. *Physiol. Behav.* 6, 65 (1971).

(5) COQUERY, J.M., VITTON, N. Altérations des potentiels évoqués sur le cortex somesthésique du chat durant un mouvement conditionné. *Physiol. Behav.* 8, 963 (1972).

(6) COQUERY, J.M., COULMANCE, M., LERON, M.C. Modifications des potentiels évoqués corticaux durant des mouvements actifs et passifs chez l'homme. *Electroenceph. clin. Neurophysiol.* 33, 269 (1972).

(7) COQUERY, J.M., MALCUIT, G. COULMANCE, M. Altérations de la perception d'un stimulus somesthésique durant un mouvement volontaire. *C.R. Soc. Biol. (Paris)* 1965, 1946, (1971).

(8) COULTER, J.D. Sensory transmission through lemniscal pathway during voluntary movement in the cat. *J. Neurophysiol.* 37, 831 (1974).

(9) EKMAN, G. Two generalized ratio scaling methods. *J. Psychol.* 45, 287 (1958).

(10) EVARTS, E.V. Pyramidal tract activity associated with a conditioned hand movement in the monkey. *J. Neurophysiol.* 29, 1011 (1966).

(11) GHEZ, C., LENZI, G.L. Modulation of sensory transmission in cat lemniscal system during voluntary movement. *Pflügers Arch.* 323, 273 (1971).

(12) GHEZ, C., PISA, M. Inhibition of afferent transmission in cuneate nucleus during voluntary movement in the cat. *Brain Res.* 40, 145 (1972).

(13) GIBLIN, D.R. Somatosensory evoked potentials in healthy subjects and in patients with lesions of the nervous system. *Ann. N.Y. Acad. Sci.* 112, 93 (1964).

(14) LEE, R.G., WHITE, D.G. Modification of the human somatosensory evoked response during voluntary movement. *Electroenceph. clin. Neurophysiol.* 36, 53 (1974).

(15) RICCI, G.F., VALASSI, F. Studio elettrofisiologico degli effetti causati dall' apprendimento e dal condizionamento nell' area sensorimotoria della scimmia. In *Problemi di neurologia e psichiatria,* il Pensiero Scientifico, Roma, 271 (1968).

(16) TROUCHE, E., SANTIBANEZ, G., LELORD, G. Effets sur l'amplitude des potentiels évoqués somatiques de la phase active d'un réflexe conditionné défensif classique. *J. Physiol. (Paris),* 55, 348 (1963).

PERCEPTION OF TACTILE STIMULI BEFORE BALLISTIC AND DURING TRACKING MOVEMENTS

Poul Dyhre-Poulsen

Institute of Neurophysiology, 2100 Ø, Copenhagen, Denmark

It has been shown that the amplitude of potentials evoked in the medial lemniscus by stimuli applied to a cutaneous nerve in cats is reduced before and during a conditioned movement. The same was observed during free movement of the limb from which the test response was elicited (3,6). The reduction started about 150 ms before the movement and persisted during the movement. The significance of this reduction is unknown; changes in surround inhibition and background activity may serve to sharpen the signal and increase the spatial discriminatory capacity in a particular part of the body. Therefore we know that the transmission is changed but the transmission of information may be either increased or reduced (2). A movement is controlled and adapted to a task, and probably utilizes information from both cutaneous and muscular receptors (7). Although the role of long loop reflexes in the control of movement is unknown it would be of interest to know whether or not the transmission of information from cutaneous receptors is changed in the cuneate nucleus during movement. There is a heavy input from pyramidal neurones to the cuneate nucleus (1) and the sensory input is probably controlled by a feedforward mechanism. Many regions in the brain control or modulate the transmission through the dorsal column nuclei. For review see A.L. Towe (9).

The aim of the study to be reported was to determine if the reduced amplitude of potentials evoked in the medial lemniscus by stimuli applied to a cutaneous nerve indicate a change in the transmission of information before movement. Since it is impossible to measure the amount of information transmitted through the nervous system directly, I have instead tried to do it indirectly by measuring the threshold of a tactile stimulus before movement. Detecting weak stimuli requires reliable transmission, thus if the ability to detect weak stimuli decreases I presume that the ability of the system to carry information is reduced.

Threshold measurements were most easily performed with human subjects. Although the anatomical situation in the monkey is similar to that of the cat, the physiology is different. In the cat most neurones may be inhibited or excited by cortical stimulation whereas in the monkey there is less, though a more varied effect of cortical stimulation (9). To avoid the generalization from cat to primate I first determined if there was a reduction of the amplitude of potentials evoked in the medial lemniscus by stimuli applied to a cutaneous nerve also in the monkey.

Physiological experiments with electrical stimuli.

The monkeys (Cercopithecus aetiops) were trained to lift their right arm, partially depress a lever, maintain a fixed position and depress the lever fully at the onset of a tone to get food reward. When the monkey reached stable performance a bipolar recording and a bipolar stimulating electrode were implanted on a peripheral nerve in the right arm. With a motordriven micromanipulator a gross electrode was placed in the contralateral medial lemniscus before every recording session. With this technique we could record for several months. With chronic electrodes the potential deteriorates within weeks.

A control stimulus was applied to the peripheral nerve about two seconds before the onset of the tone and a test stimulus was applied at different times before the movement. The velocity of the movement, the potential recorded in the medial lemniscus and the ingoing volley recorded from the bipolar electrode on the nerve were recorded on magnetic tape. Later the evoked potentials were recorded with the time scale expanded by a transient recorder on an inkwriter. The time from test stimulus to onset of movement was measured by the first visible change in velocity and the amplitude of the potentials were measured. The reduction in the test potential was calculated and the data were grouped on the basis of time from test stimulation to movement. In some experiments the onset of movement was recorded by electromyography. The time from the onset of EMG to the first detectable movement was rather constant about 40 ms and the first change in velocity was therefore used as onset of movement. The results from two monkeys are shown in Fig. 1. The amplitude reduction is plotted against time before movement. There was a reduction of about 20% in the amplitude of the evoked potential beginning about 200 ms before the movement. As in the cat there was no correlation between the force of the movement and amplitude of the lemniscus response.

Physiological experiments with tactile stimuli.

In these experiments electrical stimuli were used. Electrical stimuli evoked a synchronous volley in the peripheral nerve, and the results obtained may only be valid for such stimuli. A vibrator was mounted on the lever and the monkey placed the hand on a plate mounted on the vibrator. The stimulus consisted of one vertical sinusoidal movement lasting 10 ms and with an amplitude of about 100 μm. This stimulus was so small that it was necessary to use electronic averaging to retrieve the data. The mean amplitude of 500 control stimuli and 500 test stimuli applied between 50 and 250 ms before the movement are shown on Fig. 2. There was a marked reduction in the amplitude of the potential evoked in the medial lemniscus by the tactile stimulus before movement, thus with both electrical and natural stimulation there is a reduction in the evoked potential.

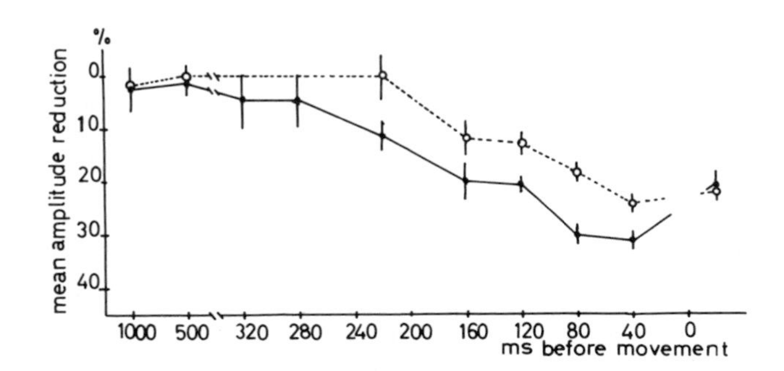

Figure 1. Relation between mean amplitude reduction and the time from stimulus to movement. For every trial the time from stimulus to movement was recorded and the trials were then grouped in time intervals from stimulus to movement. The mean of the reduction in test response in percent of control response amplitude was calculated in each group. Vertical bar indicates standard error of the mean. Data from two monkeys.

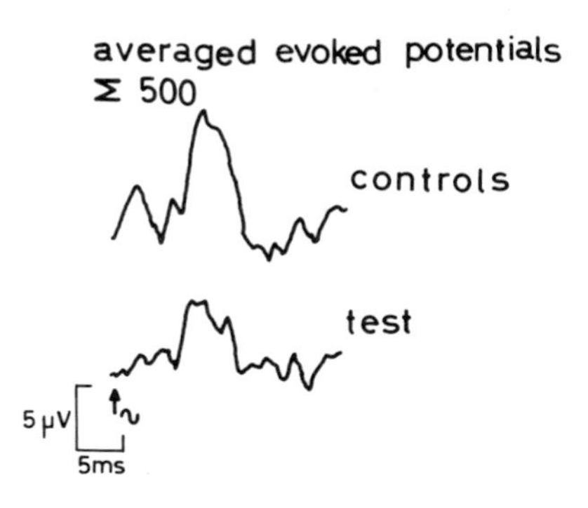

Figure 2. Average of 500 responses evoked in the medial lemniscus by tactile stimuli. Upper trace: Stimuli applied more than 1 sec before movement. Lower trace: Stimuli applied between 50 and 250 ms before movement. Arrow indicates onset of stimulus.

Psychophysical experiments with ballistic movement.

The threshold of a vibratory stimulus was measured in a two alternative forced choice task with two stimulus intervals and a response interval (2-AFC). One of the stimulus intervals contained the stimulus. The method gives a fast, simple and reliable measure of a person's performance independent of his response criterion (8). The person sat in a chair with the right arm supported from the elbow to the wrist. The tip of the right index finger rested on a knob (1/3 sphere 6 mm in diameter) mounted on a vibrator. The vibrator was suspended with a counterweight and the pressure against the index finger was constant (0.5 Newton). The amplitude of the stimulus, a sinusoidal movement (eight cycles of a 200 Hz sine wave), was controlled by a feedback circuit. The left index finger tip rested on another lever with two small response switches attached. The persons were trained to flex the right index finger 1.6 sec after onset of a light indicating the stimulus interval then return to the initial position and repeat the same movement in the next stimulus interval. The stimulus was applied in one of the stimulus intervals at different times before flexion of the finger. After two stimulus intervals and two flexions of the right index finger the person indicated the interval he thought contained the stimulus by pressing one of the switches with the left finger. Before the experiment the vibratory threshold was measured and a stimulus amplitude giving about 85% correct responses was selected. Trials were grouped according to time from stimulus to movement. A trial was discarded if there were any movement before the stimulus and if the movement occurred at an inappropriate time. The percentage of correct responses was calculated in each group of trials and was plotted against the mean of the time interval (Fig. 3). From about 200 ms before the movement the percentage of correct responses decreased from 80 to 60% indicating an increase in threshold. When the person held the stimulated finger immobile and flexed the contralateral index finger there was a decrease from about 80 to 70% beginning about 25 ms before the movement. The decrease was significant ($P < 0.05$; χ^2 test) but it was smaller and it occurred later than in trials with flexion of the stimulated finger. When movement was absent there was no change in the percentage of correct response (4).

It is possible that a rise in threshold occurs only before ballistic movements. Active explorative movement enhances tactile recognition and it was therefore decided to investigate the threshold during movement.

Psychophysical experiment with a tracking movement.

The threshold of a vibratory stimulus (four cycles of a 100 Hz sine wave) was determined with a yes-no procedure while the person performed pursuit tracking movements. The person sat in a chair, the right arm rested on a horizontal bar and the tip of the right index finger rested on a force transducer mounted on the shaft of a torque motor. The axis of rotation of the motor was aligned with the axis of the proximal interphalangeal joint of the right

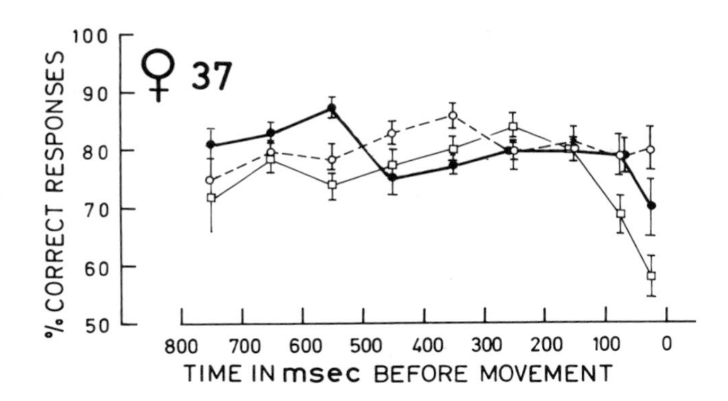

Figure 3. Relation between the percentage of correct responses and the time from stimulus to movement. For every trial the time from stimulus to movement was recorded and the trials were then grouped in time intervals from stimulus to movement. The percentage of correct responses was calculated in each group. In the trials without movement the trials were grouped in time intervals from stimulus to 1.6 sec after onset of the observation interval. Vertical bar indicates the standard error of the mean. Open square: Flexion of the stimulated index finger. Filled circle: Flexion of the index finger contralateral to the stimulated finger. Open circle: No movement in the observation interval.

index finger. A servo system maintained constant force (0.5 Newton) against the finger while the finger could move freely in the joint. The position of the finger controlled the tracking signal, a spot on an oscilloscope. The target, an out of focus spot on the same oscilloscope screen followed a pseudo random signal. The stimulus was applied by the torque motor randomly in 50% of the stimulus intervals indicated by a light. In every other stimulus interval the target was immobile. The percentage of correct responses was 75 when the stimulated finger was immobile and fell to about 65% while the person performed tracking movements. The threshold of the vibratory stimulus therefore increased during movement.

Conclusion.

The amplitude of potentials evoked in the medial lemniscus in a primate by either electrical or tactile stimuli was decreased before movement of the limb from which the test response was elicited. Movements of other limbs had no effect. In a psychophysical experiment in humans the threshold of a vibratory stimulus was increased before ballistic and during tracking movement of the stimulated finger. If the finger did not move or the person moved the contralateral finger there was no change in threshold. The increase in threshold began about 200 ms before the ballistic movement. Thus, when there is inhibition in the cuneate nucleus in the monkey there is a rise in the vibratory threshold in an experiment with humans using the same apparatus. Although input from cutaneous and connective tissue afferents interferes with motor control and increase tracking performance (5) the threshold of a vibratory stimulus was increased during movement. The cutaneous input may be inhibited during movement to facilitate transmission of information from other receptors more important for the control of movement. Furthermore it has not been investigated if recognition of objects is changed during movement. A discussion of masking and recognition problems can be found in the papers by J.M. Coquery and J.C. Craig in this book.

REFERENCES

(1) P. Andersen, J.C. Eccles, T. Oshima and R.F. Schmidt, Mechanisms of synaptic transmission in the cuneate nucleus, J. Neurophysiol. 27, 1096 (1964).
(2) Cole, J.D. and G. Gordon. Differences in timing of corticocuneate and corticogracile actions, In: Y. Zotterman (Ed.) Sensory Functions ot the Skin in Primates, Pergamon Press, Oxford (1976).
(3) J.D. Coulter, Sensory transmission through lemniscal pathway during voluntary movement in the cat, J. Neurophysiol. 37, 831 (1974).
(4) P. Dyhre-Poulsen, Increased vibration threshold before movements in human subjects, Exp. Neurology, 47, 516 (1975).
(5) P. Dyhre-Poulsen and A. Djørup, The effect of sensory input on reflex load compensation, Acta Physiol. Scand. suppl. 440, 58 (1976).
(6) C. Ghez and G.L. Lenzi, Modulation of sensory transmission in cat lemniscal system during voluntary movement, Pflügers Arch. Gesamte Physiol. Menschen Tiere, 323, 273, (1971).
(7) C.D. Marsden, P.A. Merton and H.B. Morton, Servo action in the human thumb, J. Physiol. 257, 1 (1976).
(8) Swets, J.A. (Ed.) Signal Detection and Recognition by Human Observers, Wiley, New York (1964).
(9) Towe, A.L. , Somatosensory cortex: Descending influence on ascending systems. In: A. Iggo (Ed.) Handbook of Sensory Physiology, II, Springer Verlag, Berlin (1973).

ROLE OF INPUTS FROM SKIN, JOINTS AND MUSCLES AND OF COROLLARY DISCHARGES, IN HUMAN DISCRIMINATORY TASKS

D. I. McCloskey and S. C. Gandevia

School of Physiology and Pharmacology, University of New South Wales, Kensington, Sydney, Australia

INTRODUCTION

When one contracts a muscle to move a part or to develop a tension the position and movement of the part can be perceived, together with the force required or tension developed. Potential sources of such information to the central nervous system are sensory nerves in the joints, muscles and skin, and the centrally-generated motor command itself which may provide "corollary discharges" to centres concerned with sensation.

SENSE OF POSITION AND MOVEMENT

Until quite recently it was widely believed that the conscious awareness of limb position and movement depends almost entirely upon the activity of sensory nerve endings in joint capsules and ligaments. The sensory nerve endings in muscles were thought to play important roles in motor control at a subconscious level, but not to contribute to perception. Recent electro-physiological studies on joint receptors in the cat, however, cast serious doubt on the ability of joint receptors to provide sufficiently detailed information for position sense (1,2,11). Also, intra-muscular receptors have been shown to be capable of providing perceived kinaesthetic signals in the form of illusions of false movement and position during muscle vibration at 100 Hz, and kinaesthetic sensation has been shown to persist after elimination of discharges from joint and cutaneous receptors (10). Conventional views on the basis of the senses of position and movement have, therefore, had to be re-appraised. In particular, the relative contributions of joint, muscle and cutaneous receptors to these sensations have had to be considered.

The distal interphalangeal joint of the middle finger provides a convenient opportunity to consider this problem because at this joint an anatomical peculiarity permits the hand to be positioned so as to disengage the muscles from the joint, or be positioned so that the muscles are again engaged (5). If the index, ring and little fingers are extended and then the middle finger alone is flexed maximally at the proximal interphalangeal joint, the terminal phalanx of the middle finger cannot be moved by voluntary effort because the long flexor and extensor muscles which usually move it are held at inappropriate lengths. In this position the joint is freed from effective muscular attachment. If anaethetized in this position the joint loses all position sense, but regains it when the adjacent fingers are aligned with the test finger, restoring effective muscular attachment. Thus, 'joint' sense (really 'joint plus cutaneous' sense) can be tested by imposing movements on the joint, while the muscles are disengaged, and can be compared with full position sense similarly tested when the muscles are re-engaged. 'Muscle' sense can be assessed in further tests in which the muscles are engaged but cutaneous and joint receptors are anaesthetized locally within the finger. In all of our tests on this joint movements into flexion and extension were imposed on the joint and subjects were scored according to the number of correct detections of movement and direction made in sets of 10 such displacements.

Three tests of kinaesthetic sensibility were used, first with the muscles disengaged and then with the muscles engaged. In all tests the displacements of the joint commenced from an initial angle of 30° into flexion. In the first test the distal interphalangeal joint was moved 10° into flexion or extension at angular velocities from 1 to 10°/sec. In the second test the joint was moved for a fixed time (1.2 sec) at velocities from 1 to 10°/sec. In the third test the joint was displaced through 2.5°, 5°, 7.5° and 10° at a fixed angular velocity (~8°/sec). In all tests the movements into flexion and extension were given in random order. The first test was then repeated in subjects with muscles engaged at the joint but in whom any contribution from cutaneous and joint receptors was abolished by local anaesthesia of the test finger.

Proprioceptive acuity for the joint ('joint plus skin') was determined in 12 subjects. These subjects were unable to detect any 10° displacements made at 1°/sec, and no subject was able to detect all 10° displacements made at less than 4°/sec. With the muscles engaged, the proprioceptive acuity of all subjects was improved: many subjects could now detect all 10° displacements made at 1°/sec, and all subjects could detect all 10° displacements made at 3°/sec. Similar results were found in all three tests. There was no significant bias towards detection of displacements into flexion or into extension. Figure 1 shows results from one subject.

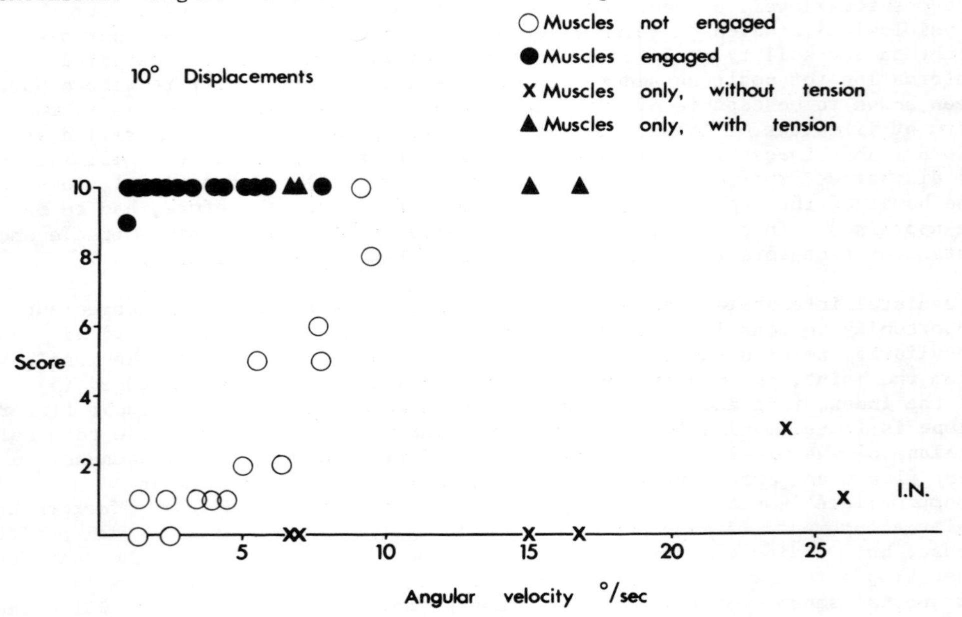

Fig. 1. Detection of joint displacements.

As muscular engagement caused no consistent change in the measured 'stiffness' of the joint it is unlikely that the improved performance resulted from increased discharge from joint receptors. Clearly, intramuscular receptors contribute to normal position sense.

When 'muscle sense' was tested alone following local anaesthesia of the joint and skin some subjects showed very poor proprioceptive acuity (see Fig. 1) while others were almost as accurate as with all the proprioceptive machinery intact. In all subjects muscle sense was greatly improved during the exertion of voluntary muscular force (see Fig. 1).

Intramuscular receptors, therefore, appear essential for full proprioceptive acuity. Receptors in joints or skin can also subserve position sense but it is difficult to anaesthetize one without the other to enable their relative contributions to be assessed. There is very little reduction in proprioceptive acuity in patients in whom the bony joints are totally replaced by prosthetic devices, a procedure which presumably destroys joint receptors (3,12). Because of this, and the electrophysiologically demonstrated inadequacy of joint receptors to signal position sense accurately, it seems likely that cutaneous receptors may be the more important complement to intramuscular receptors in subserving normal position sense. In this respect it is of interest that Knibestöl (15) has recorded in the human finger from cutaneous receptors whose discharges are closely related to the angle of the underlying joint.

No sense of movement or of altered position accompanies voluntary motor commands despatched to paralysed muscles (10,16,20). Corollary discharges do not, therefore, provide perceived sensations of movement or position.

SENSE OF MUSCULAR FORCE OR TENSION

Three mechanisms are usually proposed for the capacity to estimate the weights of lifted objects. First, cutaneous receptors may signal the pressure of the object on the skin; second, receptors in the contracting muscles or in their tendons or joints may signal the forces exerted in the task; and, third, corollary discharges arising from the motor command itself may give an estimate of the effort required to support the object.

It is a common experience that a weight lifted by a muscle until it becomes fatigued feels heavier than it felt before the fatigue. In such a situation peripheral sensory nerves in the skin of the supporting part and in the muscles, tendons and joints employed in the lift presumably continue to provide signals of the true tensions and pressures involved and so seem unlikely to be the basis of the erroneous judgements. The increase in perceived heaviness could be explained, however, if the preferred signal available derives from the centrally-generated voluntary motor command, as this command would have to increase as the muscle fatigues. This common experience has been tested objectively by having subjects match weights lifted by a fatigued muscle with weights similarly lifted on the unaffected side, and it has been demonstrated that subjects prefer to disregard any peripherally-generated sensory signals related to tensions and pressures in favour of alternative signals related to the effort required (19).

An increase in the perceived heaviness of a lifted object occurs whenever the lifting muscle is weakened. It occurs whether the weakness is caused by muscular fatigue, by partial curarization, by activation of muscle spindle afferents in its antagonist leading to inhibition of its motoneurones, or by neurological disorders (such as simple motor 'strokes' or unilateral cerebellar dysfunction) in which weakness without sensory loss is a feature (7,14,19). In all of these situations subjects prefer to judge the weight of a lifted object according to the effort involved in lifting it and to disregard any available, but conflicting peripheral signals. Figure 2 gives a diagrammatic representation of the motor pathways delivering command signals to a contracting muscle, and shows the points in the pathways at which a lesion or an experimental intervention can cause weakness of the muscle and an increase in the perceived heaviness of objects lifted by it.

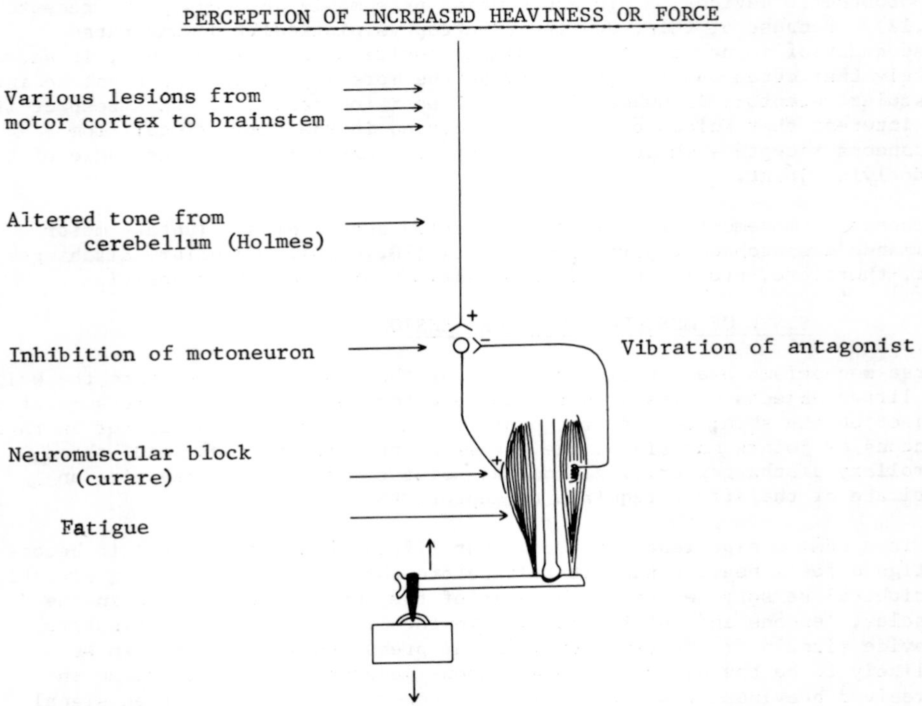

PERCEPTION OF INCREASED HEAVINESS OR FORCE

Various lesions from motor cortex to brainstem

Altered tone from cerebellum (Holmes)

Inhibition of motoneuron

Vibration of antagonist

Neuromuscular block (curare)

Fatigue

Fig. 2. Factors increasing perceived heaviness or force.

While subjects prefer to rely on centrally-generated command signals as the basis of their judgements on the heaviness of lifted objects, it can be shown that alternative signals do exist. For example, it can be shown that subjects perceive signals related to cutaneous distortion (18) or to intra-muscular tension (19) when lifting objects - such alternative signals, however, seem usually to be disregarded.

The specific possibility should be considered that, because of parallel activation of skeleto-motor and fusimotor fibres (α-γ co-activation), there is always an increase in the discharge from muscle spindles when effort is increased. Such activity of the muscle spindles might be thought to provide the perceived signals used to estimate heaviness. This is not so. When the muscle spindles in a muscle are activated by vibration of the muscle at 100Hz, no feeling of increased heaviness or weight is induced. Instead, "the subject gets a feeling of relief or lessening of tension" (13), and chooses smaller tensions than before in unaffected muscle to match any tension achieved by the vibrated muscle (19).

The sensation of heaviness, therefore, depends upon sensing the effort or motor command required to lift or support an object, and is preferred to other available signals.

EFFECTS OF SENSORY INPUTS UPON MOTOR COMMANDS

Because the perception of heaviness of a lifted object derives from the centrally-generated voluntary motor command used to lift it, it follows that the central motor command delivered to the motoneurons can be studied through observations on perceived heaviness.

We have investigated the perception of the heaviness of lifted objects when sensory inputs from parts related to the lifting task were altered. In all tests subjects were given a reference weight to lift with a muscle group on the left side and they matched it with weights similarly presented to the right (indicator) side. The weights chosen on the right side thus gave an objective indication of the perceived heaviness of the reference weight lifted on the left side. Matches were made as controls and during disturbances of sensory input on the left side.

Our results indicate that the perception of heaviness is altered by sensory inputs arising peripherally. For example, Fig. 3 shows how sensory inputs from the left index finger and thumb influence the perceived heaviness of objects lifted by flexion of either of these digits. Electrical stimulation of the index finger (90V, 40-100Hz, 2 ms) over the digital nerves, so as to produce a pressing sensation over the whole of the finger, causes weights lifted by flexion of the distal joint of the thumb to feel lighter. Local anaesthesia of the index finger, however, causes weights lifted by thumb flexion to feel heavier. Anaesthesia of the thumb itself, which blocks cutaneous and joint receptors within the thumb but leaves the innervation of its long flexor muscle unaffected, also causes weights lifted by thumb flexion to feel heavier. All these findings, for 6 subjects, are shown on the left of Fig. 3. Similar results are found for weights lifted by flexion of the index finger. A weight lifted by flexion of the index finger feels lighter when the thumb is stimulated electrically, and heavier when the thumb is anaesthetized. (Fig. 3, at right).

These changes in perceived heaviness give some insight into how motor commands must alter when inputs arising peripherally are altered. A weight lifted by thumb or index finger flexion must feel heavier during anaesthesia of either

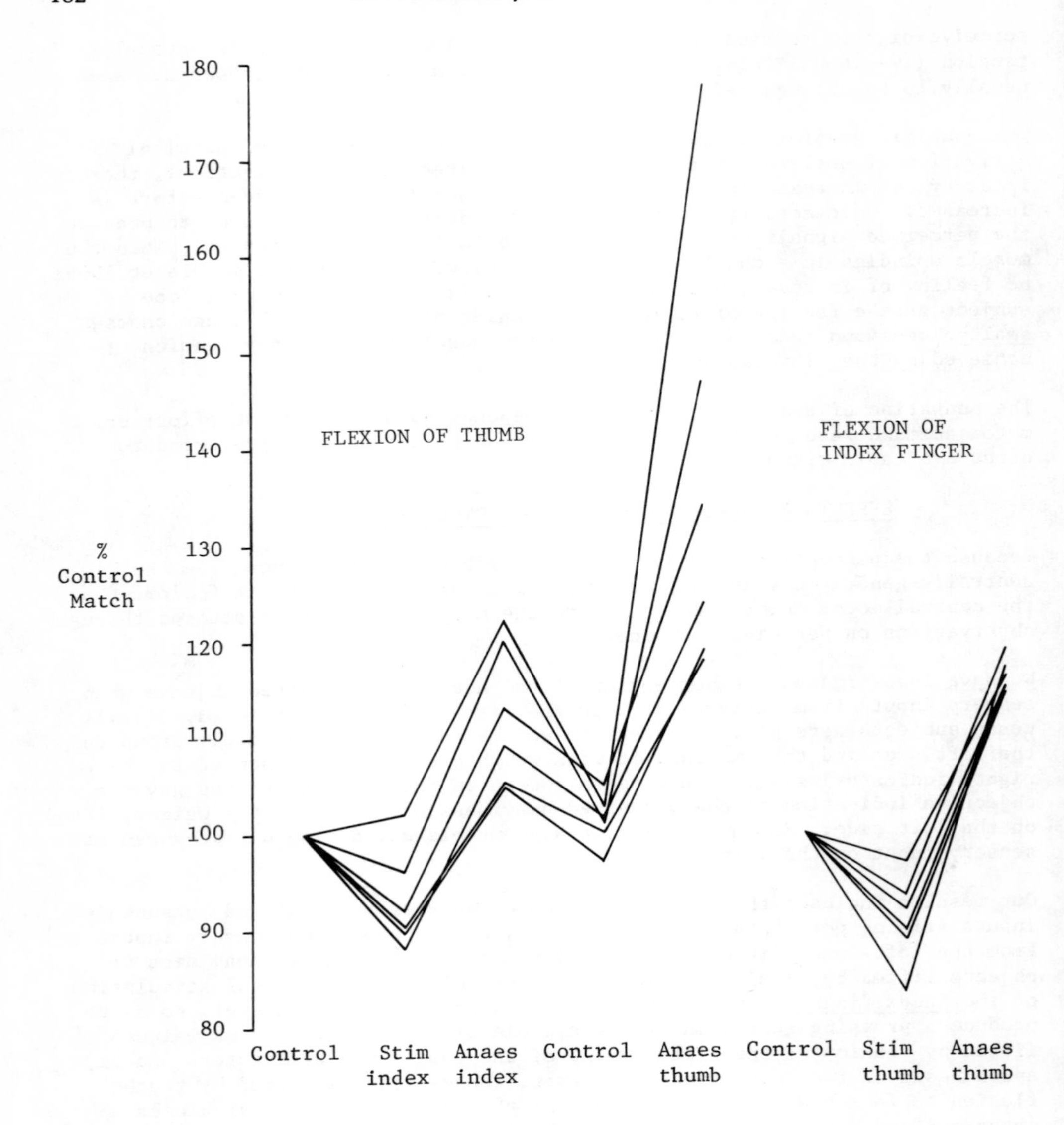

Fig. 3. Effects of alterations of sensory inputs on the perceived heaviness of objects lifted by flexion of the distal joint of the thumb (at left), or by flexion of the index finger (at right). Each line joins the means of 10 estimates of weights chosen by an individual subject in each condition. Paired t-tests show that each experimental intervention (stimulation or anaesthesia) caused a significant departure from the control choice of matching weight. See text.

the thumb or index finger because a greater central command or effort is employed in a lift. It follows that some tonic source of facilitation of the muscular contraction has been removed by the local anaesthesia. Presumably such facilitation arises in cutaneous or joint receptors. When the facilitatory influence is enhanced by electrical stimulation the central command required to perform the lift is decreased and the reduced command is perceived as a reduced heaviness.

A most interesting aspect of our results is that sensory inputs arising from both index finger and thumb (but not from the little finger, which we have also tested) can assist the motor command signals descending to either the index finger flexors or the thumb flexor. The flexors of the index finger and thumb are used more commonly in co-operation than in isolation - indeed, this functional co-operation constitutes the highly evolved 'precision grip' of man. What we have shown is that the motor commands to the flexors of either thumb or index finger flexors are assisted by sensory inputs arising from a wide sensory field usually involved in co-operative motor performances carried out by both muscle groups together. The loop between sensory input and motor output is thus closed in the periphery within the sensory field of the total co-operative performance.

The experiments of Marsden, Merton & Morton (17) and Dyhre-Poulsen & Djørup (4) suggest the mechanisms by which peripheral inputs assist the descending motor commands. Marsden *et al.* applied sudden stretches to the contracting long flexor of the human thumb and found that the principal reflex facilitation to the contraction, presumably caused by excitation of its muscle spindle afferents, is not exerted through a monosynaptic spinal arc onto its motoneurons as occurs with the tendon jerk. Instead, load compensation begins after a longer latency which is consistent with a supraspinal path for the reflex. This reflex assistance has a unique property - it is abolished by anaesthesia of the thumb. Presumably, therefore, afferents from the skin or joint of the thumb co-operate with afferents from the long flexor of the thumb to permit the "long-latency load-compensating" reflex to operate. In our experiments anaesthesia of the thumb or index finger may abolish or greatly reduce the reflex assistance coming from this reflex, thereby requiring a greater centrally-generated effort to lift or support any weight. This increased effort would be perceived as increased heaviness. The findings of Dyhre-Poulsen & Djørup (4) are especially interesting in this respect. These authors found that the load-compensating reflex of the long flexor of the thumb, which could be abolished by anaesthesia of the thumb, could be restored in the thumb by electrical stimulation of the adjacent index finger. Clearly, this fits well with our findings and with the interpretation we have placed upon them (see also 8).

If the reason for the increase in perceived heaviness during thumb anaesthesia is the removal of the long-latency stretch reflex, as we have suggested, then our observations permit us to calculate the contribution made by this reflex in normal lifting contractions of the long flexor of the left thumb. For the subjects shown in fig. 3 the average increase in heaviness during thumb anaesthesia was about 30%. The contribution of the stretch reflex was, therefore, about 23% $(\frac{30}{100+30} \times 100)$ (see also 9).

The peripheral interaction between sensory inputs and motor outflow which we have described for the finger and thumb can be seen in other areas as well.

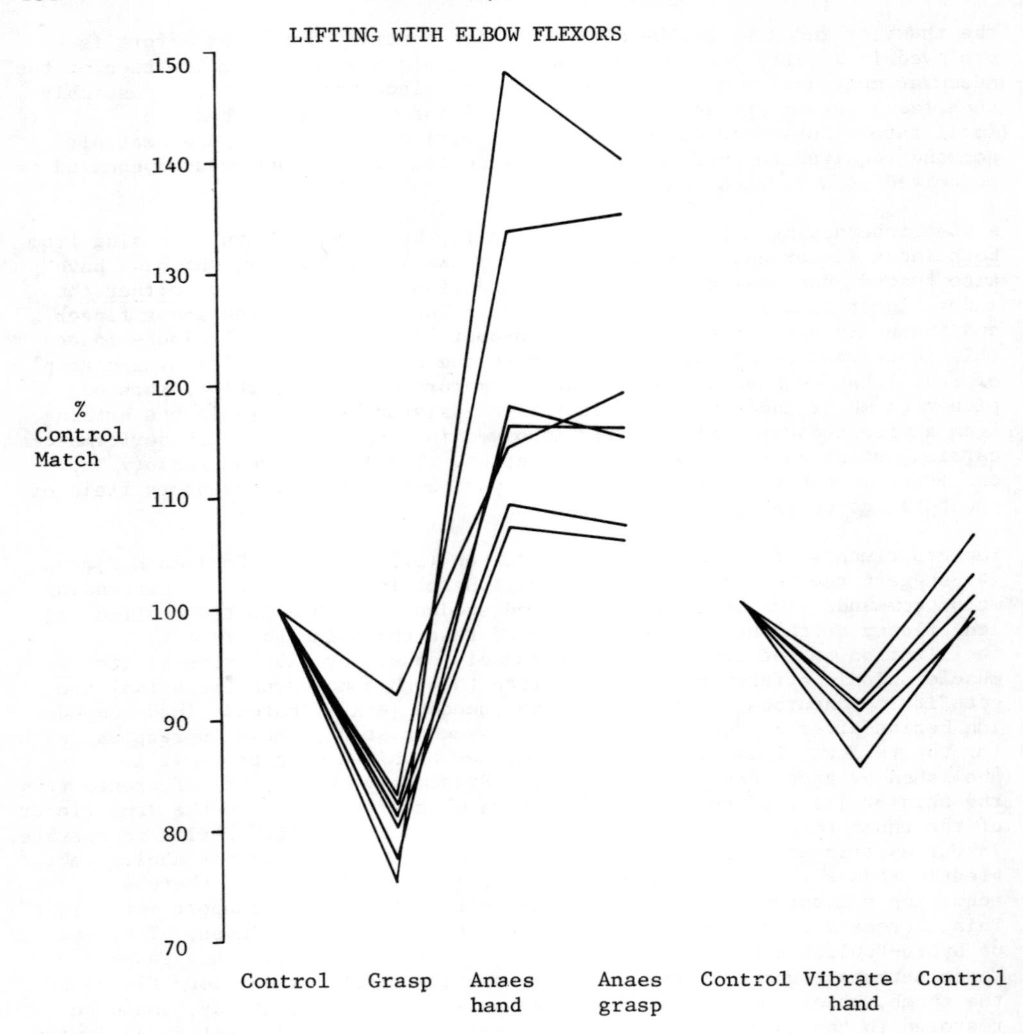

Fig. 4. Effects of alterations of sensory inputs from the hand on the perceived heaviness of objects lifted by flexion of the elbow. Each line joins the means of 10 estimates of weights chosen by an individual subject in each condition. Paired t-tests show that grasping with the hand, anaesthesia of the hand and vibration of the hand caused significant departures from the control choice of matching weight. Grasping did not significantly alter the effect of anaesthetizing the hand. See text.

Closure of the hand and flexion of the elbow are individual movements which often occur together as co-operative motor performances. In another series of weight-matching tasks (6) we looked at the perceived heaviness of objects lifted by flexion of the elbow: in the subjects in these experiments the weights were attached and lifted through a band around the wrist, so that on neither side did the hand participate in the lift. Figure 4 shows the results of these experiments. When the hand on the side (left) that lifts the reference weight at the wrist firmly grasps a piece of rubber tubing, the reference weight is perceived to be lighter than it was when the same tubing simply rested lightly in the hand. When the left hand is anaesthetized by cutting off its blood supply with a sphygmomanometer cuff, the reference weight feels heavier. In this case the sensory inputs from the skin or joints of the hand assist the contraction of the elbow flexors. When the sensory inputs are enhanced, by grasping (Fig. 4 at left), or by vibrating the palm of the hand at 100 Hz (Fig 4., at right), the motor command to the elbow flexors is assisted and so less command is required. Thus, the object lifted by elbow flexion feels light. The sensory inputs involved in this assistance appear to arise in the hand rather than in the forearm, because anaesthesia of the hand causes perceived heaviness to increase whether the anaesthetized hand is involved in grasping or not (Fig. 4). Moreover, vibration of the forearm instead of the hand causes no significant change in perceived heaviness.

The index finger and thumb, and the elbow flexors and the hand, give two examples in which the afferent input from the sensory field of a total co-operative motor performance assists the total performance. This assistance might well arise, as we have suggested, through modulation of the functional stretch reflex assistance available to the muscle groups involved. Whatever the mechanism, the phenomenon provides a means by which the central nervous system can treat common composites of motor performances as wholes, and by which motor activity can be focussed by peripheral feedback from within quite wide, but related, sensory fields.

ACKNOWLEDGEMENTS

This work was supported by a grant from the National Health and Medical Research Council of Australia. Miss Diane Madden provided expert technical assistance. Figs. 1 and 3 are reproduced by permission of the editors of The Journal of Physiology.

REFERENCES

(1) P.R. Burgess and F.J. Clark, Characteristics of knee joint receptors in the cat. J. Physiol, 203, 317-335 (1969).

(2) F.J. Clark and P.R. Burgess, Slowly adapting receptors in cat knee joint: can they signal joint angle? J. Neurophysiol. 38, 1448-1463 (1975).

(3) M.J. Cross and D.I. McCloskey, Position sense following surgical removal of joints in man, Brain Research 55, 443-445 (1973).

(4) P. Dyhre-Poulsen and A. Djørup, The effect of sensory input on reflex load compensation, Acta physiol. scand. suppl 440, 58 (1976).

(5) S.C. Gandevia and D.I. McCloskey, Joint sense, muscle sense, and their combination as position sense, measured at the distal interphalangeal joint of the middle finger, J. Physiol. 260, 387-407 (1976).

(6) S.C. Gandevia and D.I. McCloskey, Perceived heaviness of lifted objects and effects of sensory inputs from related, non-lifting parts, Brain Research 109, 399-401 (1976).

(7) S.C. Gandevia and D.I. McCloskey, Sensations of heaviness, Brain 100, 345-354 (1977).

(8) S.C. Gandevia and D.I. McCloskey, Effects of related sensory inputs on motor performances in man studied through changes in perceived heaviness. J. Physiol in press.

(9) S.C. Gandevia and D.I. McCloskey, Changes in motor commands, as shown by changes in perceived heaviness, during partial curarization and peripheral anaesthesia in man, J. Physiol, in press.

(10) G.M. Goodwin, D.I. McCloskey and P.B.C. Matthews, The contribution of muscle afferents to kinaesthesia shown by vibration induced illusions of movement and by the effects of paralysing joint afferents, Brain 95, 705-748 (1972).

(11) P. Grigg, Mechanical factors influencing response of joint afferent neurons from cat knee, J. Neurophysiol 38, 1473-1484 (1975).

(12) P. Grigg, G.A. Finerman and L.H. Riley, Joint position sense after total hip replacement, J. Bone Joint Surg 55-A, 1016-1025 (1973).

(13) K.-E Hagbarth and G. Eklund, Motor effects of vibratory muscle stimuli in man, in Nobel Symposium I: Muscular afferents and motor control, ed. by R. Granit: Wiley, New York (1966).

(14) G. Holmes, The symptoms of acute cerebellar injuries due to gunshot wounds, Brain 50, 413-427 (1917).

(15) M. Knibestöl, Stimulus-response functions of slowly adapting mechano-receptors in the human glabrous skin area, J. Physiol 245, 63-80 (1975).

(16) J. Laszlo, The performances of a simple motor task with kinaesthetic sense loss, Q. Jl. exp. Psychol 18, 1-8 (1966).

(17) C.D. Marsden, P.A. Merton and H.B. Morton, Servo action in the human thumb, J. Physiol 257, 1-44 (1976).

(18) D.I. McCloskey, Muscular and cutaneous mechanisms in the estimation of the weights of grasped objects, Neuropsychologia 12, 513-520 (1974).

(19) D.I. McCloskey, P. Ebeling and G.M. Goodwin, Estimation of weights and tensions and apparent involvement of a "sense of effort", Exptl. Neurol 42, 220-232 (1974).

(20) D.I. McCloskey and T.A. G. Torda, Corollary motor discharges and kinaesthesia, Brain Research 100, 467-470 (1975).

DIFFERENTIAL ENCODING OF LOCATION CUES BY ACTIVE AND PASSIVE TOUCH

Jacques Paillard, Michèle Brouchon-Viton and Pierre Jordan

C.N.R.S., Institut de Neurophysiologie & Psychophysiologie, 13274 Marseille, Cedex 2, France

ABSTRACT

The position of the index fingertip was used as a target for the reaching movement of the other hand in man. Systematic differences in the accuracy of pointing were interpreted as reflecting differential encoding of location cues involved in generating the reaching motor programmes. Results showed that tactile information was differently encoded as location cues when combined with movement information of proprioceptive origin involved in active touching as contrasted with proprioceptive positional information about the resting position of a limb passively touched.

INTRODUCTION

The present study was initiated as a direct extension of a previous set of experiments on fingertip position sense that showed locating errors were significantly smaller when the target hand is actively moved than when passively displaced by external forces (ref. 8.10). The results were interpreted in terms of a differential contribution of articular and muscular proprioceptive afferents to the performance. This interpretation stems from the assumption that systematic differences in the accuracy of reaching may reflect some differences in the encoding of location cues which affects the central command to generate the motor program.

Tactile cues were used in these new experiments on localization pointing in which correcting feedback was prevented. Tactile stimulation of the fingertip was associated with either active or passive positioning of the target hand or with its stable position actively maintained or passively sustained.

Results confirm and extend those already presented in an earlier publication (ref. 1). They show that tactile information was differently encoded as location cues when combined with movement cues as contrasted with positional cues of proprioceptive origin.

METHOD

As in the preceding studies (ref. 1.8.10) the position of the right index fingertip is used as a target for the reaching movement of the left index finger. The subject is sitting, with head and trunk fixed before a sheet of plexiglass set in the frontal plane about 40 cm away from him and carrying two parallel tracks 60 cm in length set vertically and symmetrically in relation to his mid-sagittal plane (see Fig. 1).

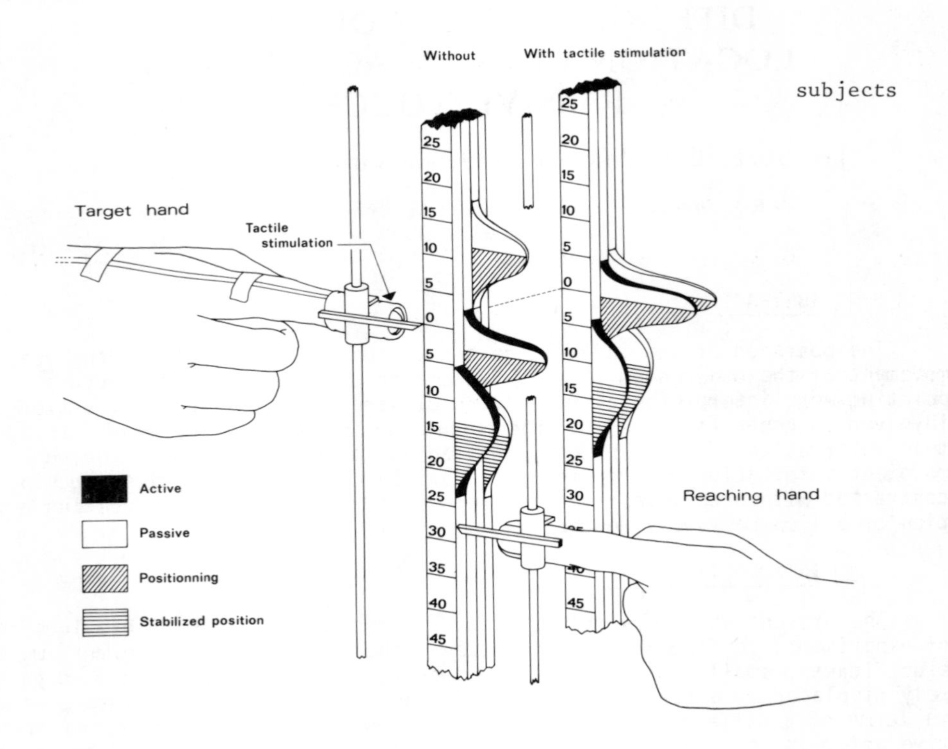

Fig. 1 Schematization of the experimental device (see text). Note that with regard to the central scale numbered in millimeters the size of the hand has been reduced for clarity of the drawing. The amount of error or reaching during each trial is indicated by the difference in position of the two slides measured on the central scale. Each distribution curve represented along the scale cumulates the results obtained on 13 subjects and representes each from 325 to 900 measures depending on the situation. The mean error of each distribution characterises a locus of reaching as defined by the combination of the afferences encoded as location cues. See comments in the text.

Each of the subjects' index finger can operate a movable slide along one of the tracks. The subject moves his right hand at a speed freely adopted by him along the right track and terminates his movement or his hand is passively displaced in a similar manner through a device controlled by the experimenter and stopped at different heights on each trial. The final position attained by the displaced limb is either actively stabilized by the subject or passively sustained.

The tip of the left index finger is surrounded with an adhesive strip putting a pair of metallic electrodes into contact with the skin surface. An electrical stimulation is delivered by an electronic stimulator as a single rectangular pulse of 1 ms wide and whose intensity is adjusted at a level sufficient for detection of a light tactile shock but without any unpleasant component.

The stimulation was either triggered by the experimenter at the end of the active or passive positioning movement of the target hand or introduced at least 15 seconds after stabilization of the target arm actively maintained in its final position or passively supported.

Nine subjects took part in the former experiment and 4 in the present experiment. Each subject performed all the experimental conditions according to a balanced order.

RESULTS

When compared with the performances of reaching observed when only proprioceptive information is available (see Fig. 2) :

1°) additional tactile information clearly improves the accuracy of pointing when cutaneous stimulation is associated with the end of the kinetic phase of positioning regardless whether it was actively or passively achieved (see Fig. 1), error in both cases was not significantly different from null-error ;

2°) additional tactile cues does not seem to be taken into account to locate the actively or passively stabilized position of the target finger. Errors in locating the fingertip in those conditions were essentially the same as those observed without tactile stimulation ;

3°) tactile stimulations of other parts of the body (arm or limb) did not alter performance in either of the other conditions.

DISCUSSION

The most striking result of the present experiment is that phasic tactile information combined with kinesthetic proprioceptive information causes a definite improvement in the accuracy of reaching in the passive positioning condition. The performance was not significantly different from null-error and there was no difference between the active and the passive conditions. In contrast tactile information combined with statesthetic information about the stabilized target position does not add to the localizing property of static proprioceptive information. This finally led to the conclusion that cutaneous afferents take a localizing value only when combined with phasic joint afferents. Moreover, muscular afferents whether phasic or static do not seem to add anything to the localizing value of cutaneous afferents.

It appears from the present study that information about static position of joints and information about joint movement can be differentially processed. This processing can take place not only in subjective judgements relative to "position sense" and "kinesthesia" as already stressed by Mc Closkey (4) but, as well, in the encoding of location cues used at a subconscious level to generate spatially oriented motor programmes.

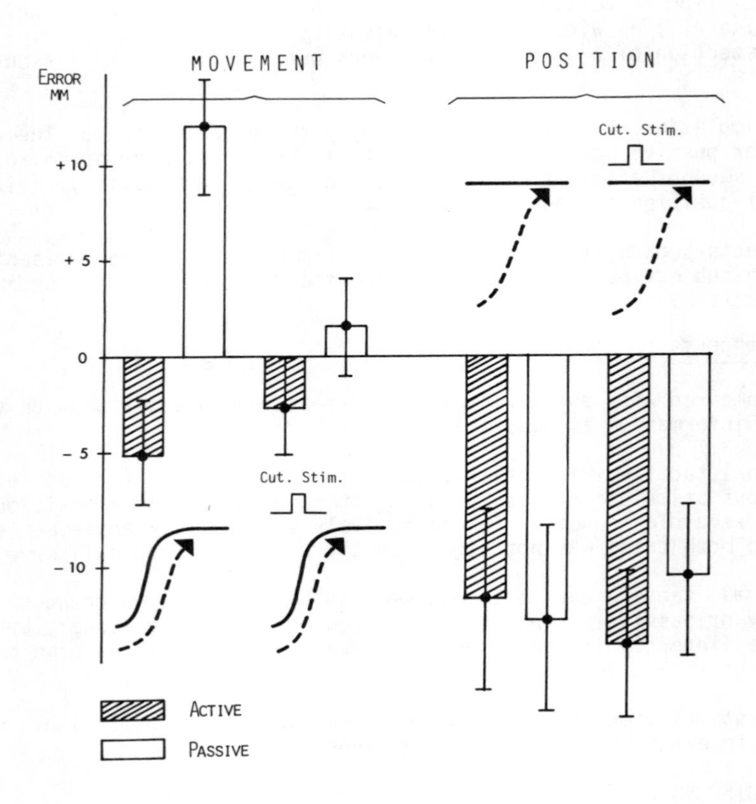

Fig. 2 Median and interquartile range of errors in locating the position of the target hand with the reaching hand. Black curves in the lower left symbolize the time course of positioning movement and black straight lines in the upper right the resting position of the target hand 15 sec after a positioning movement. The stripped curves indicated the timing of the pointing performance and the square wave those of the cutaneous stimulation.

The following table tentatively summarizes the different sources of information involved in our 8 experimental conditions.

	Positioning Movement				Stabilized Position			
	Active		Passive		Active		Passive	
Joint phasic	+		+		0		0	
Muscul. phasic	+		0		0		0	
Joint static	0		0		+		+	
Muscul. static	0		0		+		0	
Motor outflow	+		0		+		0	
Skin	0	+	0	+	0	+	0	+
Mean errors (in mm)	-5 *	-2,5 NS	+12 ***	+1,5 NS	-11,5 ***	-14 ***	-12,5 ***	-10,5 ***
	NS		***		NS		NS	

Motor outflow which could be involved as a possible explanation for the difference between active and passive condition of positioning does not seem to interfere in any of the conditions where cutaneous cues are available.

Recent observations by Millar (5) in cat and Grigg and Greenspan (3) claims that many joint afferent neurones, specially those measuring to intermediate angles which are silent in passive movement, could be excited by activation of muscles which, by virtue of their coupling to the joint, could directly stretch the joint capsule.

Grigg and Greenspan (3) suggest that our previous observation on the distinction between active and passive positioning (ref. 10) could be explained by such an effect. However, we formerly discussed this point (ref. 8) and discarded such an explanation on two grounds : first, active and passive resting posture of the target hand might likely be differentiated for the same reason. Active maintenance of a position involves a muscular load assumed to stretch the joint capsule that is not occuring in a passively supported limb. The performance is nevertheless identical in both conditions. However it could be argued here that phasic joint receptors could be more sensitive than the static one to this effect.

A second argument is more convincing : cooling or vibrating the muscles of the target limb, without changing the muscular load, dramatically suppresses the superiority of active positioning over passive positioning in our experimental conditions. Such a result led us to conclude that there is a critical contribution of spindle afferents in the phenomena (ref. 10).

The addition of tactile cues to proprioceptive one revealed another experimental condition in which the distinctive effects of passive and active positioning is suppressed. Mountcastle and Powell (6) found in the sensori-motor cortex convergence of joint and cutaneous afferents in similar cells which were inhibited by afferent cutaneous volleys. Convergence of articular, muscular and cutaneous afferents in area 3 of the precentral cortex are now well documented as well as in the posterior parietal association cortex where they seem to be involved in operations within extrapersonal space (ref. 7). It has been recently found that relay cells of the thalamic ventrobasal complex transmitting kinesthesia (identified as "joint movement" cells) were inhibited by cutaneous electrical stimulation in the medial lemnis in the cat (ref. 13). Moreover Rosén (11) showed that in ventrobasal relay cells activated by Group I afferents IPSP's were evoked by cutaneous nerve stimulation. This could explain the prevalence taken by cutaneous information in locating the target when associated with proprioceptive movement cues in passive or active positioning of the target hand.

In contrast, following the electrical stimulation of fast pyramidal tract fibers, facilitations of all joint movement cells and inhibition of cutaneous relay cells in the ventrobasal complex,via collaterals projecting in the thalamic nuclei, have been reported (ref. 13). This led the authors to suggest that "when voluntary movements of limbs are initiated by pyramidal tract impulses, preference of transmission in the ventrobasal relay cells may shift from the cutaneous sense to kinesthesia as a consequence of pyramidal tract induced facilitation". They also conclude that "during the resting state of animals cutaneous sensory transmission in ventrobasal complex is given preference over kinesthetic transmission".

This sounds in opposition to our results showing that positional proprioceptive cues seem to prevail over tactual cues in the resting state of the target hand. Several possible explanations might explain this discrepancy.

First a more precise distinction between movement and positional articular afferents needs to be made to identify their interference with phasic or static muscular afferents ; correspondingly the same distinction needs to be made when studying their interference with cutaneous afferents. Secondly some special "hair units" corresponding to the distal end of the forearm-forepaw have been described in the cat (ref. 13) as activated and not inhibited during active movement, thus indicating some functional differences among cutaneous afferents. Moreover, gating of cutaneous afferents through the efferent command (ref. 2) is likely to occur when it is possible to anticipate the skin contact which generally terminates an intended reaching act.

Finally (as already suggested by Brouchon and Hay (1)) our results point, by analogy to that has been found in the visual system (ref. 12), to the evidence of a double somesthetic system respectively for identifying and locating function : first, identification of patterns could be based on the analysis of the relative position of the points stimulated on the skin surface ; second, location of a stimulated point could be based on a space coordinated system relative to the body postural reference (ref. 9). The separability of the first system is basically dependent upon the "acuity" of the receptive surface (as testable by a two point discrimination procedure). The second system is dependent on the encoding of cutaneous afferents as locations cues in a body-centered reference system (as testable by a pointing localization procedure). Thus, the mapping of the tactual space may proceed differently in each system. Active touch may carry out different functions when exploring objects by digital palpation compared to when locating them. Further studies along these lines could help to elucidate the specific nervous mechanisms involved in these dual sensori-motor process.

REFERENCES

(1) M. Brouchon, L. Hay, Analyse des interférences entre les informations proprioceptives et cutanées dans l'appréciation des positions du corps propre, Psychologie Française 17, 135 (1972).

(2) J. M. Coquery, Fonctions motrices et contrôle des messages sensoriels d'origine somatique. J. Physiol. (Paris) 64, 533 (1972).

(3) P. Grigg, B. J. Greenspan, Response of primate joint afferent neurons to mechanical stimulation of knee joint, J. Neurophysiol. 40, 1 (1977).

(4) D. I. Mc Closkey, Differences between the senses of movement and position shown by the effects of loading and vibration of muscles in man, Brain Research 63, 119 (1973).

(5) J. Millar, Joint afferent discharche following muscle contraction in the absence of joint movement, J. Physiol. (London) 226, 72P (1972).

(6) V. B. Mountcastle, T. P. S. Powell, Central mechanisms subserving position sense and kinesthesis, Bull. Johns Hopkins Hosp. 105, 173 (1959).

(7) V. B. Mountcastle, J. C. Lynch, A. Georgeopoulos, H. Sakata, C. Acuna, Posterior parietal association cortex of the monkey : command functions for operations within extrapersonal space, J. Neurophysiol. 38, 871 (1975).

(8) J. Paillard, M. Brouchon, Active and passive movement in the calibration of position sense, in Freedman S.J. (1968) Neurophysiology of spatially oriented behavior, Dorsey Press, Homewood, Ill, p. 37.

(9) J. Paillard, Les déterminants moteurs de l'organisation de l'espace, Cahiers de Psychologie 14, 261 (1971).

(10) J. Paillard, M. Brouchon, A proprioceptive contribution to the spatial encoding of position cues for ballistic movements, Brain Research 71, 273 (1974).

(11) I. Rosén, Excitation of Group I activated thalamocortical relay neurons in the cat, J. Physiol (London) 205, 237 (1969).

(12) G. E. Schneider, Two visual systems, Science 163, 895 (1969).

(13) T. Tsumoto, S. Nakamura, K. Iwana, Pyramidal tract control over cutaneous and kinesthetic sensory transmission in the cat thalamus, Exp. Brain Res. 22, 281 (1975).

TONIC FINGER FLEXION REFLEX INDUCED BY VIBRATORY ACTIVATION OF DIGITAL MECHANORECEPTORS

H. Erik Torebjörk, Karl-Erik Hagbarth and Göran Eklund

Department of Clinical Neurophysiology, University Hospital, Uppsala, Sweden

INTRODUCTION

It is an old observation that grasping a vibrating object may cause a "magnet reaction"; the fingers tend to adhere to the object and difficulty may be experienced in attempting to loosen the grip (Rood 1860). The nature of this vibration-induced, forced grasping has been investigated in the present study. Attempts were made to elucidate whether it is a proprioceptive tonic vibration reflex (TVR), resulting from spread of the vibratory stimulus to the muscle spindles in the finger flexor muscles (Hagbarth and Eklund, 1966; Lance et al. 1966) or whether it is an exteroceptive reflex depending on excitation of mechanoreceptors in the fingers.

METHODS

Two mechanical testing procedures were used to measure finger flexion force during vibration:

I: A small cylindrical DC-motor with an eccentric load on the axis was attached to one of the fingers, usually on the volar side of the index finger. When the finger was unsupported the running motor produced finger oscillations of about 0.5 mm at the finger tip. The vibration frequency could be varied from 50 to 200 Hz. During the tests the hand was fixed to a support holding the wrist and fingers extended and the thumb abducted. The pulp of the vibrated finger rested on a fixed plate connected to a strain gauge measuring finger flexion force. The standard instruction to the subjects was that they should maintain a weak constant effort in pressing the finger pulp against the plate. They were told not to counteract any force or positional changes which they might perceive during the periods of vibration, which usually lasted for 30 seconds.

II: The subject held the pulp of his thumb against a strain gauge plate with a whole in the centre through which a blunt motor driven probe oscillated at a rate of 100 Hz causing skin indentations of about 0.5 mm. Also in this test the standard instruction was to maintain a weak constant voluntary effort in holding the grip.

Twenty healthy volunteers of both sexes, aged 25 to 53 years, were investigated.

RESULTS

Reflex response to finger vibration.

Given the standard instruction to maintain a weak finger flexion effort, all subjects responded to the index finger vibration with an involuntary slow increase in the flexion force (Fig. 1B). During prolonged vibration at a frequency of 100 Hz, it usually took 20-30 seconds until a comparatively stable plateau level was reached. At the end of vibration, the finger flexion force returned to the control level within a few seconds. On instruction to remain passive and relaxed, 5 of the 20 subjects responded with a finger flexion reflex of a similar time course but of a lower strength than in the standard tests (Fig. 1A). In the remaining 15 subjects, no torque change occurred in the relaxed state.

The finger flexion force induced by vibration was frequency-dependent; the height of the plateau level increased with increasing frequencies up to about 130 Hz. No apparent increase occurred with higher frequencies. 100 Hz was choosen as the standard frequency in the experiments to be described.

The strength of the reflex, as measured in the standard tests, varied from one subject to the next, but was relatively constant

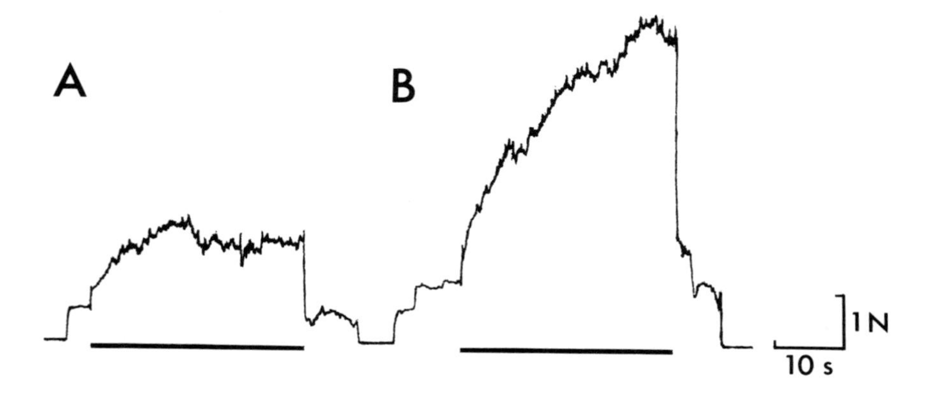

Fig. 1. Index finger flexion force induced by 100 Hz vibration (bars) when the subject is told (A) to remain passive and (B) to maintain a weak finger flexion effort during vibration. The first increase of the strain gauge signal from the zero level indicates that the finger tip is passively put on the plate. The second increase (in B) indicates the intentional finger flexion force. Note the slow unintentional increase in force during vibration, which is much more pronounced when there is a central command to flex the finger (B) than in the passive state (A).

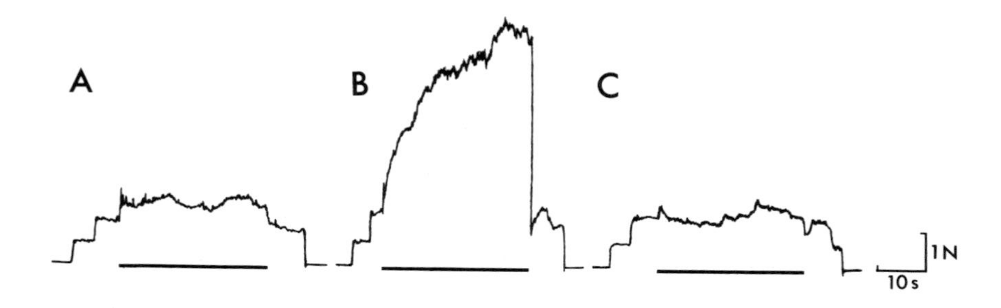

Fig. 2. Index finger flexion force induced by vibration of (A) the third finger, (B) the index finger and (C) the thumb. Note that the flexion reflex is fairly restricted to the finger exposed to vibration.

on repeated trials with the same subject. Subjects with strong reflexes plateaued at a level reaching 4-5 N, corresponding to about 1/5 of their maximal voluntary finger flexion power.

The vibration-induced finger flexion reflex was fairly restricted to the particular finger vibrated. As shown in Fig. 2, the index finger response to index finger vibration was much greater than the index finger responses to vibration of the thumb or the middle finger.

Application of the vibrator to the tendons of the finger flexor muscles at the wrist induced TVR responses (Fig. 3A) which were always much weaker than when the vibrator was attached to the volar side of the index finger (Fig. 3B). Furthermore, the weak tonic vibration reflex occurring in response to vibration over the flexor muscle tendons did not show the slow build-up typical for the motor response to finger vibration. Although a component of TVR responses from muscle spindles cannot be excluded in the records, it was evident that the muscle TVR alone could not account for the vibration-induced finger flexion.

Effect of local anaesthesia.

When the index finger was anaesthetized by a lidocaine block around the base of the finger, vibration of the anaesthetic finger produced almost no flexor response at all. A marked reduction of the motor response to the vibratory stimulus in the index finger was also noted when only the distal phalanx of the finger was made anaesthetic by a lidocaine block (Fig. 4). The remaining weak response was reminiscent of that produced by vibration over the finger flexor tendons at the wrist. On recovery from the anaesthesia, the motor response was temporarily enhanced, as compared to the controls prior to the anaesthesia. During that period paraesthesias were present in the distal phalanx of the finger, with marked tingling sensations arising during vibration and also in response to local taps on the finger tip. At this stage the perception of temperature changes and pain had not returned. With increasing time after the block

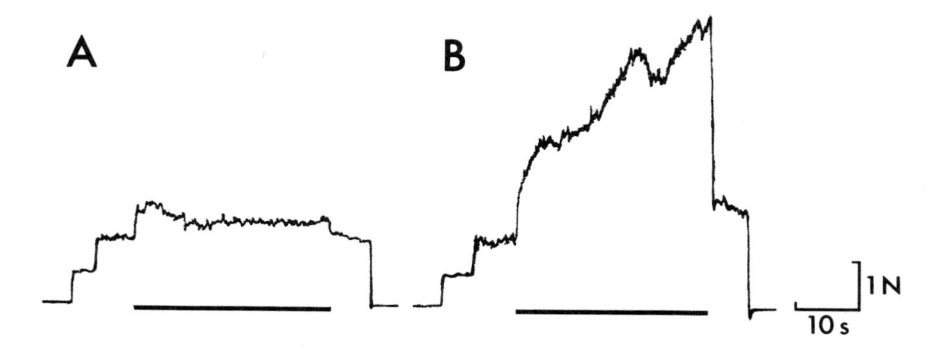

Fig. 3. Index finger flexion induced by vibration of (A) the tendons of the finger flexor muscles at the wrist and (B) the index finger. The TVR response in A is lower and does not show the slow build-up typical of the vibration induced finger flexion reflex (B).

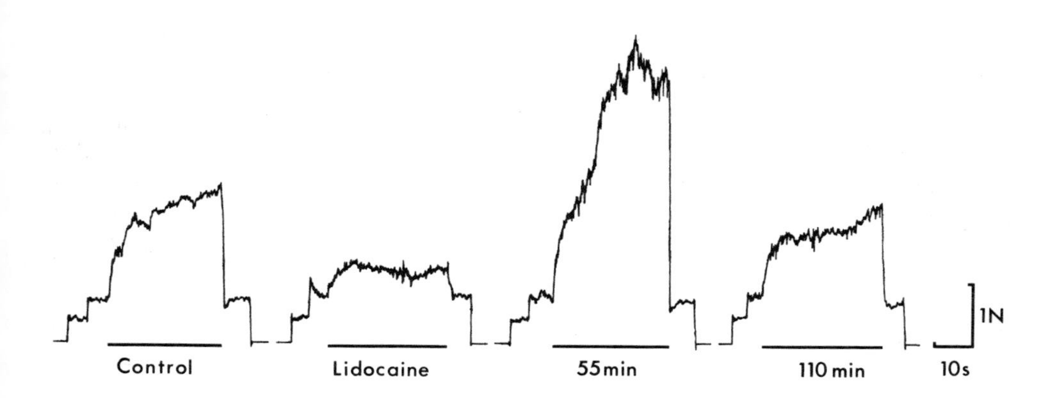

Fig. 4. Vibration-induced index finger flexion before, during and 55-110 min after lidocaine anaesthesia of the distal phalanx. During anaesthesia (second graph) the reflex response has decreased. During a period of paraesthesias in the finger tip on recovery from anaesthesia (third graph), the reflex response has increased as compared to controls before and after the anaesthesia (first and fourth graphs).

the paraesthesias disappeared and the perceptions of different types of stimuli reappeared as the motor response returned to control levels.

In further attempts to localize the receptors responsible for the reflex, the skin of the thumb pulp was anaesthetized with Chlorethyl spray. Localized vibration of the anaesthetic skin in the pulp produced a weaker flexor response than controls prior and after the anaesthesia.

DISCUSSION

The results indicate that the described reflex is an exteroceptive tonic vibration reflex, distinct from the TVR of muscle spindle origin. As judged by the blocking experiments, mechanoreceptors with myelinated axons in the distal phalanx constitute the major afferent source. The receptors probably have small receptive fields, since the end organs excited by a vibrating probe can be blocked by localized skin cooling. As shown by Vallbo and Johansson (this volume), rapidly adapting (RA) receptors and also slowly adapting (SA I) receptors are numerous in the finger tips. With their high sensitivity to mechanical stimulation combined with small receptive fields, the signals from these receptors are likely to be important for the sensory discrimination of textures. The present results indicate that afferent signals from such receptors, by a feedback mechanism, also might influence the motor performance of the exploring fingers.

Our results are compatible with those obtained with a different technique by Marsden, Merton and Morton (1977) who found that anaesthesia of the thumb may abolish the compensatory increase in active force normally arising when a voluntary thumb flexion movement is suddenly opposed by an external force. They also noted that it needed a greater effort from the subject to overcome a resistance and move the thumb when it was anaesthetic. Their suggestion that long loop cortical reflex arcs are involved may well hold true also for the tonic tactile reflex we have been studying, although we certainly do not exclude participation of segmental reflex mechanisms.

REFERENCES

Hagbarth, K.-E. and G. Eklund, Motor effects of vibratory muscle stimuli in man. In R. Granit (Ed.): Proceedings of the First Nobel Symposium: Muscular afferents and motor control, Almqvist and Wiksell, Stockholm, pp 177-186, 1966.

Lance, J.W., P. de Gail and P.D. Nelson, Tonic and phasic spinal cord mechanisms in man. J. Neurol. Neurosurg. Psychiat.29, 535-544, 1966.

Marsden, C.D., P.A. Merton and H.B. Morton, The sensory mechanism of servo action in human muscle, J. Physiol. 265, 521-535, 1977.

Rood, O.N., On contraction of the muscles induced by contact with bodies in vibration. Am. J. Sci. Arts 24, 449, 1860.

Vallbo, Å.B. and R.F. Johansson, The tactile sensory innervation of the glabrous skin of the human hand. (This volume).

HEIGHTENING TACTILE IMPRESSIONS OF SURFACE TEXTURE

Susan J. Lederman

Queen's University, Kingston, Ontario, Canada

ABSTRACT

In 1975, Gordon and Cooper described an interesting phenomenon. The orientation of gentle undulations on the surface of an object could be detected more accurately when a person moved an intermediate paper over the surface, than when the bare fingers were used alone. We have found that surface *roughness* can be exaggerated by using a similar manner of touching. In my paper, I discuss several experiments which explore this alteration in texture perception. A "reduced-shear" interpretation of both tactile phenomena is offered, and possible receptor mechanisms considered. Two alternative explanations of the data are suggested briefly. Finally, the potential relevance of this work to the facilitation of tactile map reading by the blind is considered.

Background

The work I am going to discuss follows directly from a simple but intriguing result published in Nature by Gordon & Cooper (1975). Let me first describe the effect under everyday circumstances, because it is even more marked outside the laboratory. Run your fingers across some surface such as a veneered door or table which appears fairly smooth to the touch; then repeat this while holding a thin piece of paper in your fingers. Make sure the paper moves *with* the fingers across the surface. What you will notice is that undulations are either revealed for the first time, or else are exaggerated when you use the paper. Apparently this phenomenon has been used for a long time by craftsmen and people in the automobile industry. From response to his publication, Gordon (personal communication) learned, for example, that the inspection of the coachwork of Aston Martin Sports cars is done by rubbing over the surface with a cotton glove on the hand. German craftsmen test the finish of molded surfaces by using silk. And polishers have similarly used cloth to detect bumps on surfaces. All the same, it is unlikely that this tactile phenomenon has ever before been subject to scientific investigation.

Gordon & Cooper demonstrated this effect quite simply under laboratory conditions. They used a plate, precision-ground to leave a raised central strip, rectangular in shape, 3 mm wide and approximately 13μ high. Covering the surface with a thin piece of cardboard produced a gentle undulation in the plate. Accuracy of identifying the orientation of this undulation was approximately 60% when examined with the bare fingers, increasing to 80% when the intermediate paper was used. The authors suggest that as skin contains more than one type of receptor responsive to mechanical deformation (Talbot *et al*, 1968), rubbing an undulating surface such as theirs may result in receptor responses to roughness (by a light touch system), and also to gradual surface changes or undulation (mediated by a deep pressure system). The response to roughness may interfere with or mask the response to undulation. The effect of using the paper would be to reduce the masking, thus permitting more accurate detection of orientation.

Heightening Tactile Impressions of Roughness

Experiment 1. If this explanation were true, might not the *perceived* roughness of the surface on which the undulations occur, or of any surface for that matter, be relatively *less* when the masking paper is used? Apparently not. Surfaces actually feel rougher when touching is performed with the intermediate paper. We presented people with textured surfaces consisting of various grades of sandpaper, each covered with a piece of writing paper to prevent particles from being dislodged. On half of the trials, the surfaces were felt with the bare fingers ("no-paper" condition), and on the other half, through a piece of paper held in the fingers ("paper" condition). Subjects estimated roughness using a procedure known as magnitude estimation. This involves assigning numbers (fractions, decimals or whole numbers) in proportion to the roughness of a stimulus relative to the one immediately preceding it. Thus, if the first stimulus was called "10", and the second felt twice as rough, it would be called "20". The data are presented in Fig. 1.

The ordinate (log scale) shows the geometric means of roughness estimates, and the abscissa (log scale) represents grit value of the sandpaper. (*Grit* refers to the number of openings per inch in the sieves used to produce the sandpapers.) The top curve displays the results of the paper condition; the bottom curve shows the no-paper condition. In all cases, surfaces felt rougher in the paper condition.

Of course, we are interested in whether roughness *discrimination* improves as well. Unfortunately, psychophysical functions which describe what S.S. Stevens calls "prothetic" continua (i.e. intensive dimensions), do not provide such information. Consequently, we plan in future to use a signal detection paradigm when we determine whether the intermediate paper increases roughness discrimination as well as sensation magnitude (now in progress).

A "Reduced-Shear" Interpretation

How might the results of Experiment 1 be explained? To judge the roughness of a surface, there must be relative motion between that surface and the hand. This means that as the hand is moved, both downward (normal) and lateral (shear) forces are applied to the skin of the fingertips. What roles might these forces play in the perception of roughness?

Martin Taylor and I (Taylor & Lederman, 1975) have proposed a model of tactile roughness based on the static deformation of the skin touching regularly-grooved metal surfaces. We found that perceived roughness could be predicted well by several aspects of skin displacement. Since our early experimental work (e.g. Lederman & Taylor, 1972) pointed to the importance of normal forces in the perception of roughness, we included them in our theoretical calculations. The latter indicated that values of these several deformation parameters would increase (as did perceived roughness) with increases in the force applied normal to the skin. Thus normal forces have played a significant part

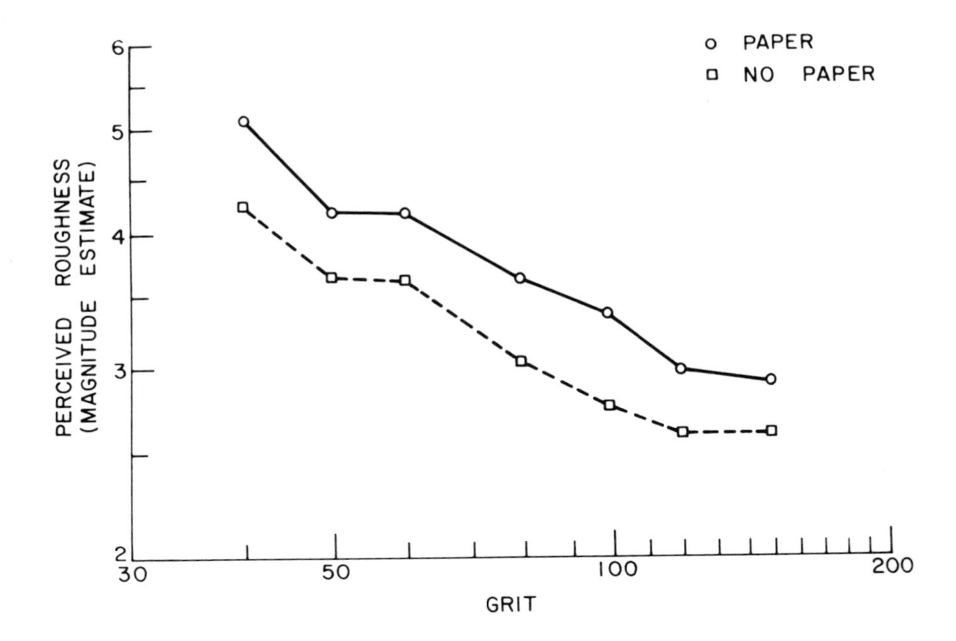

Fig. 1. Perceived roughness (magnitude estimates) as a function of sandpaper grit in paper and no-paper conditions. Each point is based on 50 observations.

in both our theoretical and experimental work on roughness perception.

What effects might the shear forces contribute to perceived roughness? This parameter was not of course considered in our original static model. I would like to suggest now that shear might interfere with or mask the effects of the relevant normal forces on perceived roughness.

According to what I will call a "reduced-shear" interpretation, the intermediate paper used in the first experiment serves to decrease shear, and thus the amount by which the roughness signal is masked. This analysis can also be applied to Gordon & Cooper's finding, since detecting the orientation of a single undulation (like a cluster of tiny elevated bumps) similarly involves the application of both normal and shear forces to the skin.

There is at least one important difference between the two experiments, however. My study deals with the magnitude of roughness sensations, while Gordon and Cooper's study addresses itself to the accuracy of orientation detection. Since the task is not simply undulation detection, but rather detection of the *orientation* of this undulation (an important distinction not made by Gordon & Cooper), we must consider the issue somewhat further. It is possible that Gordon & Cooper's subjects were performing close to threshold when they used their bare fingers. Errors in orientation might in fact be due to errors in detecting the presence of the undulation. Reduced masking in the paper condition might serve to increase the accuracy of detecting the orientation of the undulation by raising the deformation signal above detection threshold. If this were correct, we would expect to find a height beyond which judgments were similar, both with and without the paper. In their article, the authors only reported per cent orientation detection of the 13μ ridge. However, all of the higher-ridged surfaces used in the practice sessions yielded similar (near perfect) performance, both with and without the paper (Gordon, personal communication). Without further measurement and experimentation, it is difficult to describe the nature of their stimulus precisely, or to estimate the appropriate threshold value for undulation detection. For now then, we can only suggest that the increased accuracy of orientation detection may be a threshold-related phenomenon. If such is the case, the reduced-shear interpretation can be applied to Gordon & Cooper's findings as well.

Experiment 2. As an alternative to the reduced-shear explanation, one could argue that our data are the result of subjects pressing harder when they moved the intermediate paper across a surface. As mentioned earlier, Lederman and Taylor (1975) have shown that perceived roughness increases with increasing force. But downward force was not controlled in the first experiment. We therefore carried out a control experiment to determine whether the phenomenon would remain when finger force was held constant in both paper and no-paper conditions. We essentially repeated the first experiment (using fewer surfaces), once with subjects pressing with ~40 gm ("light" condition), once with ~120 gm ("medium" condition), and finally with the finger force maintained at ~290 gm ("heavy" condition). These values were chosen to represent a range of forces which are "comfortable" to apply when touching with and without the paper. Force was controlled by an apparatus designed along the lines of a classical balance scale. The results are shown in Fig. 2. Perceived roughness is plotted as a function of sandpaper grit on log scales, for paper and no-paper conditions. The data are shown separately for the light, medium and heavy force conditions (top, middle, and bottom, respectively). Briefly, the results showed that the heightened roughness effect remained when subjects used ~120 and ~290 gm, and tended in this direction (though not significantly) when they used ~40 gm. It would appear then that the increased roughness which occurs when an intermediate paper is used cannot be explained by the suggestion that people press harder. But there does seem to be a lower limit to the phenomenon.

There are two additional tests of the reduced-shear explanation which you may try for yourself. Place a thin piece of paper over a surface. Move your fingers across the paper, but without displacing it. The surface will feel smoother than when you move both paper *and* fingers across the surface. The perception is to be expected because this time shear has been increased and, thus we suggest, masking as well. Next, move your fingers back and forth across the surface, again with the piece of paper underneath. This time, move the intermediate paper, but in some direction other than the one in which you are moving your fingers. The surface feels a little smoother than when you feel the surface through the paper moving *with* the fingers. This experience would also be predicted by the reduced-shear explanation, since shearing of the skin has been increased when the finger moves relative to the paper.

Experiment 3. In Experiment 3, we attempted to manipulate shear directly. Three levels of skin shear were used. In all conditions, each of the four surfaces was produced by placing a single layer of spherical, glass beads (of a given diameter range) on a relatively smooth base. In the "high shear" condition, the particles were firmly fixed to the base, and the "surfaces"

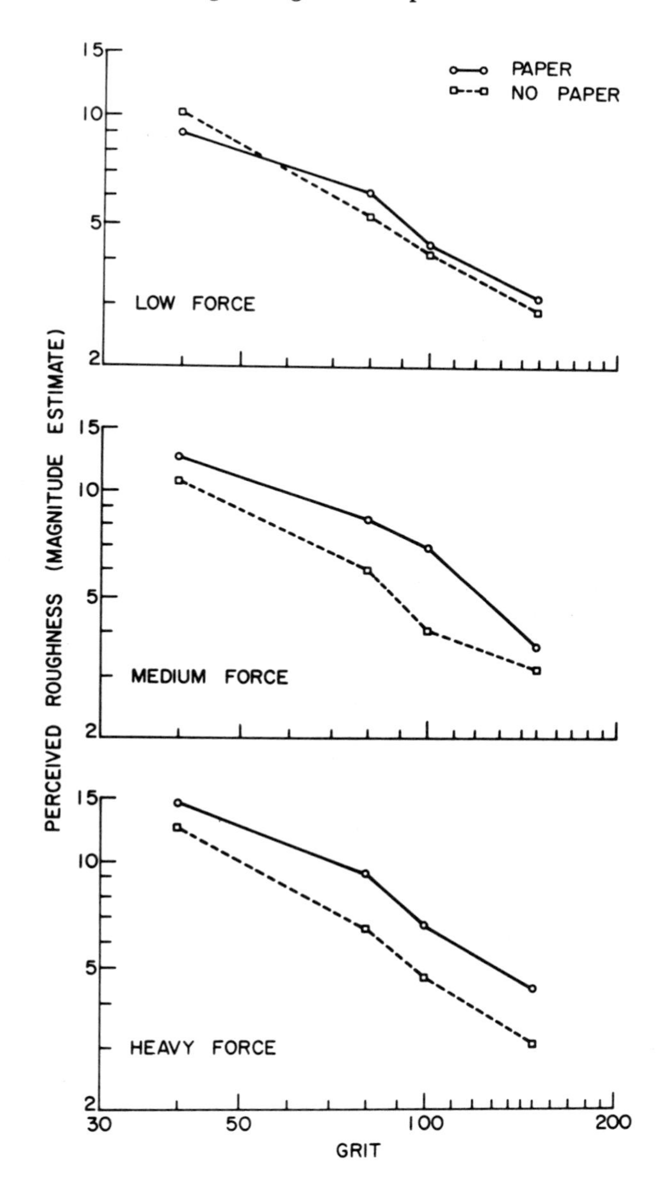

Fig. 2. Perceived roughness (magnitude estimates) as a function of sandpaper grit in paper and no-paper conditions. *Top* light force condition; *middle* medium force condition; *bottom* heavy force condition. Each point is based on 60 observations.

thus produced were tightly covered with a double layer of tissue paper. During a "high shear" trial, the subject moved his fingers *over* the (double-) papered, beaded surface. In the "medium shear" condition, the subject felt the same four surfaces, but this time by moving the top layer of tissue paper *with* the fingers across the (single-) papered, beaded surface. The medium and high shear conditions are much like the paper and no-paper conditions, respectively, of Experiments 1 and 2. In the "low shear" condition, we used the same beaded surfaces, but this time allowed the beads to roll freely. The subject moved a double layer of tissue paper over the beads. However, due to the spherical nature of the particles, only forces normal to the paper can be applied. Therefore, there can (theoretically) be no shear on the paper moving over the beads, and hence none on the skin which is stationary relative to the paper. Almost none of our blindfolded subjects realized the particles were rolling, or seemed to have any difficulty judging the roughness of our somewhat unusual "surfaces".

The reduced shear explanation predicts that the four stimulus surfaces presented to the subjects should be judged (by magnitude estimation) as roughest in the low shear condition, followed by the medium, and finally the high shear conditions. The predictions were clearly confirmed, as shown in Fig. 3. Geometric means of the roughness estimates are plotted as a function of the nominal aperture width of the sieves used in the production of the four beaded surfaces (log scales). This experiment offers further support for the reduced-shear interpretation.

Receptor Mechanisms

Is it possible to translate the explanation above into sensory physiological terms? Vallbo and Johansson have pointed out in their paper that there are four distinct types of low threshold mechanoreceptive units in the glabrous skin of the human hand, PC, RA, SAI, and SAII. Any of these receptive units, singly or in some combination, could be candidates for coding skin deformation caused by shear and normal forces. Such discussion as follows must of course remain purely speculative until the response characteristics of these afferent units to both kinds of force can be determined.

Given that the SAI and RA units occur most frequently in the fingertip (Vallbo and Johansson), they might be likely candidates to consider first. One intriguing notion implicating the SAI units (ending in the Merkel cell neurite complex) may be derived from work by Cauna (1954).

In the digital skin of the hand, two kinds of epidermal pegs, called limiting and intermediate or sweat ridges, project into the dermis. These are shown schematically in Fig. 4a. The limiting ridges are found directly below the surface grooves, and are firmly attached to the dermis. The intermediate ridge is embedded in loose connective tissue, and may follow the movements of the surface ridge above when force is applied. Cauna has suggested that the intermediate ridge may improve tactile acuity by functioning as a primitive lever mechanism, its movements serving to magnify the tactile stimulation at the surface for the receptors below. These receptors are the Merkel's corpuscles, which are associated with the deep surface of the intermediate ridge (see Fig. 4a).

A similar micromechanical system might be involved in tactile texture perception, and in the heightened impressions of roughness reported here. Force

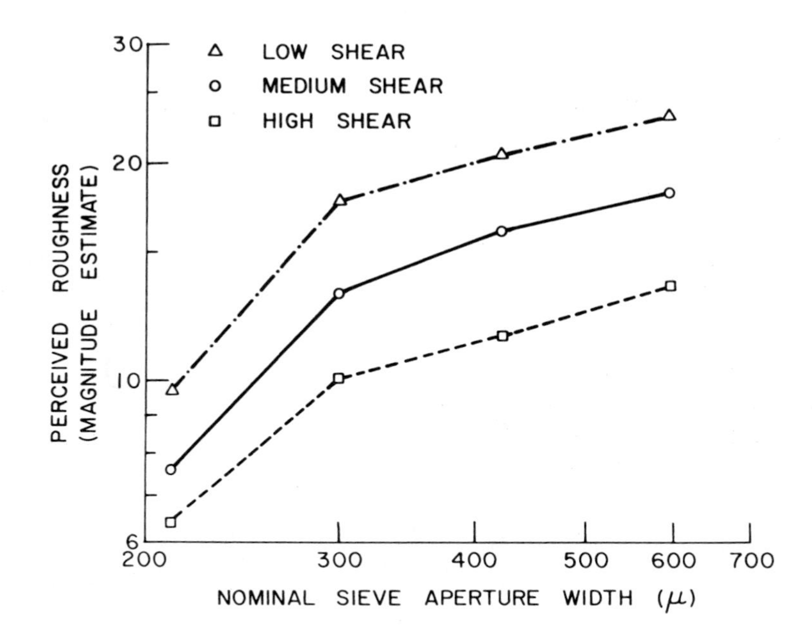

Fig 3. Perceived roughness (magnitude estimates) as a function of nominal width of sieve aperture and shear. Each point is based on 70 observations.

exerted normal to the surface ridge might provide the signal for "roughness" by deforming the intermediate ridge downward (assuming that the distribution of forces along the narrow surface ridge is equivalent to a single downward force in the centre of the ridge). The suggested movement is shown in Fig. 4b. The dotted lines represent the resting position of the intermediate ridge and the solid lines, the position the intermediate ridge might assume when a normal force is applied to the skin as shown. But to perceive roughness, as mentioned earlier, there must be relative motion between hand and surface. Hence, the effects of shear must be considered as well. Applying a lateral force might add a sideways bias to the motion of the intermediate ridge, as shown in Figs. 4c. and 4d. The precise details concerning extent and direction of sideways and downward motion are not known. Calculations are further complicated by the fact that non-rigid bodies are involved. Figs. 4b.-4d. are therefore intended only as visual aids to understanding the more general proposition that intermediate ridge motion involving lateral and downward components takes place when the hand moves across a surface. Might the afferent fibres ending in Merkel's corpuscles code both normal and shear forces, the effects of shear interfering with the simultaneous coding of normal force, and hence with perceived roughness? According to this analysis, when shear is decreased by using the paper, the sideways bias of the movements of the inter-

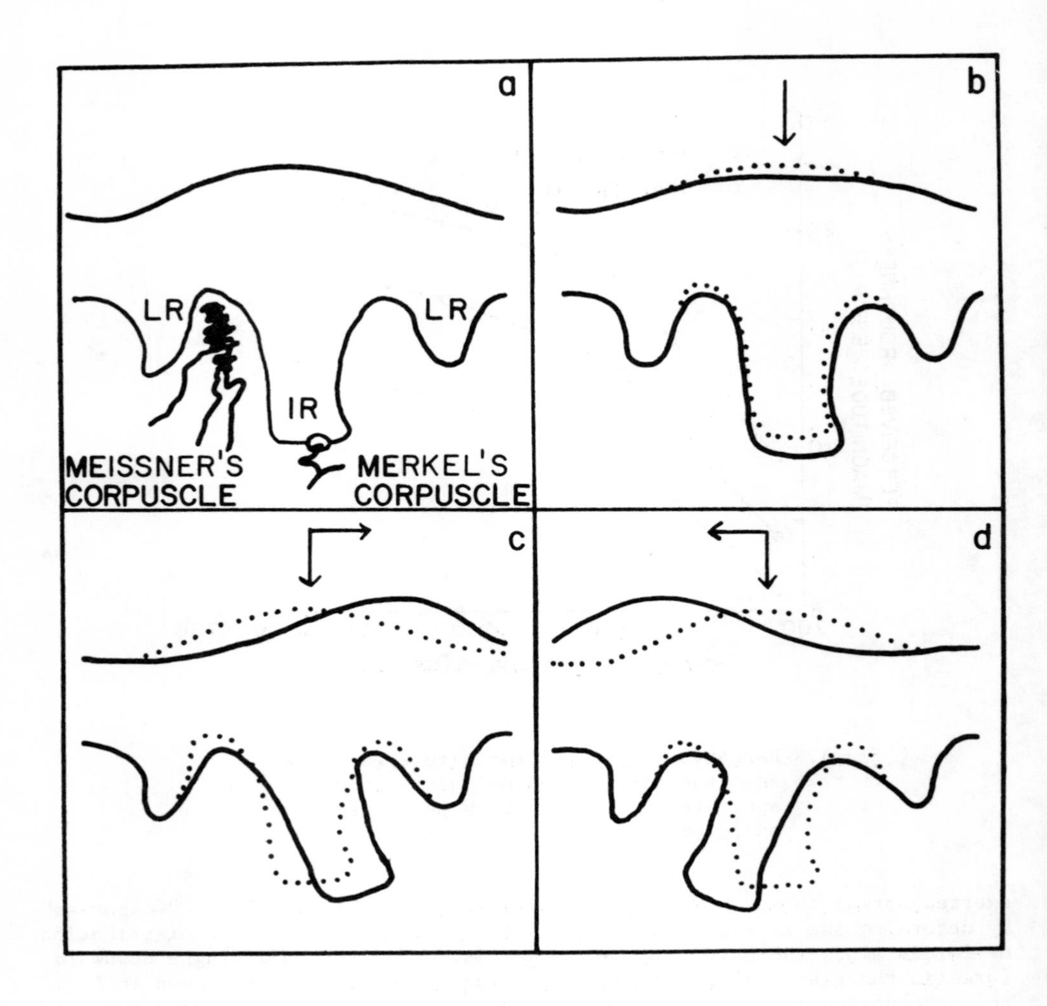

Fig. 4. *a* Schematic representation of human digital skin (adapted from Cauna, 1968) showing intermediate ridge (IR) and limiting ridges (LR) projecting into the dermis. The IR "floats" freely in loose connective tissue, while the limiting ridges are attached firmly to the dermis. Also shown is Merkel's corpuscle, attached to the deep surface of the IR, and Meissner's corpuscle, attached to the papillary ridge. *b Suggested* movement of the IR when a normal force is applied to the centre of a surface ridge. *c* and *d Suggested* movement of the IR when normal and shear forces are applied to the surface ridge as shown (dotted lines represent the resting position of the IR). Note the downward and sideways motions of the IR.

mediate ridge would be lessened. This in turn would result in a relatively larger roughness signal.

Also shown in Fig 4a. are Meissner's corpuscles, found in the dermal ridges, and presumed to be the end organs of the RA units. Deformation of the surface ridges due to normal and shear forces might therefore be transmitted directly to and coded by Meissner's rather than Merkel's corpuscles.

Of course, there are numerous hypotheses one could suggest. For example, both RA and SAI units might be involved. What is needed now are single-unit recordings of the various candidate receptive units when lateral and normal forces are applied. I would like to point out, however, that in discussing the possible functions served by epidermal ridges and dermal receptors in acuity and texture perception, we have, of course, been addressing issues arising from Quilliam's paper.

Other Interpretations

People have informally suggested two other ways in which an intermediate paper may improve detection of the orientation of an undulation and increase the magnitude of roughness sensations. One possibility is that the paper serves as an amplifier. It might produce travelling waves along its surface (caused by the edges catching on surface irregularities beneath), the waves increasing roughness by adding to those vibrations occurring directly under the fingertips. Another possibility is that the paper serves as a low-pass filter, letting through energy only at the lower frequencies caused by the larger, more-widely spaced bumps. This last interpretation, depends upon a frequency-dependent notion of tactile roughness. Further work is being carried out to evaluate the relative merits of these alternatives.

Applications to Tactile Maps for the Blind

Quite apart from understanding the mechanisms which underly this tactile phenomenon, it may serve a useful function for the blind. Earlier in this paper, I mentioned that we wished to determine whether texture discrimination as well as sensation magnitude is improved by using a very thin hand covering. Should this prove to be so, it would be worth considering whether this technique could improve the discrimination of (textured) areal symbols used in tactile maps. Of course, there must be no concomitant interference with the proper encoding of the line and point symbols also used in this communication system. Such work relates directly to the concerns of Armstrong and the Blind Mobility Research Unit in Nottingham.

Summary

In summary, this paper has considered a new perceptual phenomenon in which the perception of tactile roughness is heightened. The work has been linked to a related phenomenon in which the accuracy of detecting the orientation of surface undulations is increased. A reduced-shear interpretation was offered for both findings, and possible receptor mechanisms considered. Two other explanations were suggested briefly. Finally, the potential value of this work for the blind in interpreting tactile maps was discussed.

REFERENCES

N. Cauna, Nature and functions of the papillary ridges of the digital skin, *Anatomical Record*, 119, 559 (1954).

N. Cauna, Light and electron microscopal structure of sensory end-organs in human skin. In D. Kenshalo, *The Skin Senses*, C. Thomas, Springfield (1968).

I. Gordon & C. Cooper, Improving one's touch, *Nature*, 256, 203 (1975).

S.J. Lederman & M.M. Taylor, Fingertip force, surface geometry, and the perception of roughness by active touch, *Perception & Psychophysics*, 12, 401 (1972).

M.M. Taylor, & S.J. Lederman, Tactile roughness of grooved surface: a model and the effect of friction, *Perception & Psychophysics*, 17 (1), 23 (1975).

W.H. Talbot, I. Darian-Smith, H. Kornhuber, and V.B. Mountcastle, The sense of flutter-vibration: comparison of the human capacity with response patterns of mechanoreceptive afferents from the monkey hand, *Journal of Neurophysiology*, 31, 301 (1968).

ASPECTS OF MEMORY FOR INFORMATION FROM TOUCH AND MOVEMENT

Susanna Millar

Department of Experimental Psychology, University of Oxford, South Parks Road, Oxford OX1 3UD, England

INTRODUCTION

My question is how shapes are coded in memory by active touch. I shall first review some published findings with blind and sighted children. These suggest that memory for familiar or easily named shapes differs little from visual and verbal memory. By contrast, memory for tactual nonsense shapes seems to have little access to attentional strategies. The results resemble findings on short-term motor memory in conditions that preclude spatial reference. I shall then present results from ongoing studies. These indicate that tactual shapes are not initially coded in spatial terms. Some findings also suggest that texture differences may be used. I shall finally consider how this may explain the characteristics of short-term tactual memory.

MEMORY FOR HAPTIC INFORMATION

Findings from a number of converging experimental designs suggest that memory for tactual shapes differs from visual and verbal memory, unless the shapes can be named quickly. Unfilled delays and delays filled with attention demanding tasks can be compared to assess how far memory depends on attentional capacity (Ref. 1,2). Children's memory for unfamiliar nonsense shapes was found to depend only on the length of delay while memory for familiarized shapes was disturbed mainly by attentional demands during delay (Ref. 3). The form of coding can also be inferred by comparing memory for lists of items that are either similar or dissimilar on specific characteristics (Ref. 4,5). Similarity in the name sounds of easily named successive Braille letters disturbed tactually probed recall. This suggests that the letters were coded verbally. But recall by children who were slow at naming the letters was only disturbed when successive letters were similar in "feel" rather than in name-sound (Ref. 6). These subjects must have coded tactual features. Further, spacing verbal or visual items in groups typically improves recall. Subjects seem to be able to "chunk" or rehearse the input better under these conditions. Spacing easily named Braille letters in groups also improved tactually probed recall by blind children. But grouping tactual nonsense shapes had adverse effects on recall by the same children (Ref. 7). Recall spans were also much lower for these shapes. Thus coding seems to differ for tactual nonsense shapes, and attentional strategies are less possible or useful. Nevertheless, coding tactual features can be an advantage. Successive Braille matches were found to be much faster if the physical format of letter pairs was identical than when it differed even when the name and shape (configuration) of the letter pairs were the same (Ref. 8). This showed not only that tactual information about the physical format was coded in memory. It also implies that such codes can be used to speed matching.

Discrepancies in the extent to which memory depends on available attentional capacity have also been found in short-term motor memory (Ref. 9,10). It has been argued cogently that memory for discrete blind movements is subject to attentional demands only if inputs can be coded verbally or in terms of spatial references (Ref. 10). When spatial cues are made unreliable, short-term motor memory deteriorates with delay rather than with attentional demands (Ref. 11). Applied to tactual memory this suggests that subjects may not initially code tactual shapes in terms of spatial features at all. Some of my previous findings are consistent with this. For instance, blind children, especially slower readers, had difficulty in recognizing enlarged Braille letters even though the overall shape and spatial relations between cells remained the same (Ref. 8). Matching identical visual shapes is typically faster than matching different shapes (Ref. 12, 13). It is assumed that identical pairs can be matched on global shape while different shapes need more exhaustive scanning (Ref. 12, 14). But for tactual shapes matching takes as long for identical as for different pairs except possibly by very experienced readers (Ref. 8, 15). This suggests that global shape is not coded initially. Finally, blind-folded sighted children who had been trained on Braille letters, had surprisingly little idea of the spatial relations between dots in a letter, let alone its global shape (Ref. 15).

Nevertheless, it is often implied that Braille letters are recognised as outline figures by touch as in vision (e.g. Nolan & Kederis, 16). Apkarian-Stielau and Loomis (17) suggest indeed that shapes are coded by touch as in very 'blurred' vision. Differences are merely due to poorer tactual spatial resolution. However, Taylor, Lederman and Gibson (18) argue that touch is specialized for texture. Extremely fine discriminations are possible. It thus seems reasonable to assume that tactual shapes may initially be coded by differences in texture, and not in spatial terms. The findings I should like to present come from ongoing studies designed to explore some of these assumptions.

TACTUAL MATCHING STUDIES

Three types of evidence were sought in the first study. It was argued that if Braille letters are coded in terms of spatial features, the time it takes to decide that two shapes differ should vary with the number of different spatial features between them. This has been reported for visual letters (Ref. 19). Also, if global shapes are recognized, identical pairs should be matched faster, and it should not matter how many dots make up a shape. Three types of spatial change between "different" Braille letter pairs were used: (1) Change in one location of the 3 x 2 Braille matrix, by omitting one dot in a given location. (2) Change in two locations by omitting a dot in one location, and adding one in a different location. (3) Change in three locations by omitting dots in two locations, and adding a third in a different location. Identical pairs were chosen from all the letters making up the "different" pairs and interspersed in the lists in Gellerman order. A group of slower and a group of faster Braille readers were tested on "same" - "different" judgments of these letter pairs.

The results, shown in Figure 1, were unequivocal. The slower group took longer over both "different" ($p < .01$) and "same" ($p < .005$) pairs. But the number of changed features between "different" pairs had no significant effect on either group. The number of dots in identical shapes did matter ($F = 13.39$; d.f. = 2,28; $p < .001$). A groups x conditions interaction ($F = 8.81$; d.f. = 2,28; $p < .005$) showed that this was due to the slower group ($p < .001$).

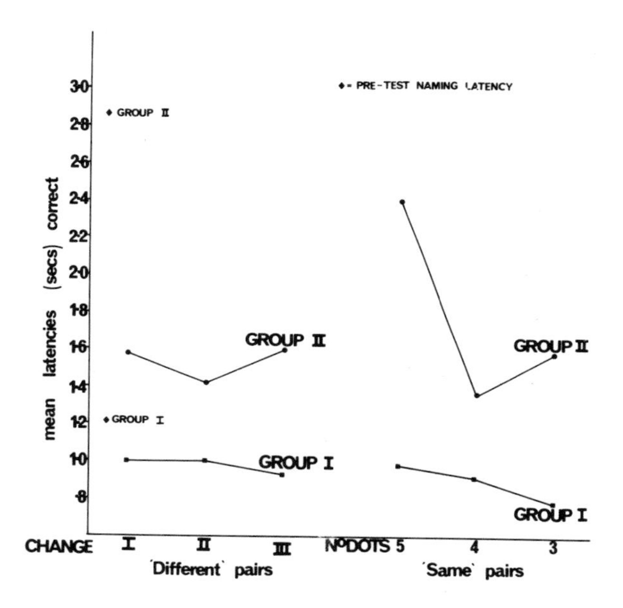

Fig. 1. Mean latencies for different pairs (three types of change) and same pairs (5,4 or 3 dot shapes) for faster (Group I) and slower (Group II) readers.

For "same" - "different" tests there was a tendency to an interaction suggesting that some of the better group were beginning to match on global shape, as shown in Fig. 2. But this did not quite reach significance level. Considering the amount of experience with Braille shapes by both groups of subjects, it is rather startling to find so little evidence that identical letters were processed as global figures, and also that "different" shapes were evidently not matched on spatial features. There is no reason to suppose that this depended on "blurred" perception. Presentation was self-paced, and accuracy was extremely high (1.8% errors "different", 3.56% errors "same").

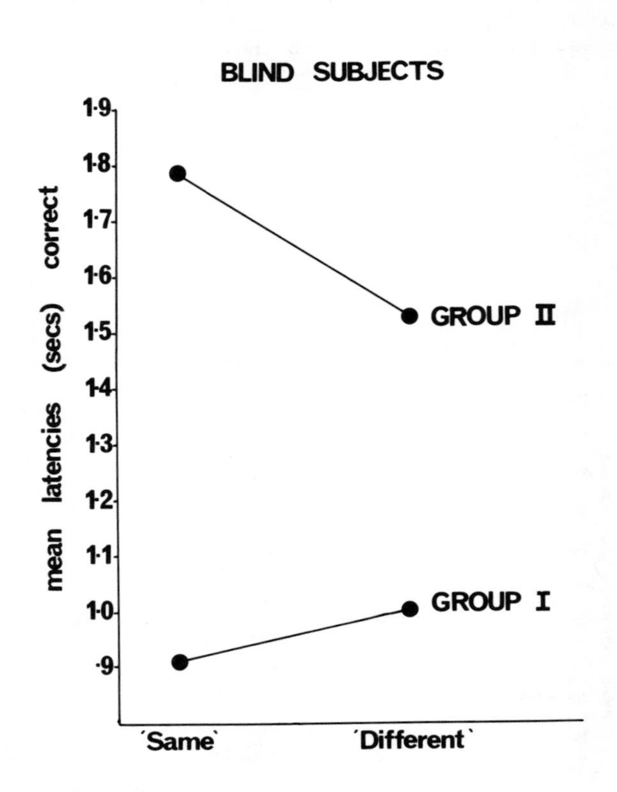

Fig. 2 Mean latencies for "same" and "different" pairs by faster (Group I) and slower (Group II) readers.

Evidence on two further parameters was sought next. Firstly, symmetry, i.e. equal distribution about the vertical or horizontal axis is an important spatial cue in recognizing visual shapes even for pre-school children (Ref. 20,21). It was argued that if tactual shapes are coded by spatial features, "different judgments for pairs of shapes differing on this cue should be easier. If subjects code texture differences, "crowding" (differences in dot number) should be an easier cue. Secondly, spatial coding would be indicated by more accurate judgments with the left hand. The minor hemisphere is specialized for spatial tasks even when shapes are tactual (Ref. 22,23). However, if the shapes are not spatially coded, there is presumably no reason to expect a left-hand advantage. Early in learning such an advantage does not seem to occur (Ref. 15,24).

Two groups of right-handed 6 and 9 year olds were tested blind-fold with either the left or right hand on "same" - "different" judgments. Symmetric and asymmetric patterns containing either 8 or 5 dots were generated from a 3 x 3 raised dot matrix (8 x 8 mm) as shown in Figure 3. There were four lists of pairs: (1) Eight-dot patterns in which "different" pairs differed in terms of symmetry (symmetric versus asymmetric); (2) Five-dot patterns in which "different" pairs differed in terms of symmetry; (3) "Different" pairs were symmetric shapes which differed in terms of crowding (8 versus 5 dots); (4) "Different" pairs were asymmetric shapes which differed in terms of crowding. "Same" pairs were 8-dot symmetric and asymmetric pairs in list 1 (SS/AA); 5-dot symmetric and asymmetric pairs in list 2 (RR/II); 8-and-5 dot symmetric pairs in list 3 (SS/RR) and 8-and-5 dot asymmetric pairs in list 4 (AA/II). These were interspersed in Gellerman type order. The order of lists was counterbalanced.

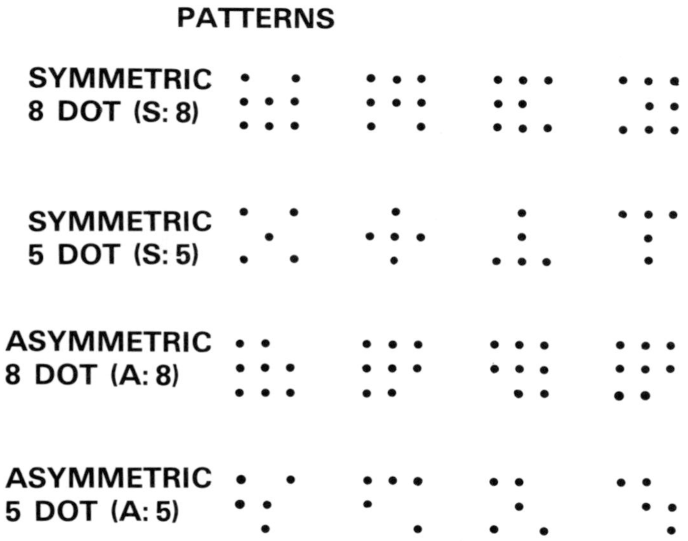

PATTERNS

SYMMETRIC 8 DOT (S:8)

SYMMETRIC 5 DOT (S:5)

ASYMMETRIC 8 DOT (A:8)

ASYMMETRIC 5 DOT (A:5)

Fig. 3. Symmetric and asymmetric shapes used in 4 lists for "same" and "different" pairs.

Results are shown in Figure 4. There was no significant advantage for the left hand. "Different" pairs were judged significantly more accurately when they differed in terms of crowding (dot number differences) than in either of the two lists based on differences in symmetry ($p < .01$ respectively on Newman

Keuls tests following a significant conditions effect in the Anova $\underline{F} = 16.5$, d.f. = 3,84; $\underline{p} < .001$).

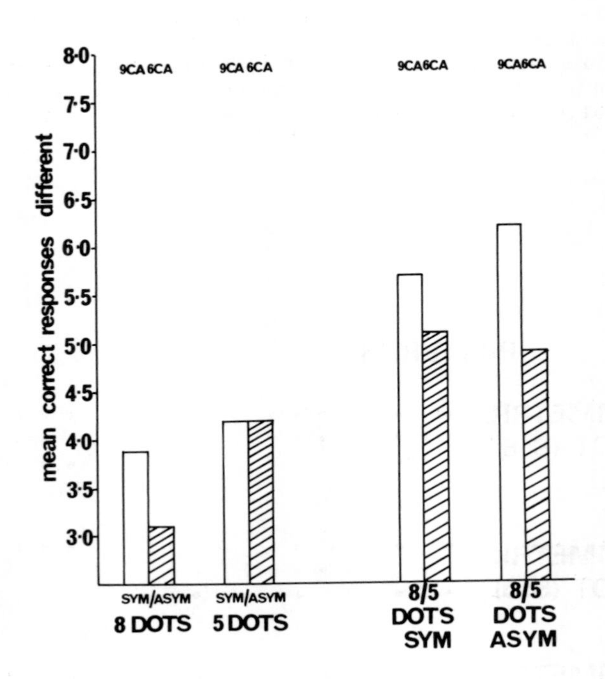

Fig. 4. Results for 8-dot pairs differing on symmetry (Sum/Asym) ; 5-dot pairs differing on symmetry (Sym/Asym) , symmetric pairs differing on crowding (8/5) , and asymmetric pairs differing on crowding (8/5) by 9 and 6 year old (9 CA, 6CA) blindfolded sighted children.

Error rates for "same" pairs were relatively high (40.5% versus 33% on "different" tests). This may have been due to response bias. On the other hand, "same" judgments were affected by the type of list in which they occurred. The effect of conditions was reduced, but significant ($\underline{F} = 4.84$; d.f. = 3,84; $p < .01$); performance was better for lists testing "crowding". There was thus a context effect for "same" judgments.Taken across conditions, symmetric and asymmetric "same" pairs did not differ. The important point for identical shapes is that symmetric pairs were judged no more accurately than asymmetric pairs even by the older group. The laterality results and the ineffectiveness of symmetry as a cue again suggest that subjects did not code the shapes as spatial patterns. The superiority of crowding cues for "different" pairs is consistent with the view that successful coding depended on texture differences.

The next question is whether these findings might be reversed for blind children who have a good deal of experience with tactual dot patterns and how delay affects this performance. Studying this exhaustively has only just begun. But I can report one result on a small group of congenitally blind children. They were tested with half of each of the 4 lists in (8 secs) delay and half in immediate test conditions in counterbalanced order. Errors and latencies were dependent measures.

The results for "different" pairs are shown in Figure 5. Errors (18%) differed significantly between conditions ($\underline{F}$ = 4.27; d.f. = 3,18; $p < .025$), and were highest for the 8-dot list with the symmetry cue than for the others ($p < .05$). The other three lists did not differ significantly. If anything, the two lists with the "crowding" cues were most accurate.

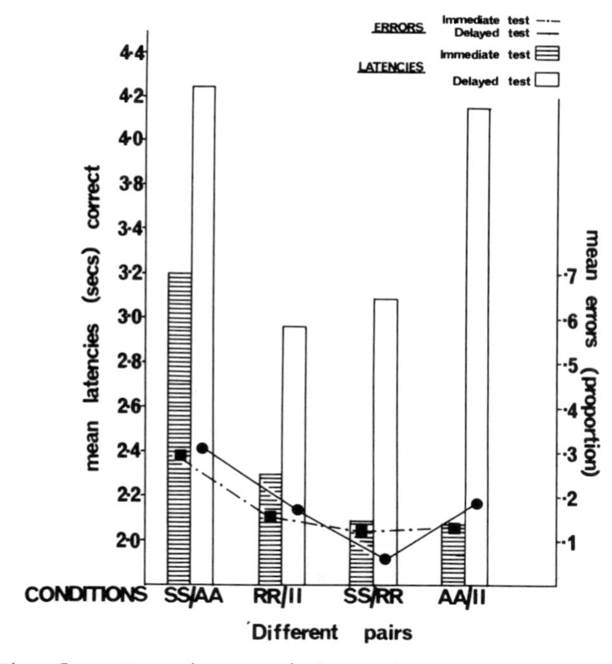

Fig. 5. Mean (correct) latencies and error proportions in immediate and delayed tests for 8-dot pairs differing on symmetry (SS/AA); 5-dot pairs differing on symmetry (RR/II); symmetric pairs differing on crowding (SS/RR); and asymmetric pairs differing on crowding (AA/II).

This was the case also for latencies of correct responses on immediate test. The fact that delay significantly increased latencies for correct responses ($\underline{F}$ = 23.04; d.f. = 1,6; p < .005) is consistent with previous results which suggest that subjects tend not to "rehearse" or actively maintain tactual inputs in memory across unfilled intervals. There was some support also in the latency data (Fig.5) for the hypothesis that maintenance across delay was less easy for discriminations based on "crowding" (texture) differences than when differences in symmetry were coded correctly. But from the present data this cannot as yet be a firm conclusion, and requires further study. The main point of interest here is that even for blind children differences in spatial symmetry seemed to be relatively ineffective as cues for discriminating between pairs of unfamiliar shapes. Particularly when delays were minimal successive discrimination tended to be faster and more accurate when gross differences in the number of dots between shapes ("crowding") could be used.

Analysis of identical matches was of interest only to test whether

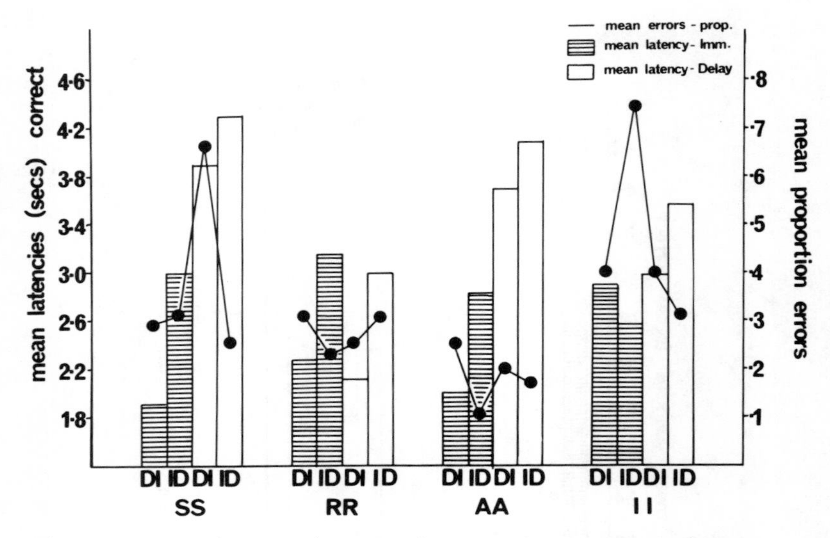

Fig. 6. Mean (correct) latencies and error proportions for immediate (hatched) and delayed (not hatched) tests for "same" SS(8-dot symmetric), RR(5-dot symmetric), AA(8-dot asymmetric) and II(5-dot asymmetric) pairs by subjects receiving delayed tests first (DI), and subjects receiving immediate tests first (ID).

symmetric shapes as such were easier or more difficult than asymmetric shapes. There was no evidence for this. Delay slowed matching ($\underline{F}$ = 10.54; d.f.=1,6; $\underline{p}$ < .05). Errors (32. 5%) produced an interaction between Shapes and Delay ($\underline{F}$=10.64;d.f.=3,18;$\underline{p}$ <.001) which depended on the order of presentation ($\underline{F}$=3.46; d.f.=3,18; p <.05). It meant that subjects who received immediate tests first (ID) had special difficulties with 5-dot asymmetric shapes on first test; subjects who received delayed tests first (DI) had special difficulties with 8-dot symmetric shapes on first test. The level of these errors as well as their dependence on task order indicated faulty response strategies rather than random responses on first test. But these difficulties involved both symmetric and asymmetric shapes, and there were no other differences in speed or accuracy.

It is of some interest that differences in symmetry about the vertical and horizontal axes should have so little effect in shape discrimination even by congenitally totally blind children who have general experience with dot patterns. At the same time, it fits in with findings on young blind children in complex spatial movement tasks in which the current relative position of body and limbs is not a reliable cue. These children seem to have difficulty in tasks which require reference to past experience of external spatial coordinates (Ref. 25). They have little difficulty in spatial tasks in which self-referent information can be used reliably. Crucial evidence on the neurophysiological mechanisms underlying this has been provided by many contributors to this symposium. We still seem to be some way from understanding the processes involved in performance without visual feedback in complex spatial problem and memory tasks. But these questions should also eventually prove capable of solution.

DISCUSSION

The main findings of the experiments presented here suggest that spatial features may not be the most salient cues for successive discrimination or coding in memory of relatively unfamiliar small tactual dot patterns by children. Thus discrimination speed in Experiment 1 was not affected by the number of spatial features that differed between shapes; and speeds for identical pairs were not consistent with an explanation of matching based on global configuration. The left hand (right hemisphere) advantage to be expected if spatial features are coded was not found for unfamiliar shapes in the second experiment. The findings showed that subjects were more accurate in discriminating successive patterns that differed in the number of dots contained within them than for patterns that differed in symmetry about the vertical and horizontal axes. A substantially similar result was found for blind children in the third experiment. These findings, may not, in fact, be unduly surprising. Spatial coding requires references in relation to which a location or direction may be coded as up/down, right/left, vertical, horizontal or oblique. In exploring Braille and similar small complex dot-patterns without visual feedback, by means of small, and - for unfamiliar shapes typically unsystematic - movements with the first segment of the index finger, such reference is not available. If I understand the interesting findings by Professor Sakata and Iwamura (this volume) correctly, feature detectors, for instance neurons selectively responsive to a given direction of movement of an edge across the skin, operate in relation to the receptive fields for a particular area of skin (segment of the phalanx). One might speculate that what may need to be learned in order to discriminate spatial features in unfamiliar dot patterns are systematic exploratory strategies that expose the finger to the stimulation in particular directions relative to the skin surface. By contrast even very young infants can discriminate rectangular from round objects by hand grasp. The report (Sakata and Iwamura, this volume) that particular combinations of inputs from joint and skin converge in single neurons, driving these when the monkey grasped objects of particular shapes, was thus especially interesting. But learning to discriminate Braille shapes is difficult. The findings here suggest that this may be because invariant spatial reference is lacking in initial explorations. Subjects therefore tend to rely on texture differences for coding successive inputs in memory. I shall suggest below how this might explain the characteristics of tactual memory for unfamiliar shapes which motivated the present studies.

The findings on tactual memory to which I referred at the beginning of the paper showed that, unlike memory for visual or for familiar tactual shapes, memory for unfamiliar tactual shapes was disturbed by similarity in the feel of successive items, by grouping items, and by unfilled delays, but not by attention demanding tasks in other modalities. Several explanations may be considered. For this it is useful to assume that memory depends on at least three experimentally distinguishable processes: 1. A very brief peripheral after-effect, persisting in uncoded form after off-set of the stimulation (Ref. 26); 2. A short-term process that uses attentional capacity to code and maintain the input across delays; and 3. Longer-term or permanent storage. One explanation of tactual memory might be that this depends only on peripheral "after-glow". This is unsatisfactory. It could not explain the fact that subjects can remember more than one, and often three of four successive inputs to the same finger (Ref. 6,7). Presumably each successive input overwrites or erases the previous sensory after-effect. Another possibility is that touch is like "blurred" vision or perception in noise. Difficult discriminations would pre-empt limited attention (Ref. 27). Only very few items would be coded therefore, and there would be little spare capacity to rehearse items. This would explain the more limited span, and why grouping did not facilitate recall. Some aspects of tactual memory seem to be accounted for in these terms. But it would not explain adverse effects of grouping, or why coding physical tactual features can speed matching more than naming. Explanations in terms of blurred perception would not account for the findings on experienced blind subjects whose accuracy in Braille recognition is high. All the same, it seems unnecessary to assume that there are a multiplicity of separate modality-specific short-term stores, each with its own decay properties.

The explanation I should like to put forward is as follows: The extent to which attention is involved depends on the type of code which subjects use. Only higher order codes that condense a good deal of information require attentional capacity at encoding. Such codes are also easier to use actively to maintain the information. Spatial codes would seem to be "summary" codes of this kind. By their nature, they condense and "tag" a great deal of spatial information by referring spatial features to the two major spatial axes. When spatial reference is unreliable, as in exploring unfamiliar dot patterns, crude differences in texture are coded instead. Texture discriminations can be acute and specific. But they provide many inputs without simple or obvious coordinate systems, or "summary" description by which the information is easily condensed. Coding by texture would thus demand little attention at encoding. But, equally, attentional strategies would be more difficult with such codes so that items cannot easily be maintained across delay. Particularly for raised dot patterns, differences in texture do not lend themselves easily to perceptual grouping. The large number of relatively specific inputs even for a single shape would thus quickly overload the system. Consequently grouping items by spacing would merely increase the overload. Grouping would thus have adverse effects. When orthogonal spatial references are provided, either because subjects are more experienced and can refer to vertical and horizontal axes by using the sides of the page or the finger, or proprioceptive information from the position of the body and limbs, or when shapes are delimited by vertical lines, coding tactual shapes spatially should be easier. This could also explain the result found by Apkarian-Spielau and Loomis (17). In that study vertical slits were placed over the shapes. This was intended by them to reduce blurring. They found that this significantly improved tactual letter recognition, but had no effect on blurred vision. On the account here, the result would be explained by the fact that placing vertical slits over the shapes provides spatial (vertical) references, and this would aid in coding spatial features. For vision these are unnecessary since spatial references are continuously available in any case.

In summary, memory for tactual shapes, like short-term motor memory, deteriorates with delay rather than with attentional demands, unless inputs are coded verbally or in terms of spatial references. The findings presented here suggest that tactual shapes are not initially coded in spatial terms, either as global configurations or by spatial features. There were some indications also that they may be coded by texture differences. I have argued that this can explain the findings on tactual memory if it is assumed: 1. that attentional demands in memory depend on the use of "summary" codes that can organize and condense large amounts of information to relatively simple or economic "descriptions"; 2. that spatial (and verbal) codes are of this order, but are applied to tactual shapes only with experience; 3. that coding only texture differences in spatial patterns does not lend itself easily to perceptual grouping or condensed "description"; 4. such codes therefore tend to overload the system in spatial tasks, and are not as easily used in attentional strategies. At the same time, texture differences are not merely uncoded after-effects of the stimulation, but can be coded and used in memory.

REFERENCES

1. Peterson, L.R. & Peterson, M.J. Short-term retention of individual verbal items. J. exp. Psychol. 58,193 (1959).

2. Dillon, R.F. & Reid, L.S. Short-term memory as a function of information processing during the retention interval. J. exp. Psychol. 81, 261 (1969).

3. Millar, S. Tactile short-term memory by blind and sighted children. Br. J. Psychol. 65, 253 (1974).

4. Conrad, R. Acoustic confusions in immediate memory. Br. J. Psychol. 55, 75 (1964).

5. Conrad, R. The chronology of the development of covert speech in children. Dev. Psychol. 5, 398 (1971).

6. Millar, S. Effects of tactual and phonological similarity on the recall of Braille letters by children. Br. J. Psychol. 66, 193 (1975).

7. Millar, S. Short-term serial tactual recall: Effects of grouping on

tactually probed recall of Braille letters and nonsense shapes by blind children. Br. J. Psychol. in press (1977).

8. Millar, S. Tactual and name matching by blind children. Br. J. Psychol. in press (1977).

9. Russell, D.G. Spatial location cues and movement production. In G.E. Stelmach (Ed.) Motor Control Issues and Trends, Academic Press, New York (1976).

10. Martenuik, R.G. Cognitive information processes in motor short-term memory. In G.E. Stelmach (Ed.) Motor Control: Issues and Trends, Academic Press, New York (1976).

11. Laabs, G.L. Retention characteristics of different reproduction cues in motor short-term memory. J. exp. Psychol. 100, 168 (1973).

12. Bamber, D. Reaction times and error rates for 'same'-'different' judgments of multidimensional stimuli. Percept. & Psychophys. 6, 169 (1969).

13. Nickerson, R.S. Response time to 'same'-'different' judgments. Percept. & Mot. Skills 20, 15 (1965).

14. Kruger, L.E. Effects of irrelevant surrounding material on speed of same-different judgments of two adjacent stimuli. J. exp. Psychol. 98, 252 (1973).

15. Millar, S. Early stages of tactual matching. Perception 6, 333 (1977).

16. Nolan, C.Y. & Kedris, C.J. Perceptual factors in Braille Word Recognition, Amer. Found. Blind, New York (1969).

17. Apkarian-Stielau, P. & Loomis, J.M. A comparison of tactile and blurred visual form perception. Percept. & Psychophys. 18, 362 (1975).

18. Taylor, M.M., Lederman, S.J. & Gibson, R.H. Tactual perception of texture. In E. Carterette & M. Friedman (Eds.) Handbook of Perception 3, Academic Press, New York (1973).

19. Taylor, D.A. Holistic and analytic processes in the comparison of letters. Percept. & Psychophys. 20, 187 (1976).

20. Gaines, R. The discriminability of form among young children. J. exp. Child Psychol. 8, 418 (1969).

21. Munsinger, H. & Forsman, R. Symmetry, development and tachistoscopic recognition. J. exp. Child Psychol. 3, 168 (1966).

22. Milner, B. & Taylor, L. Right hemisphere superiority in tactile pattern recognition after cerebral commissurotomy. Neuropsychologica 10, 1 (1972).

23. Kumar, S. Short-term memory for a nonverbal tactual task after cerebral commissurotomy. Cortex 13, 55 (1977).

24. Rudel, R.G., Denkla, M.B. & Spalten, E. The functional asymmetry of Braille letter learning in normal, sighted children. Neurology 24, 733 (1974).

25. Millar, S. Spatial representation by blind and sighted children. In G. Butterworth (Ed.) The Child's Representation of the World, Plenum Press, New York (1977).

26. Melzack, R. & Eisenberg, H. Skin sensory afterglows. Science 159, 445 (1968).

27. Rabbitt, P.M.A. Channel-capacity, intelligibility and immediate memory. Qu. J. exp. Psychol. 20, 241 (1968).

VIBROTACTILE PATTERN RECOGNITION AND MASKING*

James C. Craig

Department of Psychology, Indiana University, Bloomington, Indiana 47401, USA

Masking has proved to be a useful tool in investigating the way in which the skin processes tactile stimuli. In the basic masking paradigm, the detectability of a test stimulus is measured in the presence of and in the absence of some additional stimulus, the masking stimulus. Using this paradigm, investigators have examined such problems of tactile sensitivity as the nature of interaction among widely disparate loci on the skin (1,2,3), the effect of psychophysical task on localization and detection (4), the effect of increasing contactor area on detectability (5), and the way in which interaction among vibrotactile stimuli may be increased by compensating for time of arrival at the central nervous system (2). These problems and many more have been investigated by simply measuring changes in detectability in the presence of masking stimuli, i.e., detection masking.

The development of tactile communication systems using vibrotactile stimuli such as the Optacon (6), TVSS (7), and Optohapt (8) has provided additional impetus for the study of masking. These systems have made it possible to present to the skin complex, spatio-temporal patterns of vibration. In doing so, these systems have generated new questions about how tactile patterns are processed and have provided the means by which some of these questions may be answered. The problems I have been concerned with over the past several years have revolved around one of these systems, the Optacon, and the way in which cutaneous patterns generated on the Optacon display are perceived by the skin. One of the ways we have been investigating cutaneous pattern perception is by studying the way in which one stimulus or pattern interferes with the recognition of another pattern. In short, we are studying what is usually referred to as "recognition masking."

The Optacon is a reading aid for the blind. It converts printed material to patterns of vibration which can be felt by the blind user's fingertip. Following a training period, many blind people are able to read ordinary print at rates of 30 to 60 words per minute (9), and in some cases even greater rates are achieved (10,11). Our previous studies have shown that a number of variables affects letter recognition and masking with the Optacon. These variables include the position of a letter in a trigram, whether the letter is subject to forward masking (a masking stimulus presented before the letter to be identified) or to backward masking (a masking stimulus presented after the letter to be identified), or to both forward and backward masking, the nature of the letter to be identified, and the time between the letter and

*This study was supported by N.I.H. Grant NS09783.

the masking stimuli (12). In addition, it has been shown that there are large differences among observers as to how much masking they show and in some cases whether any masking is demonstrated or not (11). The present study extends some of our earlier findings by examining separately the role of temporal intervals in forward and backward masking, by comparing detection and recognition masking, and by measuring the effectiveness of various patterns in producing interference in letter recognition.

FORWARD AND BACKWARD MASKING

A previous study using the Optacon (12) examined the ability of observers to identify target letters when masking stimuli consisting of black rectangles, the same size and shape as the letter "H," were presented right before and right after the target letter. The sequential presentation of black rectangle, letter, black rectangle produced considerable interference in letter recognition. The percent correct recognition dropped about 40% when comparing letter recognition in the absence of the masking stimuli to letter recognition in the presence of the masking stimuli. It would probably not surprise anyone to learn that, as the time between the target letter and the masking stimuli was increased, the amount of interference produced decreased considerably. The masking stimuli were presented both before and after the letter. Thus, there were two temporal gaps, one between the offset of the first masking stimulus and the onset of the letter and the second between the offset of the letter and the onset of the second masking stimulus. The two gaps were always set at the same duration. When the gaps exceeded 150 msec, there was little interference in letter recognition. The presentation of a masking stimulus both before and after the letter made it impossible to evaluate separately the contribution of forward and backward masking, the topic of our first study.

The particular time course of forward as compared to backward masking is of some interest because of the use such time courses have been put to in general descriptions of pattern recognition processes. There are two general processes which have been used to explain how forward and backward masking might interfere with visual pattern recognition (13,14,15). The two processes are integration and interruption. In the integration process, a masking stimulus is thought to interfere with a target by adding its energy to the target and making it difficult for the observer to separate the two. Thus the closer in time the masker is to the target, the more difficult it is for the observer to separate the two, and interference is evident. If integration is the sole explanation for masking, the amounts of forward and backward masking should, according to several investigators (13,14), be equal. Studies of visual masking have, however, found that there is more backward than forward masking, i.e., for the same temporal separation between target and mask, a masker presented after the target produces more interference than does a masker presented before the target. To account for finding greater backward than forward masking, it has been hypothesized that the stimulus pattern begins a neural process which is interrupted by the masking stimulus. If interruption were the sole process operating in recognition masking, there would be presumably no forward masking evident. It is assumed that finding both forward and backward masking but somewhat more of the latter than the former is the result of both integration and interruption (13,14).

As in our previous studies, the Optacon was used to generate the letters and the masking stimuli. The Optacon contains a camera with an array of photosensitive elements. The array contains 144 elements, 6 columns by 24 rows.

As the camera is passed over a letter of the alphabet, the pattern of light and dark produced by the letter is registered by the camera. The camera is connected to a 6 x 24 array of vibratory stimulators. The pattern of light and dark produced by the letter is reproduced on the tactile array. The letter "O," for example, produces a circular pattern of vibration that moves from right to left across the user's fingertip as the camera is moved across the letter.

Uppercase letters of the alphabet were used as the patterns to be identified. The camera of the Optacon was adjusted so that a single letter directly beneath the camera would occupy the top 18 rows of the tactile display and all 6 columns. A single letter would occupy a space on the observer's fingertip 1.1 by 2.0 cm. Horizontal strokes of the letters would, in general, activate two rows, and vertical strokes, one column. The amplitude of vibration on the tactile array was set at a comfortable level for the observers and was left unchanged across observers and conditions.

As the Optacon is ordinarily used, the camera is moved by the blind person across the material to be read. In the present studies, however, the camera was moved across the letters by the automatic page scanner, a modified x-y plotter. The automatic page scanner produces smooth and controlled rates of movement of the vibrotactile pattern across the user's fingertip.

The first set of measurements involved all 26 uppercase letters of the alphabet. The letters were seriph letters generated by I.B.M. type, prestige elite. Sighted observers received training in identifying the letters presented singly, i.e., without any masking stimuli. The scan rate, the time it took for a particular point on a letter to pass across the tactile array, was 90 msec. The total time for a letter to pass across the array was 150 msec. This rate corresponds roughly to a reading rate of 100 words per minute. Observers were presented randomly generated lists of 40 letters to identify. A particular list would involve only one condition of the experiment. The conditions tested were single-letter recognition (no masking stimulus), forward masking with the times between the offset of the masker and the onset of the target set at either 45, 90, 165, or 220 msec and backward masking with the times between the offset of the target and the onset of the masker set at 45, 90, 165, or 220 msec. The masking stimulus was generated as in previous studies, by a black rectangle the same size as the letter "H." Figure 1A shows the type of

Fig. 1. Examples of patterns used with Optacon. Set A was used in forward-backward masking measurements. Set B shows the various backward maskers tested.

letter and masking stimuli used. After each trial, the observer attempted to identify the letter presented and was then told what letter had been presented.

The results from these measurements are shown in Fig. 2. Each point in the masking conditions represents the mean of 800 trials, 200 from each of four observers. The point at $-\infty$ represents single-letter recognition, no masking

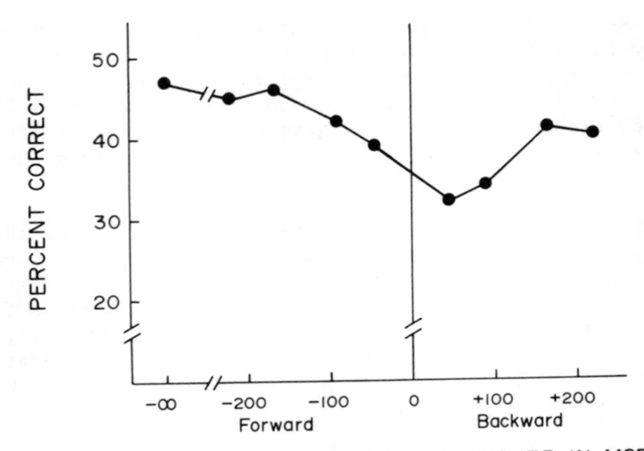

Fig. 2. Percent correct letter recognition as a function of the time between the masking stimulus and target letter. Point at -∞ represents letter recognition with no masking stimulus present. All 26 letters tested.

stimulus, and is the mean result from 1600 trials. The results show rather clearly that a backward masker is more effective than a forward masker and that this increased effectiveness extends to 150 to 200 msec following the offset of the target.

The previous work (12) also showed that a backward masker is more effective than a forward masker and that this effectiveness is, in part, the result of the nature of the letters used and the manner in which the Optacon displays information. The letters move across the tactile array of the Optacon from right to left. In the case of a backward masker, the trailing edge of the letter, the right-hand side, is closer both temporally and spatially to the backward masker than to the forward masker. A number of the uppercase letters of the alphabet has more information required for their recognition on the right-hand side of the letters. Craig demonstrated that these right-hand letters, "B, C, D, E, F, G, K, L, P, Q," and "R," were interfered with significantly more by a backward masker than were the remaining 15 letters of the alphabet. These 15 letters were called symmetrical letters because the information required for their identification could be obtained from both the left and the right sides of the letter. Thus, it is not surprising that more backward masking might be evident under the conditions of the present experiment. Indeed, an analysis of the present results shows that there was a significantly greater decrement in recognition in going from the forward to the backward masking condition for the right-hand letters as compared to the symmetrical letters ($p < .002$, two-tail, Mann-Whitney U test).

It might be argued that, when the results are thus biased by the nature of the patterns, attempting to analyze the time course of masking is difficult. On the other hand, the results shown in Fig. 2 are indicative of what might be expected with the full alphabet, or at least the uppercase letters. However, to eliminate this bias, the measurements of forward and backward masking were repeated with a subset of the symmetrical letters. The subset of symmetrical letters tested consisted of 11 letters, all of which were approximately the same width. The 11 letters were "A, H, N, O, S, T, U, V, X, Y" and "Z." The same set of 11 letters was used throughout the remaining studies I will be discussing. Letter recognition was tested in the same way as described before, and the results are shown in Fig. 3.

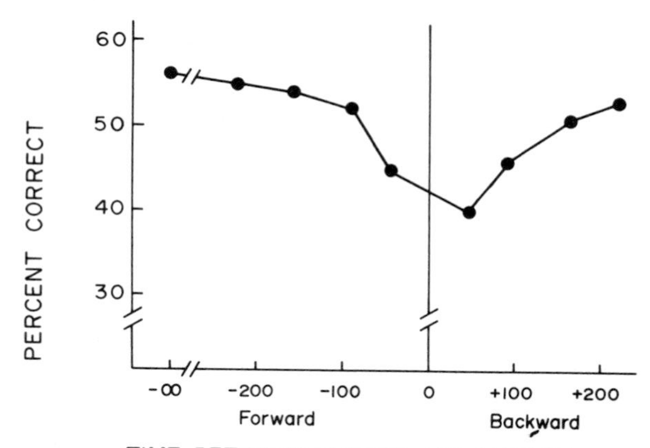

Fig. 3. Percent correct letter recognition as a function of the time between the masking stimulus and target letter. Point at -∞ represents letter recognition with no masking stimulus present. The 11 symmetrical letters were tested.

The results are the mean results from four observers. Two of the observers were tested for 400 trials on each of the masking conditions and for 800 trials on the no-masking conditions, the point at -∞. The remaining two observers were tested for half as many trials under each condition. The results show that, even when the patterns to be identified are symmetrical, there is more backward than forward masking present.

The function shown in Fig. 3 is similar to those which have been generated by using letters presented visually (14). Such visual results have been used to argue that both the processes of integration and interruption are necessary to account for the effect that a masking stimulus has on visual pattern recognition. It would, however, clearly require a considerable number of additional experiments to establish that these two processes are the explanations for cutaneous pattern recognition and masking effects. The results also show that the representation of the cutaneous pattern in the nervous system can be interfered with for as long as 150 msec after the offset of the pattern. In cognitive terms, such a result indicates that it takes at least 150 msec for the cutaneous "icon" to be stored in more permanent memory.

Additional caution should be exercised in comparing these data to visual data, particularly with regard to the time scale shown in both Figs. 2 and 3. It should be remembered that the display time for the letter is 90 msec and that the time from the onset of the leading edge of an average letter on the right side of the display to the offset of the trailing edge of the letter on the left side of the display is 150 msec. The masker follows a similar time course. Therefore, if, rather than plotting time between target and masker, the interstimulus interval, one were to plot the time between the onset of the target and the onset of the masking stimulus, stimulus onset asynchrony, there would be an additional 150 msec inserted between the forward and the backward masking functions. In many of the visual experiments, the display of the target stimuli for less than 5 msec makes the difference between presenting the data as a function of interstimulus interval and stimulus onset asynchrony negligible. The difference in the way in which the present data are displayed is more than merely one of convenience. The difference involves the way in which one views the processes of pattern recognition and of masking. Questions may be raised as to what portion of the pattern is necessary for accurate recognition and what portion of the masking stimulus is effective in interfering with pattern recognition. To answer these questions would require further research using different modes of presenting the target letters and the masking stimuli.

DETECTION MASKING

A second series of measurements examined forward and backward detection masking under conditions comparable to those already examined in recognition masking. Several previous studies have examined forward and backward detection masking, although not under conditions comparable to those under which recognition masking was examined in the present study (2,3). There are several reasons for wanting to examine detection processes and to compare the results with recognition processes. One reason is to see to what extent common processes are being measured. If recognition and detection masking were to prove to be very similar processes, then it might be possible to use the results from the more numerous detection studies to make predictions about the outcome of recognition measurements. There might, indeed, be some reason to suppose that the results of such a comparison would be similar inasmuch as most theories of pattern recognition assume detection to be necessary prior to recognition (13).

A second reason for examining detection masking is that previous studies of tactile recognition have proposed explanations that bear little relation to explanations of detection. Using the TVSS Loomis and Apkarian-Stielau (16) studied the ability of an observer to recognize letters of the alphabet presented through an array of vibrators placed against the observer's back. They found that presenting masking stimuli on columns of the array adjacent to the columns on which the letters were displayed resulted in considerable interference in letter recognition. They concluded that this kind of recognition masking resulted from a reduction in the ability of the skin to resolve patterns spatially. Such an explanation is not usually used to explain how detection might be interfered with. In another study an Optacon-like display was employed to study rapid letter recognition (17). The investigators in this latter study concluded that many of the errors in letter recognition are the result of responding with correct letters in the incorrect order. Again such an explanation is not likely to serve for the kind of masking observed in a detection process.

In any comparison of detection and recognition processes in the present context, it should be made clear that the observers in our letter-recognition studies do not have any difficulty detecting that something has occurred. The observers can feel the target letters move across their fingertips. The difficulty lies in identifying which particular letters passed across their fingers.

Detection masking was measured under conditions which were similar to those under which recognition masking was measured. The measurements were made twice, with the same observers using two different psychophysical procedures, a tracking procedure and a two-interval, forced-choice procedure (2IFC).

Several recent studies using tactile stimuli have found that differences in detection-masking functions reflect the psychophysical methods used in measuring the amount of masking. Gescheider et al. (18) measured tactile thresholds when a masking stimulus was presented to the same side of the body as the test stimulus, ipsilateral masking, and to the opposite side of the body, contralateral masking. Using a tracking procedure, they found that there was considerable ipsilateral and contralateral masking. With a 2IFC procedure they found approximately the same amount of ipsilateral masking as they did with a tracking procedure but very little contralateral masking. Gescheider et al. interpreted their results as reflecting perhaps changes in the observer's task in the two conditions or perhaps changes in the observer's criterion. Rita Snyder, working in our laboratory, extended the analysis of the task differences resulting from differences in psychophysical procedures by setting up two different tasks for the observers to perform on test stimuli in the presence of a masking stimulus. One task required observers to detect a stimulus. The second task required them to indicate at which one of two test sites the stimulus had been presented, i.e., to localize the test stimulus. From the results she obtained, she concluded that a 2IFC procedure probably reflects the observer's ability to detect any change in the overall pattern of stimulation produced by the addition of the test stimulus to the masking stimulus, whereas classical psychophysical procedures such as tracking or method of limits reflect the observer's ability to localize the test stimulus in the presence of the masker (4).

It may be that if a tracking procedure does measure the ability of an observer to localize a test stimulus, then using such a procedure might produce results which more nearly parallel the results produced in a recognition masking task in which localization of pattern features is important. Alternatively, it may be that observers are using changes in the overall pattern of stimulation to help them identify letters. Such considerations of task requirements aside, when there are known to be such differences in detection results produced by different psychophysical procedures, it is probably best to use both procedures to see if the results are radically altered.

The measurements with both the 2IFC and tracking procedures were made with the same test and masking stimuli. The test stimulus was a 230-Hz, sinusoidal vibration delivered to the observer's left index fingertip. The test stimulus was 150 msec in duration with a rise-fall time of 10 msec. The masking stimulus was of the same frequency and duration and had the same rise-fall time as the test stimulus. The masking stimulus was set at a fixed amplitude of 58 microns peak-to-peak and delivered through the same vibrator as was the test stimulus.

The vibrator was a Goodmans V-47 vibration generator fitted with a circular contactor, 6 mm in diameter. The contactor protruded through a circular hole, 8 mm in diameter, in a metal plate. The observer rested his or her left arm on the metal plate and placed the left index finger over the hole in the metal plate in order to touch the contactor. The vibrator rested on a balance pan to maintain a constant static force of 30 grams against the observer's fingertip.

Thresholds were first measured with a tracking procedure. The electrical signal which generated the test stimulus was passed through a recording attenuator. By means of a hand switch connected to the recording attenuator, the observer could gradually increase or decrease the amplitude of the test stimulus. The observer was instructed to press the hand switch until he or she could no longer feel the test signal and then release the hand switch to allow the signal to become stronger until he or she could feel it. By alternately pressing and releasing the switch, the observer could "track" his or her threshold and leave a permanent record of his or her tracks on the recording attenuator chart. The midpoint of the observer's tracks was used as the estimate of the threshold.

The test stimuli were presented once every 1.5 sec. Thresholds were measured for the test stimulus in the presence of and in the absence of the masking stimulus. The time intervals between the test and the masking stimuli were set so that the masker preceded the test stimulus by 220, 165, 90, or 45 msec, forward-masking condition, or so that the masker followed the test stimulus by 45, 90, 165, or 220 msec, backward-masking condition.

Measurements with the 2IFC procedure used the same test and masking stimuli presented to the same loci on the skin. On each trial, two observation intervals were defined by lights flashing in front of the observer. During both observation intervals, the masking stimulus was presented, whereas the test stimulus was presented in only one of the two intervals. The observer's task was to select the interval which contained the test stimulus. Trial-by-trial feedback was provided. The intensity of the test stimulus was changed in accordance with the observer's responses to establish a level at which the observer was correct 75% of the time. This intensity was used as the threshold estimate. Approximately 100 trials were used to determine the threshold. Further details of the 2IFC procedure and method of establishing the threshold level may be found in Craig (19). When quiet thresholds were measured, the procedure was the same, with the exception that the masking stimulus was omitted. The same time intervals between test and masker were used as with the tracking procedure, with the exception that the 220-msec interval in both the forward- and backward-masking conditions was not tested.

Figure 4 shows the results from both the tracking and the 2IFC measurements. Both sets of measurements were obtained from the same four observers. Each point in the tracking condition represents the mean of approximately 40 measurements with the exception of the point at $-\infty$, no masking stimulus, which is the mean of 80 measurements. Each point in the 2IFC condition represents the mean of 14 or 15 measurements (100 trials per measurement) with the exception of the point at $-\infty$ which is the mean of 42 measurements.

Apparently, the only effect that the choice of psychophysical procedure has on the functions generated is the overall level of sensitivity. It is not surprising that observers show more sensitivity with a 2IFC procedure than with a tracking procedure. The threshold measured with a 2IFC procedure is roughly half what it is with the tracking procedure. The main conclusion is that there does not appear to be any functional change resulting from the selection of the psychophysical technique.

The most important point that Fig. 4 makes is that, in measuring detection masking, as is not the case in measuring recognition masking, there is more forward than backward masking evident. Although it may be too strong a statement to say that these results show that different processes are involved when

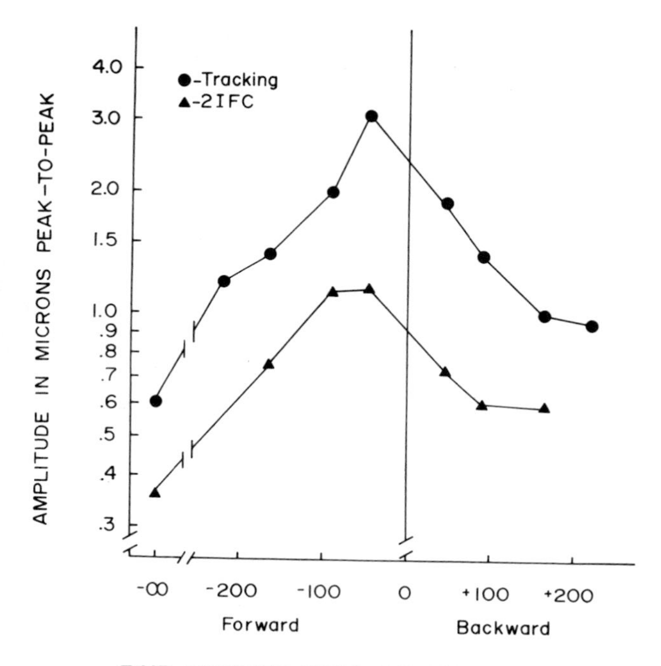

Fig. 4. The amplitude required to detect a test stimulus in the presence of a masking stimulus as a function of the time between the test and masking stimulus. Point at −∞ represents amplitude required to detect test stimulus with no masking stimulus presented. Tracking refers to measurements obtained using tracking procedure. 2IFC refers to measurements obtained using two-interval, forced-choice procedure.

detection as opposed to recognition masking is being measured, the results at least demonstrate the difficulty in generalizing from one set of measurements to another.

One additional experiment was carried out in our laboratory to compare detection and recognition masking. In this experiment, performed by Ben Mannix, the masking stimulus was a fairly intense vibratory signal presented to the observer's left palm. For one set of measurements, observers were required to detect a vibratory stimulus presented to the left index fingertip. For a second set of measurements, observers were required to identify letters generated by the Optacon also presented to the left index fingertip. Whereas the threshold for detecting test stimuli was raised considerably in the presence of the masking stimulus, there was little interference with letter recognition. In fact, only one of the four observers tested showed any decrease at all in percent correct letter recognition with the masker. With the same masker, the threshold for detecting stimuli was raised an average of 16 dB for all observers. It should be added that even the one observer who showed some recognition masking showed no recognition masking with a contralateral masking stimulus (20). In addition to pointing out another difference between our measurements of detection and of recognition masking, the results also suggest that the locus of at least part of the interaction between a masking stimulus and a target letter in a recognition task may be peripheral rather than central. The question of locus interaction will be discussed below.

One ought, however, to be cautious in making direct comparisons between detection and recognition processes based on these measurements. Although we attempted to keep conditions between the two kinds of measurements comparable, there are obviously a large number of differences between measuring detection of and measuring recognition of vibrotactile stimuli. It may be that to obtain recognition masking from a masking stimulus presented to a nearby site, it is necessary to use a masking stimulus more similar to the target. For example, had it been possible to use a second Optacon to generate masking patterns on the palm, some interference with letter recognition on the fingertips might have been obtained. Further, in the Loomis and Apkarian-Stielau study mentioned above (16), considerable interference in letter recognition was obtained with masking stimuli presented at a locus near the locus where the letters were presented.

THE NATURE OF THE MASKING STIMULUS

As the Optacon is ordinarily used, the masking stimulus for a given letter is usually some other letter. Craig (12) has demonstrated that placing target letters between two black rectangles produces considerable interference in letter recognition and further that placing the targets between two letters produces interference in the recognition of the letter to nearly the same degree. Such a result suggests that tactile letter recognition under these conditions is relatively insensitive to the type of masking stimulus used. To see to what extent recognition masking is insensitive to the type of pattern used as the masker, several types of masking stimuli were examined to see if there might be any large differences in pattern recognition that reflect the nature of the masking stimulus.

Using the same procedure described in the first series of measurements, trained sighted observers were asked to identify which one of 11 symmetrical letters, had been presented. The letters were presented at the same display rate as before, 90 msec. The time between the offset of the target letter and the onset of the masking stimulus was approximately 25 msec, somewhat shorter than it had been before. The letters were uppercase letters typed with a Dual Gothic typing ball which produces a block-like, sans-serif letter. In the first series of measurements, the ability of the observers to recognize single letters in the absence of a masking stimulus was measured and compared with their ability to recognize letters followed by another letter or by a black rectangle, i.e., backward masking. The results of these measurements are shown in Table 1. Each entry for each observer is the mean of 260 trials.

TABLE 1 Percent Correct Letter Recognition Under Various Masking Conditions

Observer	No Masker	Letter Masker	Rectangle Masker
T.B.	66.8	30.0	31.9
J.B.	87.3	60.7	66.2
M.D.	64.6	38.8	37.7

The table reveals several points worth noting. As the previous results would lead one to expect, the letter masker and the rectangle masker produce considerable, and approximately equal, amounts of interference. The results also show that the equivalent amounts of interference produced under the two conditions are not the result of averaging across observers but rather that each observer shows about the same amount of masking independent of the type of masker. What Table 1 does not show, but what a more detailed analysis of the observers' responses does show, is a similarity in the patterns of responses produced in the two masking conditions. A rank-order correlation of the percent correct for each letter in the letter-mask condition and in the rectangle-mask condition shows a positive correlation of .71. Such a correlation indicates that the way in which one letter interferes with the perception of another letter must be somewhat similar to the way in which a non-letter, and presumably meaningless, pattern interferes with recognition.

In a second set of measurements, we tested some additional hypotheses about what factors determined the masker's effectiveness in interfering with letter recognition. One of the factors which might be responsible for the interference with letter recognition is the onset of the masking stimulus. To test this possibility, the backward-masking stimulus was reduced to a single vertical line presented at the same point in time, 25 msec after the offset of the target letter, as the black rectangle was presented. If onset phenomena were important, then increasing the number of such events might very well result in a substantial increase in masking. To do this we also used two vertical lines as masking stimuli. To generate a third masking stimulus, the two vertical lines were connected at the top and the bottom by horizontal lines to form a rectangle. A fourth pattern was generated by filling in the open rectangle of the third pattern and thus creating the standard black rectangle which had been used previously. These patterns are shown in Fig. 1B. The four masking stimuli also formed a progression in terms of intensity of stimulation from the single vertical line to the black rectangle. If onset properties were foremost in determining masking effectiveness, the two vertical lines should produce the smallest percent correct. If intensity of the masking stimulus were more important, the black rectangle should be the most effective masker.

The percent correct recognition of the 11 symmetrical letters was measured in the absence of any masking stimuli and in the presence of each of the four masking stimuli described above. The four stimuli were used as backward maskers only, and the onset of each masker was 25 msec after the offset of the target letter.

The results from four observers are shown in Table 2. Each point represents

TABLE 2 Percent Correct Letter Recognition Under Various Masking Conditions

No Masker	One Line	Two Lines	Open Rectangle	Closed Rectangle
77.1	58.9	57.3	54.8	50.4

the mean of 800 trials. The general trend of the results supports the idea that changes in the intensity of the masker produce changes in letter recognition. Comparing the results obtained using a single line as a masker with the results obtained using two lines as a masker suggests that very little additional masking is produced by the addition of a second parallel line. The onset of the second parallel line is approximately 85 msec after the offset of the target which, as Fig. 3 suggests, is a short enough temporal delay to allow the presence of the second line to exert some influence on the target. The results indicate that the onset of the masker by itself is not particularly critical in interfering with letter recognition.

The most striking result from these measurements is the large amount of interference produced by a single vertical line. Although the intensity of the masker does have some effect on the amount of masking produced, it appears that a stimulus less intense than the target letter can produce considerable interference in the recognition of the letter. It may be that one critical feature of the masker is that it produce a vibrotactile pattern which sweeps across the same area of skin as the target letter occupied.

GENERAL DISCUSSION

What do the results from the present studies reveal about tactile pattern recognition and masking? First, the results suggest, following the analysis of visual recognition processes, that both interruption of processing of tactile patterns and integration with tactile patterns may be important in producing interference. Second, tactile pattern recognition can be interfered with by a masking stimulus presented as much as 150 msec after the offset of the pattern. This result suggests that processing of the pattern is still not complete for some time following the presentation of the pattern. Further, that the total durations of the letters were 150 msec made the time from the onset of the pattern to the onset of a masker, which was still effective in producing interference, 300 msec. Further research might well be directed to the effects of changing target duration. If the duration of the target were reduced, would the effectiveness of a backward masker be increased because it would be closer in time to the onset and middle portions of the target or would the effectiveness remain the same because the time between the offset of the target and masker would remain the same?

The results of the present studies may also serve as a reminder that the term "masking" may be used to cover a broad category of phenomena which, upon closer examination, differ from one another considerably. The fact that more forward masking than backward masking was evident in a detection task but that the reverse was the case in a recognition task is one indication of the difference between types of masking. Another indication of the difference between types of masking is that, in measuring the effect that various patterns had on letter recognition, patterns less intense than the target were able to interfere with the perception of the target. Such is not the case in detection tasks where the masker is more intense than the test stimulus.

A third indication of the difference between recognition and detection is the apparent difficulty of interfering with recognition by means of a masker located at a nearby site on the skin under conditions in which interfering with detection is relatively easy. This third observation also raises the question of central and peripheral interaction. As noted before, the failure to find significant masking with a strong stimulus located at a nearby site favors a peripheral explanation for recognition masking. Also favoring a peripheral explanation is the similarity of results produced by a letter masker and a non-letter masker. In support of a more central site for interaction is the relatively long period of time, 150 msec, over which backward masking may occur and the production of significant interference by relatively weak stimuli. It may very well be that recognition masking is the result of both central and peripheral interaction as detection masking seems to be (3). If the example set by studies of visual pattern recognition and masking holds true for tactile perception, it is likely that a number of additional measurements will be necessary to pin down the relative contribution of peripheral and central factors in recognition masking.

The present results also have some implications for cutaneous communication systems. If the users of such systems are required to identify patterns presented in rapid sequence, then the greater the time interval between patterns, the more likely the observer is to identify the pattern. Obviously, increasing the time interval between patterns will mean slower information transmission. Additional measurements would be necessary to determine the precise trade-off between time intervals and pattern identification needed to maximize information transmission. The present results also suggest that the kind of patterns surrounding the pattern to be identified has a smaller effect on correct identification than might be expected. Thus one might predict that, in reading with the Optacon, for example, there would be a more or less constant effect that letters presented sequentially would have on one another and that this effect would be independent of the particular letters which served as masking stimuli. There are, of course, consistent differences in the ease with which certain target letters are identified but, in general, the letters show a similar decrement when presented with adjacent masking stimuli, that is, all the letters from the easiest to identify to the most difficult to identify show a roughly comparable decrease in percent correct in the presence of a masking stimulus. Evidence for this observation is the consistently high, positive correlation between the rank orderings of the percents correct for each letter presented without a masking stimulus and the rank orderings resulting from the measurements under the various masking conditions.

Some caution should be exercised in assuming that any manipulation which reduces recognition masking will result in improved performance in such tasks as reading with the Optacon. The fact that it has already been shown that it is difficult to predict recognition masking results from detection masking results should prevent us from being too bold in predicting reading performance. A number of variables which affect reading performance have not been considered in the measurements of recognition masking. Such factors as context cues or repeated exposure to certain letter combinations may lead to a reading performance quite different from the performance which might be predicted from recognition masking results. Future research might well investigate some of the variables which affect recognition masking as these variables affect performance in tactile reading tasks.

REFERENCES

(1) R. D. Gilson, Vibrotactile masking: Effects of multiple maskers, Perception & Psychophysics 5, 181 (1968).

(2) R. D. Gilson, Vibrotactile masking: Some spatial and temporal aspects. Perception & Psychophysics 5, 176 (1969).

(3) C. E. Sherrick, Effects of double simultaneous stimulation of the skin, Amer. J. of Psych. 77, 42 (1964).

(4) R. E. Snyder, Vibrotactile masking: A comparison of psychophysical procedures, Unpublished doctoral dissertation, Indiana University, 1973.

(5) J. C. Craig, Attenuation of vibrotactile spatial summation, Sens. Processes 1, 40 (1976).

(6) J. C. Bliss, M. H. Katcher, C. H. Rogers, & R. P. Shepard, Optical-to-tactile image conversion for the blind, IEEE Trans. Man-Mach. Syst. 11, 58 (1970).

(7) Bach-y-Rita, P. (1972) Brain Mechanisms in Sensory Substitution, Academic Press, New York.

(8) F. A. Geldard, Cutaneous coding of optical signals: The optohapt, Perception & Psychophysics 1, 377 (1966).

(9) L. H. Goldish, & H. E. Taylor, The Optacon: A valuable device for blind persons, New Outlook for the Blind 68, 49 (1974).

(10) M. W. Moore, & J. C. Bliss, The Optacon reading system, Educ. Visually Handicap. 7, 15 (1975).

(11) J. C. Craig, Vibrotactile pattern perception: Extraordinary observers, Science 196, 450 (1977).

(12) J. C. Craig, Vibrotactile letter recognition: The effects of a masking stimulus, Perception & Psychophysics 20, 317 (1976).

(13) Massaro, D. W. (1975) Experimental Psychology and Information Processing, Rand McNally, Chicago.

(14) T. J. Spencer, & R. Shuntich, Evidence for an interruption theory of backward masking, J. of Exp. Psych. 85, 198 (1970).

(15) M. T. Turvey, On peripheral and central processes in vision: Inferences from an information-processing analysis of masking with patterned stimuli, Psych. Review 80, 1 (1973).

(16) J. M. Loomis, & P. Apkarian-Stielau, A lateral masking effect in tactile and blurred visual letter recognition, Perception & Psychophysics 20, 221 (1976).

(17) U. Schindler, & Å. Knapp, Ursachen der gegenseitigen Verdickung von taktil dargebotenen. Buchstaben: Unterbrechung, Summation, oder Verzogerung? Psych. Research 38, 303 (1976).

(18) G. A. Gescheider, D. D. Herman, & J. N. Phillips, Criterion shifts in the measurements of tactile masking, Perception & Psychophysics 8, 433 (1970).

(19) J. C. Craig. Difference thresholds for intensity of tactile stimuli. Perception & Psychophysics 11, 150 (1972).

(20) B. Mannix, Spatial and temporal factors in vibrotactile recognition and detection masking, Unpublished honors thesis, Indiana University, 1977.

READING MACHINES FOR THE BLIND

James C. Bliss

Telesensory Systems, Inc., 3408 Hillview Ave., Palo Alto, CA 94304, USA

The concept of a machine to convert optical images from the printed page into tactile form comprehensible to a blind person is very old. In 1880, Camille Grin (1) in France invented a machine in which

> *"an optical system threw the image of each successive letter on the back wall of a camera obscura divided into 64 compartments every 1/7th of a second and induced an electrical current whose amperage varied according to whether the selenium cells were lit (white portions of letters) or not (black portions of letters). The current was distributed to 64 electromagnets which, according to its force, caused little pins to be extruded or not, which represented a relief image of the letters, which the blind person detected by placing his finger on the display device, called a 'tactilator'."*

After almost a century of developmental efforts in reading aids for the blind, the Optacon has emerged as a widely used and successful reading device. Today, over 3,500 Optacons are in use in over 42 countries and the number of Optacons is increasing at the rate of about 1,000 per year.

The Optacon is a small, portable, electronic system which enables a blind person to read using the sense of touch. The Optacon is a kind of closed-circuit tactual television system in which the blind reader senses images with his finger produced on a rectangular array of vibrating pins. These images are acquired by a small, hand-held camera, about the size of a pocket-knife, which the blind reader moves along each line of print. In the normal reading situation, the Optacon displays tactile images on an array of 144 vibrating stimulators with a resolution roughly equivalent 20/50 vision with a one-half degree field of view. (The vibratory stimulators are arranged in 24 rows and 6 columns within an area of about one inch by one-half inch. The frequency of stimulator vibration is 230Hz and the entire array is curved to roughly fit the index finger.) The performance of some trained Optacon users indicates that this level of tactile "vision" can be actually achieved in this task.

Over 350 blind students have been taught to read with the Optacon in a nine-day course at TSI. In this course, tactile recognition of the alphabet is often taught in one and one-half days, and the remainder of the course is devoted to language skill development, text reading, and format exploration.

This emphasis on language skills in the course, rather than tactile recognition of alphabetic patterns, has been increased in the last year and a half. Figure 1 shows how the average reading rates achieved by students has improved as more emphasis has been placed on language skills. This figure also shows the effect of age on end-of-training reading rates and that, in recent years, proportionately greater increases in reading rates have been obtained by the older students.

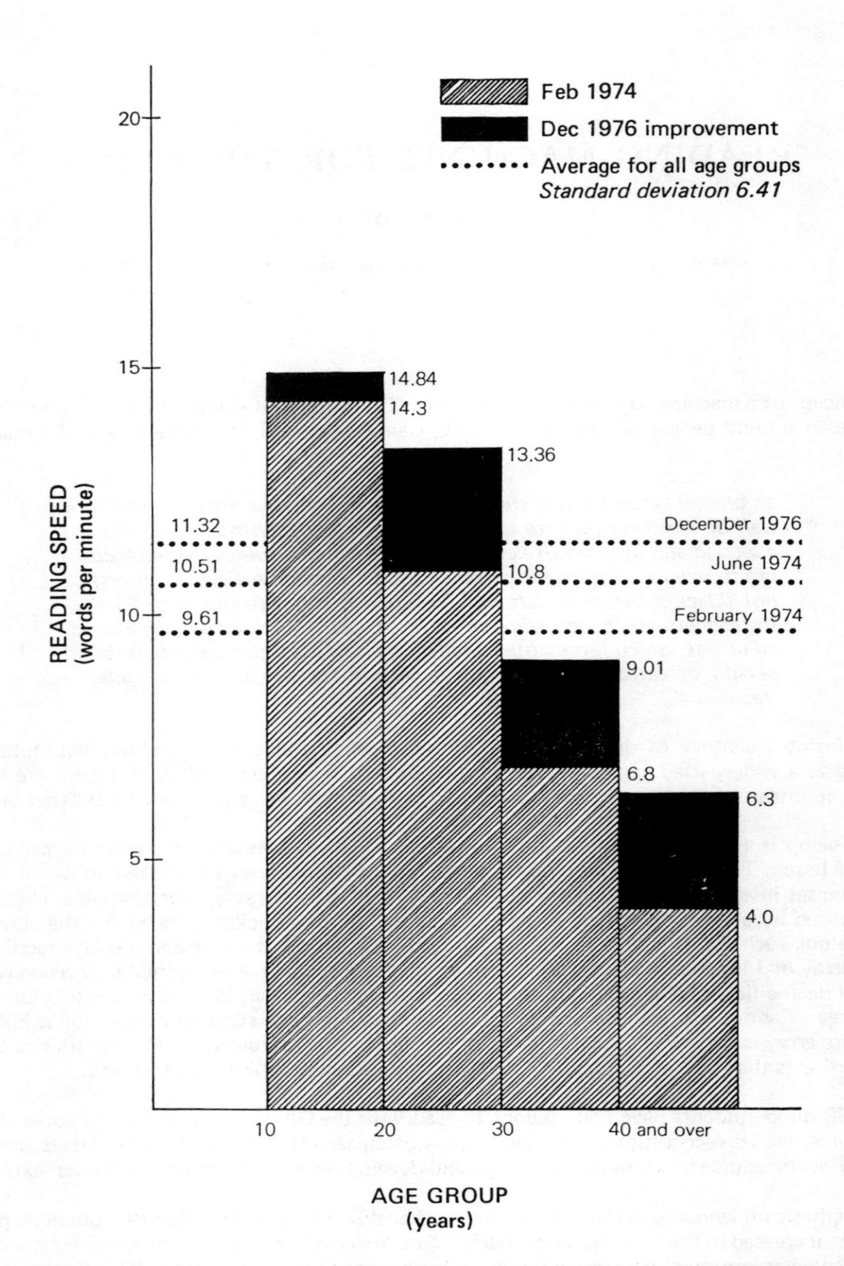

Fig. 1. Optacon average reading speed at the end of the TSI training course as a function of age group. (Based on data from 300 students between 1973 and 1976.)

The Optacon has provided access to a very wide range of printed materials for many blind people. However, two factors limit its usefulness: learning to master Optacon reading requires considerable effort, and reading rates are much slower than visual reading. Attempts to improve reading performance, both with European languages and languages with more complex graphemes (such as Kanji), by increasing the number of tactile stimulators have not been successful. Some of these attempts are described below.

Taenzer (2) developed a theoretical model to describe the tactile reading process with the Optacon. Taenzer's model is based on experimental results from two types of reading tasks. In the constant rate task, the tactile array was programmed so that connected English text appeared to move past the fingertip at a fixed rate. In the free-rate task, the subject had complete control over the rate at which letters were presented to him by tracking a small hand-held camera over each line of print on the page.

Taenzer varied the width of the field of view (window width, W, measured in letterspaces) and the rate at which text was moved past the window (display speed measured in letterspaces per second) in the constant rate experiments. He measured reading accuracy for various combinations of these parameters. In summarizing the results, Taenzer found it is useful to define a quantity called display time, t_d, which is the time duration any point on a letter is in the window.

Taenzer's results showed that for a fixed reading accuracy,

$$S_d = \frac{W}{t_d} \tag{1}$$

for window widths up to 1.5 letterspaces for tactile reading and 5 to 7 letterspaces for visual reading. Thus, for window widths less than 1.5 letterspaces, visual and tactile reading speeds are comparable and both increase proprotionately with window width.

If reading accuracy is plotted versus display time, various window widths and display times combine to produce the single psychometric curve shown in Fig. 2. This result indicates that reading accuracy is virtually zero unless the display time is 50 msec or longer. If the display time is 150 msec or longer, the reading accuracy is greater than 95%.

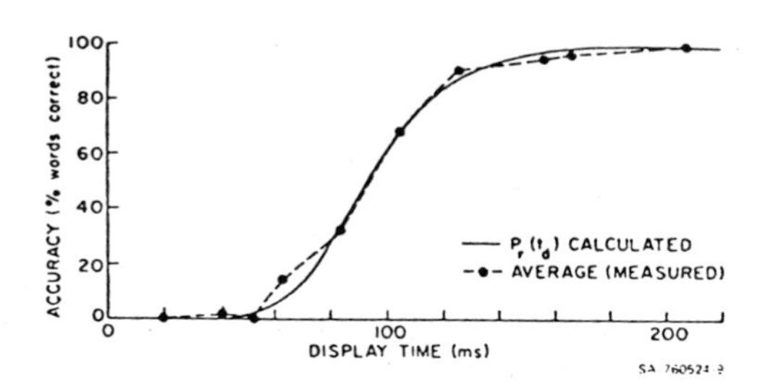

Fig. 2. Comparison of the constant-rate reading results with the theoretical result derived from the decision mechanism model. (From 2)

However, in free-rate reading, the subject has two tasks to perform: tracking the printed line with the camera and interpreting the textual information. When window width is varied in free-rate reading, speeds not directly proportional to the window width are obtained. Taenzer found that free-rate data could be accurately fitted to a function of the form

$$S_r = \frac{W}{t_p + W\tau_t} \tag{2}$$

where S_r is the reading (scanning) speed (letterspaces per second), t_p the perception and processing time (seconds), and τ_t a tracking constraint.

This equation differs from the fixed rate Eq. 1 in that a tracking term has been added, which represents the processing time needed to fulfill the task of keeping the hand-held camera moving so that new information is scanned whenever it is needed. The values of $t_p = 90$ msec and $\tau_t = 28$ msec/letterspace were found to fit these data, and these values correspond to an average scanning rate of 140 msec/letterspace.

Figure 3 shows how Equations 1 and 2 compare for constant-rate and free-rate reading, respectively. This figure suggests that if the Optacon window-width could be increased to 7 letterspaces without perceptual limitations, then constant-rate reading at a fixed accuracy should permit a linearly increasing reading speed with window width. This result led to a series of experiments by Hill (3) in an attempt to increase Optacon reading rate by increasing the tactile field of view.

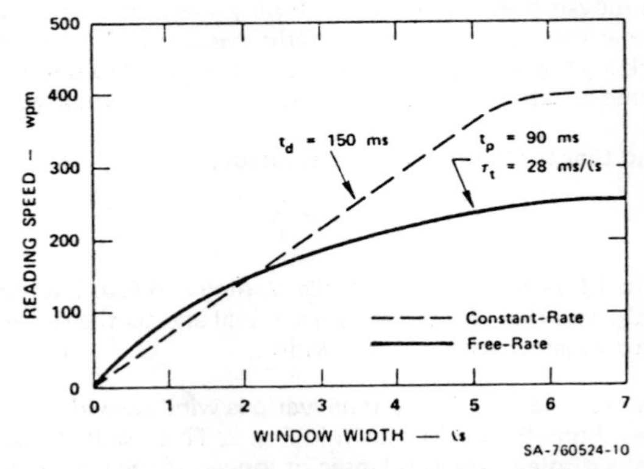

Fig. 3. Expected tactile reading speeds for both the fixed-rate and the free-rate case. (From 2)

In the series of experiments conducted by Hill, the following displays were tried to obtain a wider field of view and a corresponding higher reading rate:

1) Text was presented on two adjacent fingers with two separate 12 by 8 arrays of vibratory stimulators. The number of columns of stimulators in the display was varied from 2 to 16 and reading accuracy was measured as a function of window width.

2) A high density, single finger tactile display consisting of 12 columns and 24 rows was used to measure the effect of window width on a single finger.

3) A two-finger display similar to (1) above was used except that the alphabetic shapes moved along the length of the fingers rather than across the fingers.

4) The high density, single finger tactile display of (3) above was used with the alphabetic shapes moving along the length of the finger rather than across the finger.

Figure 4 illustrates the general type of result found in all of these experiments. In every experiment, tactile reading speed did not increase with increasingly wider windows that display more than one letter. Hill suggest that this might be due to two limitations: (1) The complexity of a single letter-space may be the maximum that can be recognized with the fingertip. This would account for the reading limitation when the wide display is used on a single finger. (2) The information on two different fingers cannot be integrated spatially. This would account for the failure to read faster using an additional neighboring finger or with an additional index finger.

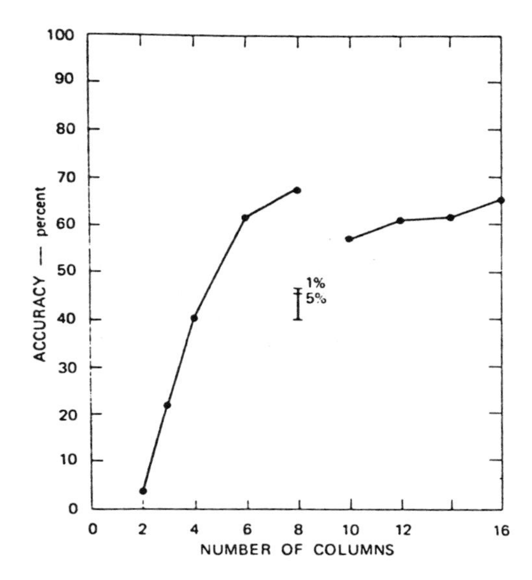

Fig. 4. Average expanding-window results of all three subjects. Each data point is the average accuracy in reading 16 blocks of text (1,600 words). The tolerance bars are the 1% and 5%, two-sided t values for comparing any pair of data points. (From 3)

The lack of spatial integration across fingers evidenced by Hill's subjects is consistent with the common impression of braille teachers that braille readers who feel the embossed dots with more than one finger at a time read no faster as a group than single-finger braille readers. Perhaps braille should be taught as a single-finger task.

These essentially unsuccessful attempts to increase tactile reading rates with the Optacon have resulted in research being directed toward augmenting the tactile display with an auditory display which provides full-word synthetic speech. Speech is an ideal output for a reading machine for the blind because it can be understood at rapid rates with a little learning.

The auditory display is being developed as an accessory to the Optaon. This accessory will receive electrical signals from the Optacon camera, recognize these as letters, and convert this text into spoken words. This spoken word output accessory, together with an Optacon, will give a blind user the ability to read limited materials at speeds up to 200 words per minute without removing any of the advantages of the Optacon, such as portability, versatility with material (such as graphs, formulae, and foreign languages), and active interaction with the printed document. The combined tactile-auditory display system places different demands on the tactile channel by shifting the tactile information processing task from a pattern recognition emphasis to a tracking control emphasis. This reading system promises to provide both high reading rates and the flexibility of handling a wide range of materials in a convenient manner.

Munson, et al, (4) have studied the human factors and systems questions of such a spoken word output/tactile display reading machine through computer simulation. These experiments have provided data on tracking tolerances, scanning strategies, and system time delays which are now being used to develop a spoken word output accessory to the Optacon. Hopefully it may soon be possible to make the over 3,500 Optacons speak which are in the field!

REFERENCES

(1) Schneider-Maunoury, Pierre. "Reading machines for the sightless", presentation for the International Congress for the 150th anniversary of the Braille System, Paris (May 22-24, 1975).

(2) Taenzer, J. C. "An information processing model for visual and tactile reading", *Perception*, 1, 147-160 (1972).

(3) Hill, J. W. "Limited field of view in reading lettershapes with the fingers", Proceedings of the Conference on Cutaneous Communications Systems and Devices, Frank A. Geldard, Chairman (April 17-18, 1973), 95-105. Published by the Psychonomic Society, Inc.

(4) Munson, J. H., R. E. Savoie, R. P. Shepard, R. W. Stearns, J. C. Bliss. "Systems study of an interactive computer-aided, text reading source for the blind available through any telephone", SRI Final Report to the National Eye Institute under Grant No. 5 R01 0450-02 (October, 1970).

ACKNOWLEDGEMENTS

The research described here was performed primarily under the following grants: Social and Rehabilitation Service Grant VRA-RD-2475-S-68; Neurological Diseases and Stroke Grant NS-08322; Office of Education Grant OEG-0-8-071112-2995; National Eye Institute Grant No. 5 R01 EY00450, and National Science Foundation Grant No. APR 76-20814.

THE DEVELOPMENT OF TACTUAL MAPS FOR THE VISUALLY HANDICAPPED

John David Armstrong

Blind Mobility Research Unit, University of Nottingham, Nottingham, England

ABSTRACT

Much of the research on mobility aids for the visually handicapped has concentrated on the development of devices and systems to protect the pedestrian from collisions with the near environment. By comparison, the development of aids to provide the blind person with information about the layout of the more distant (urban) environment has been limited. This paper describes the development and application of raised maps as an aid to the independent mobility of the blind.

THE MOBILITY OF THE VISUALLY HANDICAPPED

Purposive and successful movement in the environment depends on two processes; orientation and navigation. Orientation information enables the pedestrian to control his movements with respect to the features of the near environment such as pavement boundaries, fixed obstacles and other pedestrians. Navigation information allows the pedestrian to make appropriate changes in the direction of his movement in order to achieve specified destinations within an area. The significant difference in the mobility performances of blind and sighted pedestrians can be attributed to the different amounts of these two types of information available to them.

For the sighted person, the process of orientation is carried out at such a low level of cognitive involvement that it is not easy to analyse how it is accomplished. The sighted person is able, without effort, to move quickly and safely along the pavement and to make all appropriate corrections to the travel path well in advance of obstructions. In an unfamiliar area the sighted person shows no decrement in this ability to orientate with respect to near environmental features. By contrast, orientation for the blind person involves substantial and sustained mental effort. In totally new environments the blind person is likely to show a substantial decrement in his orientation ability.

With respect to navigation, the blind and sighted persons are both likely to experience difficulty whilst following a new route but, with increasing familiarity, their performances will diverge substantially. In a totally new area, the sighted person will need advanced information about the correct action to take at critical choice points. He will work to a sequence of instructions which give the relevant landmarks or locations where change is necessary and, in order to confirm his decisions, additional information about environmental features between actual choice points. For the sighted person, the amount of information which can be used in a description of a route is extensive to say the least; ranging from a list of street names to detailed descriptions of features to be encountered. Successive passes of the same route will lead to the development of a cognitive representation of increasing detail until the point will be reached when the route is no longer simply a series of directions but a collection of interrelated features which can be viewed and manipulated and viewed in a number of ways (e.g. birds eye view, street level view). In a totally new area, the blind person will also work on a sequence of instructions which specify points where change of direction will occur and may also have additional supportive information available. However, the blind person has to cope with two serious problems. First, because he is unfamiliar with the area, orientation is difficult and second, the range of reliable landmarks which can be used for navigation is usually extremely limited. Unlike the sighted person, subsequent experience of the route does not lead to the development of any detailed cognitive representation of the route; simply because data collected is rarely sufficiently unique to form points around which the schema can be constructed. Performance on subsequent passes may improve but this will be due mainly to an improvement in ability to detect those important landmarks and a limited development of other more subtle, though reliable, cues (for example, point sound sources, olfactory stimulation, changes in pavement texture and camber).

The difference between the mobility of the sighted and blind person becomes most obvious when complex behaviours other than straightforward route following is required. At a simple level, the sighted person taken blindfold to an intermediate point on one of a number of learned routes and then asked to complete the route sighted will have little or no difficulty in doing so. The blind person might be able to do this with difficulty but, unless some very discriminable and precise landmark was obvious, probably not at all. Here, the sighted person is able to call on a facility which might be described as 'landmark constancy'. If he approaches a known landmark from a new direction (for example, at an intersection) or is shown the landmark out of that sequence of landmarks associated with a particular route then he is able, by an appropriate manipulation of his cognitive representation of the area, to realise his position. In this way, the sighted person can be shown a series of routes within a given area which have common points and will soon be able to construct a single cognitive map which relates the features of the individual routes. The completion of a novel route made up from parts of the individual routes should present no difficulties. For the blind person, his inability to detect more than a very few unique useable landmarks means that he is able to distinguish one route from another only in terms of the sequence of infrequently occuring events (usually the end of the pavement at a road intersection). Only rarely does that blind pedestrian encounter a landmark or situation which he can reliably discriminate. The blind person could be systematically exposed to a whole series of interlocking routes without ever being able to incorporate these into a single schema. Thus, the blind person, because of his inability to make fine discrimination between environmental features, is unable to develop the "landmark constancy" which allows multiple views of the environment to be incorporated into a single spatial representation. For a simple linear route the blind person need not be too severely disadvantaged providing that landmarks are chosen judiciously. For whole urban areas, the blind person is unlikely to develop the capability to operate over a defined area without additional help.

AIDS TO NAVIGATION

Whereas the range of aids available to provide the visually handicapped with orientation information is substantial and still increasing (e.g. guide dogs, canes and many electronic devices), only a limited amount of research has been devoted to aids for navigation purposes. For the sighted person, such information is usually provided via the printed map and it would seem reasonable to suppose that a raised version of the printed map would be of use to the visually handicapped. In fact some early research (Leonard & Newman, 1970) suggested that three types of map (the raised spatial map, the written sequential map and the spoken sequential map) offered possibilities which should be explored further.

Sequential maps which provide only a series of instructions on how to complete a specified route have a number of obvious disadvantages. In particular, they provide no information about the relative positions in the environment of those locations which may be of interest to the pedestrian. In short, they do not provide the means of building up that spatial representation of the environment which the blind person is unable to develop on his own.

Spatial Conceptualisation in the Blind

Before embarking on the development of a spatial map it seems reasonable to question whether or not the visually handicapped, particularly the congenitally blind, are able to conceptualise space. Patently, a successful map needs to present information in such a way that the blind person is able to use this to determine movements and a map which provided a simple one-to-one spatial representation of the environment would be meaningless to a person who was unable to conceptualise the environment in these terms.

The question as to whether blind people are capable of conceptualising space is not new. The literature is well littered with reports of studies which arrive at conflicting conclusions with the crucial issue seeming to be whether or not prior visual experience, and consequently the capacity for visual imagery, is necessary. The majority of reported studies are concerned with the recognition of small forms and it is apparent that the type of result obtained is a function of the task involved. However, this research apart, there are very few studies which throw light on the conceptualisation of space at the macro level. Leonard & Newman (1967) showed that congenitally blind children were able not only to use spatial maps for simple route following but also to extrapolate an alternative route to the desired goal when the most direct route was made impassable. It could be argued that this study does not resolve the question of macro spatial conceptualisation in the blind but simply indicates that a blind person can learn appropriate responses to a map display without actually having a developed cognitive model of the area represented. However, the selection of a detour and successful execution of that detour does seem to indicate an appreciation of spatial layout.

Designing a Spatial Map

In terms of basic information, a tactual map for blind people is very similar to the conventional printed map. The overall structure is provided by the street layout and the locations of important features are indicated by appropriate symbols. However, on the tactual map, the amount of information relating to location of features must be limited; first, because many such features would be undetectable and, second, because the poor resolution of the skin of the finger, compared to that of the eye, requires that the associated symbols used should be quite large and reasonably well spaced.

Our earliest attempts at designing tactual maps were somewhat pragmatic with the major part of the development being devoted to the choice of materials and techniques for reproducing permanent raised lines on a surface. However, in training blind people to use the early maps a number of crucial issues emerged. Because mobility maps need to be portable, there was an obvious restriction in size and hence the area available for mapping. Since the main purpose of the map is to provide the user with an overall spatial framework, it seemed advantageous to keep the size of the map to a width which could be totally encompassed by two outspread hands (Bentzen 1971). In this way the major area of the map could be encompassed simultaneously.

Our first task was not to determine a symbology by which to represent the various structural and locational features but rather to determine a list of those environmental features which needed to be represented. This list, which is given later, was assembled on the basis of a questionnaire sent to professional mobility instructors (James, Armstrong & Campbell, 1973) who were asked to report on the various environmental features which they would mention when providing reliable verbal maps of a series of defined routes.

Choosing the Symbols

Environmental features classified as point (representing single locations), linear (representing continuous links between two places) and areal (representing a defined space). For example, a bus stop is a point feature, a road is a linear feature and a park is an areal feature. Our first series of experimental studies was to derive a vocabulary of symbols which could be used to identify these three categories of feature. We assembled a potential set of some 8 areal symbols, 17 line symbols and 30 point symbols. The eight areal symbols (see Fig. 1.) were similar, though not identical, to those used by Nolan & Morris (1971). The seventeen line symbols were also closely related to those used by Nolan & Morris but included some lines which had already been introduced pragmatically into maps being used by blind people in the United Kingdom (see Fig. 2.) In the first experiment involving sixty-two blind school children (age range from 11 to 19 years), subjects were presented in random order with pairs of areal symbols and asked to determine tactually whether they seemed to be the same or different. Each symbol was presented once in combination with itself and with the other seven symbols. In the second study, the same subjects were presented with pairs of line symbols and asked to say whether they seemed to be same or different. Again, each of the seventeen symbols was presented once in combination with itself and with each of the other symbols.

Nolan & Morris (1971) proposed that the criteria for adequate discriminability should be:

(i) that the average confusion between a symbol and other acceptable symbols should be 5% or less.

(ii) that the confusion with itself or with any single symbol acceptable under criterion (i) should be 10% or less.

Using these quite severe criteria, five of the eight areal symbols and ten of

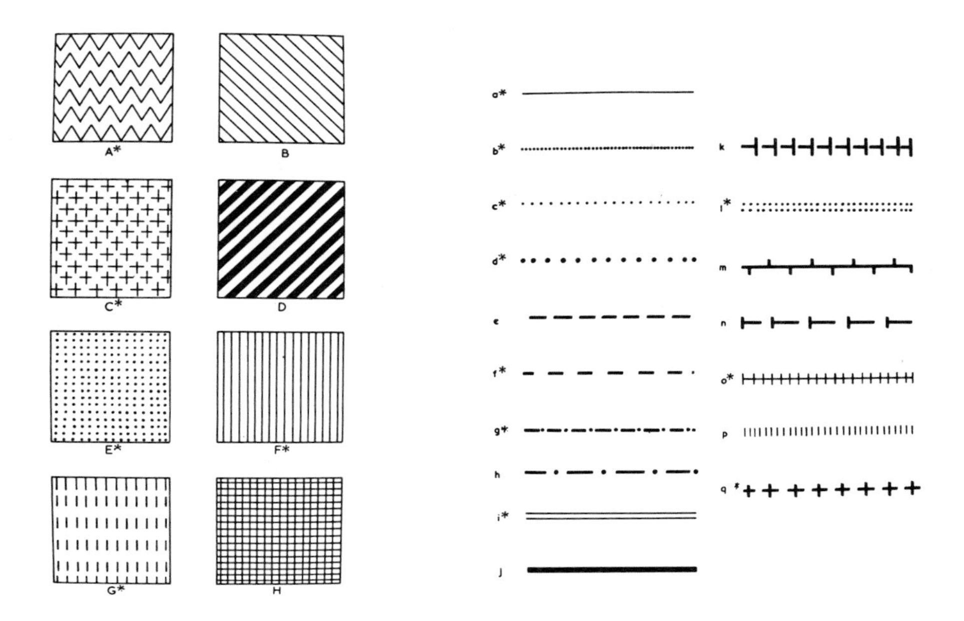

Fig. 1. Experimental areal symbols.

Fig. 2. Experimental line symbols.

the seventeen linear symbols were deemed to be discriminable. These symbols are marked with an asterisk in Figs. 1 and 2. The results are described in full by James & Gill (1975).

The potential set of 30 point symbols were more extensive than those used by Nolan & Morris and also included raised those upper-case letters found to be discriminable by Schiff (1966). In this next experiment some 97 blind school children and 59 blind adults were the subjects. For ease of presentation, each symbol was allocated to one of three groups (see Fig. 3.) and care was taken to ensure that some potentially confusable symbols appeared in the same group. The results of the paired comparison procedure indicated that 13 of the 30 symbols met the Nolan & Morris discriminability criteria. These 13 symbols are marked with an asterisk in Fig. 3. and the detailed results are reported in Gill & James (1973).

These studies of areal, line and point symbols have both extended the range of symbols introduced by Nolan & Morris and have confirmed a number of earlier findings on tactual discrimination. For areal symbols, the highest confusion was between B and D. This supports the Nolan & Morris finding that a simple change in orientation of the pattern does not lead to a discriminably different array. In this case, the thickening of the line in D was insufficient to make the pattern discriminable from B. In the case of line symbols, the highest degree of discriminability was between <u>dotted</u> lines of various spacings whereas

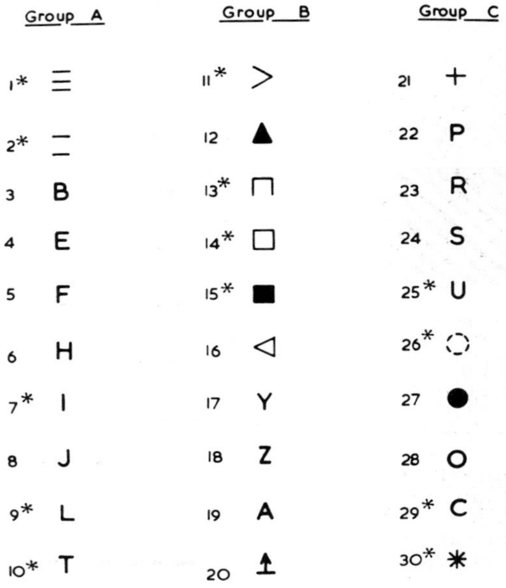

Group A		Group B		Group C	
1*	☰	11*	>	21	+
2*	=	12	▲	22	P
3	B	13*	⊓	23	R
4	E	14*	□	24	S
5	F	15*	■	25*	U
6	H	16	◁	26*	◌
7*	I	17	Y	27	●
8	J	18	Z	28	O
9*	L	19	A	29*	C
10*	T	20	↥	30*	✱

Fig. 3. Experimental point symbols.

the variation in the spacing of <u>dashed</u> lines (e.g. e and f) was not a good cue for discrimination. This suggests that the mark/space ratio is a critical factor in discrimination of discontinuous lines. The set of discriminable point symbols are, with the exception of 15, open contoured as opposed to solid symbols and this seems to be in line with the findings of Austin & Sleight (1952) who reported that outline rather than solid symbols were most discriminable. These findings add considerable weight to the hypothesis that the rate of deformation of the skin correlates with the perceived clarity or sharpness of the material being explored. In this case, rate of change of deformation is highest for the dotted lines and open contour symbols therefore discriminability is high. In this study, as in the Austin & Sleight experiments, the most successful subjects were those who made extensive though very fine exploratory finger movements.

Matching Features and Symbols

Having established a working vocabulary of discriminable symbols, the next stage was to assign these symbols to the environmental features which the map designer might want to represent. We also needed to know if the selected symbols could be used together on the same map since it was realised that high discriminability in the paired comparison situation might not necessarily be a good predictor for inter-symbol confusion on a whole map. The matching of symbol and feature (Fig. 4.) was not entirely random. Where possible,

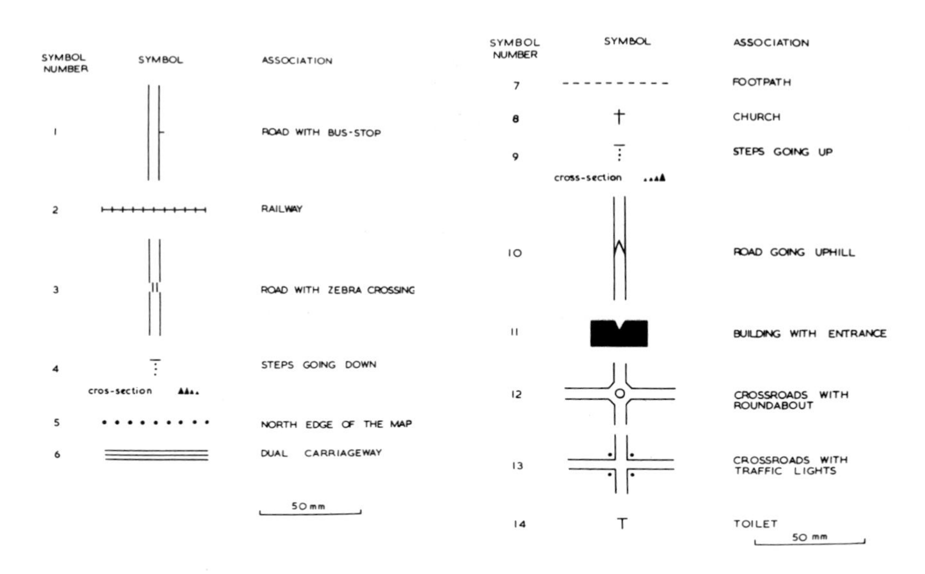

Fig. 4. The experimental symbol vocabulary.

meaningful connections between symbols and features was maintained (e.g. a cross and a church).

A small number of new symbols were introduced at this time and these included a multiple height directional symbol for steps. This symbol was inspired by the directional tactual arrow suggested by Schiff (1966) and adapted from a similar symbol used by Wiedel & Groves (1972). This type of symbol, which incorporates a physical cue to direction, tends to be more effective than the more abstract symbol for gradient (see symbol 10, Fig. 4.)

Twenty five visually handicapped school children (range 8 - 17 years) were required to learn the symbol/feature relationships and to go through a relearning phase after 3 weeks. At this second stage, a measure of the extent of necessary relearning was established (i.e. how well the symbol/ feature relationship had been remembered). After a further three weeks, the subjects were asked to identify the features represented by the symbols on a pseudomap (Fig. 5.) In general, the need for relearning the relationships was quite low and identification of the symbols on the pseudomap very high. Symbols 2, 4, 6, 9, 11 and 12 were more readily learned and subsequently identified. Symbols 3 and 10 were highly confused, probably because of the masking effect of the enclosing lines.

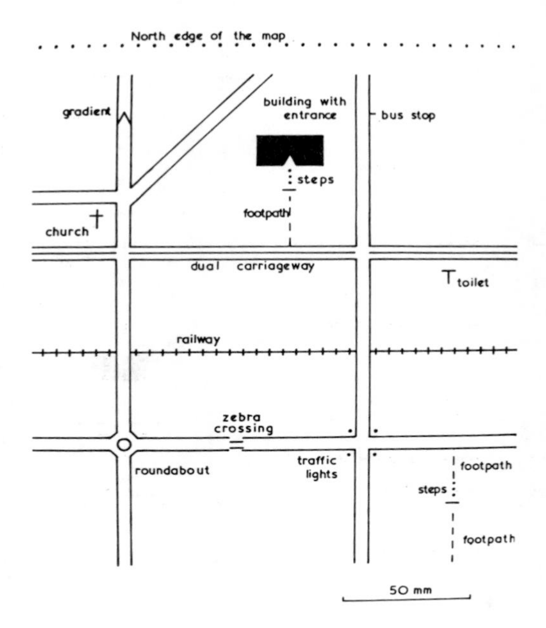

Fig. 5. A pseudo map using some of the experimental point and line symbols.

Symbol Heights

Nolan & Morris (1971) showed that the recognition of individual symbols improved when point, line and areal symbols differed in height above the background material. Some of our own unpublished pilot studies have confirmed this. In our own experiments, reported above, we used heights above base of 1.14 mm. and 0.89 mm. for point and line symbols respectively and these seemed to be optimum in terms of recognition, comfort and ease of scanning.

PRODUCTION AND EVALUATION

Standardisation

It is somewhat regrettable that, in spite of the considerable number of studies on map symbology, no standardisation of symbols and symbol sets has yet been achieved. One of the consequences of this lack of progress has been that the introduction of tactile maps for visually handicapped people has been slow. On the basis of our research, we have standardised on a set of point and line symbols (James and Armstrong, 1976) and these are shown in Fig. 6. By making relief copies of these symbols available as part of a map making kit (Fig. 7.) we have achieved a high degree of standardisation within the United Kingdom and in some other parts of the world.

FEATURE	DOUBLE-LINE SYMBOL	SINGLE-LINE SYMBOL
Main Road		
Secondary Road		
Discouragement Line		
Dual Carriageway		
Footpath		
Railway		
North Edge		
Zebra Crossing		
Signal Controlled Crossing		
Bus Stop		
Footbridge		
Subway		
Traffic Lights		
Roundabout		

FEATURE	SYMBOL
Telephone Box	
Pillar	
Ladies Toilet	
Gentlemens Toilet	*
Non-specified Symbol	•
Steps	
Building with entrance	

Fig. 6. The standardised line and point symbol vocabulary.

Although we accept that the symbology may not be perfect it is certainly adequate to allow maps to be made and used widely. It is only by making this tentative and, perhaps, arbitrary step towards standardisation of symbology that map design will eventually move towards the optimum.

Two Evaluation Studies

The final stage of the mobility map project has been a series of experimental studies using maps which have been constructed using standardised 'symbology' and have followed strict design principles with respect to size, scale and other layout considerations (see Armstrong, 1973; James & Armstrong, 1976). These studies have not been restricted to maps showing simple street layouts but have included the mapping of a range of complex situations which would normally be avoided by blind people or, at best, handled with considerable difficulty. Only the two most important studies will be mentioned here.

In the first study (James & Armstrong, 1975) the area to be mapped was a very large two storey indoor shopping centre. This centre is particularly attractive to the locally visually handicapped population because public transport to it is good, it contains a large and varied concentration of shops within a restricted area and it is totally free from normal traffic hazards. However, the problem which makes independent mobility difficult is the lack of reliable landmarks and orientation cues; the centre is of a uniform construction and pedestrian flow is multi-directional.

The first problem, therefore, was to differentiate the various parts of the centre and this was achieved by attaching braille labels to the roof support pillars which occured regularly at 6 metre spacing down both sides of the central pedestrian area and on both upper and lower floors (see Fig. 8.) Each pillar label indicated first, the floor (upper or lower) second, the side of the shopping area (East or West) and third, the pillar number. In this way, a detailed landmark system was established. In addition to a brief written description of the general layout of the centre, the map users were provided with a comprehensive reference system which listed all of the shops, the number of the pillars nearest to the shop entrances and the various products sold.

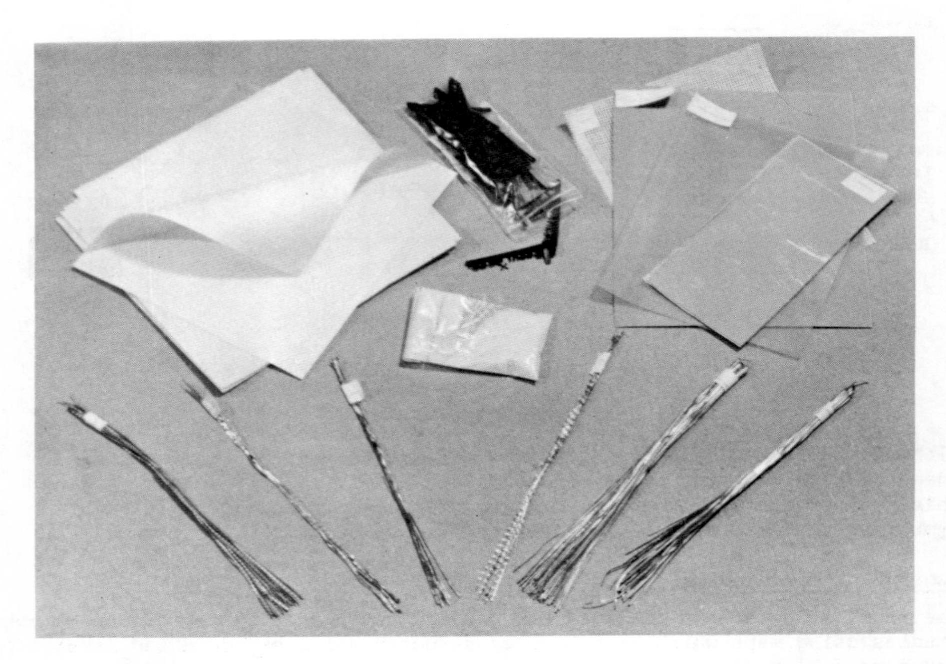

Fig. 7. Map making materials provided in the commercially available kit.

In the evaluation, a number of congenitally blind people were given a short training introduction to the map and were then allowed to study the map at leisure overnight. The following day, each subject was asked to execute 36 route tasks in the centre. The average success rate for these tasks was just over 75% with a range of 25% to 90%. Actual map reading errors were few and most problems arose from inadequacies in other mobility skills. In a subsequent experiment, subjects were asked to plan and execute a shopping expedition using the map. All subjects returned with a full complement of items.

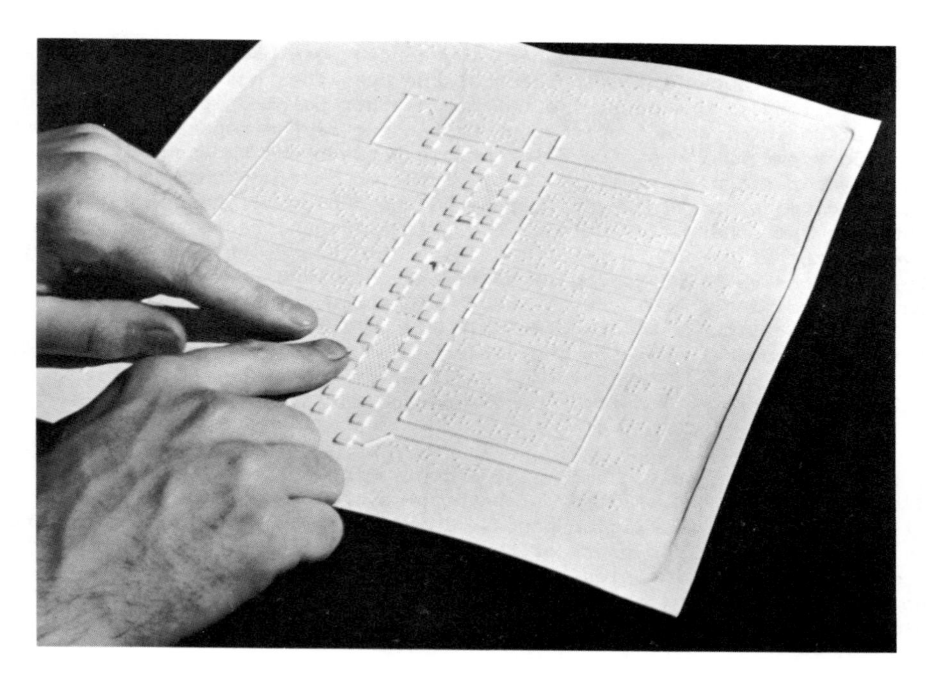

Fig. 8. Part of the shopping centre map.

The most significant finding from this study was that blind people can successfully incorporate information from a number of maps (there were two maps for each floor) and to utilise this information to plan and follow routes which involved movement in three dimensions. Although the full range of symbols was not used on this map, no discrimination errors were reported for those symbols used. The multi-level step symbol (see Fig. 8.) was reported by the subjects to be exceptionally easy to locate and recognise.

The second major study (James & Swain, 1975) explored the possibility of increasing information density on a map without increasing the level of confusability. This was achieved by designing a multilevel map which consisted of four interlocking surfaces each of which displayed a particular category of information. The basic level was a standard tactile street map of a city centre using double line representation. The second level was the underside of this map on which abbreviations for street and building names were given in reverse braille (Kidwell & Greer, 1972). The other two levels were thin plastic overlays which could be located over the main street map. These two overlays contained different sets of information relating to the city centre. Although only two overlays were used in this study, there is no limit to the number of possible overlays which could be used individually in conjunction with the base map. For example, the basic street layout could be overlayed with maps of different bus routes, locations of public buildings, public utilities and so on without the possibility of overcrowding.

For the field trial of this particular design, two overlays, showing two interlocking bus routes in Nottingham city centre were used. Subjects were asked to execute routes involving each of the bus lines and then to travel between locations on one of the routes and another point on the other route. This operation involved not only an understanding of the separate routes but also a good appreciation of the relationship between the two routes. The task was made particularly difficult by virtue of the fact that the two bus routes coincided at more than one point and the choice of changeover from one route to the other was crucial. Four visually handicapped people were involved in the study and three of them were able to answer, without error, thirty questions relating to choice of route, individual bus stops and manoeuvres involving changeovers. The fourth subject made only one error. Subsequently, all four subjects were able to complete three complex routes involving changeovers. None of the subjects had any difficulty in coping with either the symbols used, the reversed braille, or the process of gathering and assimilation of information from three separate surfaces.

These two studies are only examples of a continuing series of experiments aimed at determining the efficacy of both the symbology and the map making materials supplied in the kit of parts. In addition, we continue to receive and collate responses from users of the maps kits relating to both the symbology and design principles. To the present time, we have no firm evidence that alterations or additions should be made to the chosen set of symbols (Fig. 6.)

REFERENCES

Armstrong, J.D. The design and production of maps for the visually handicapped. Mobility Monograph, 1, University of Nottingham (1973).

Austin, T.R. & Sleight, R.B. Factors relating to speed and accuracy of tactual discrimination. Journal of Experimental Psychology, 44, 283-287 (1952).

Bentzen, B.L. An orientation and travel map for the visually handicapped; hand production,,testing and commercial reproduction. Unpublished Masters Thesis, Boston College (1971).

Gill, J.M. & James, G.A. A study on the discriminability of tactual point symbols. American Foundation for the Blind Research Bulletin, 26, 19-34 (1973).

James, G.A. & Armstrong, J.D. An evaluation of a shopping centre map for the visually handicapped. Journal of Occupational Psychology, 48, 125-128 (1975).

James, G.A. & Armstrong, J.D. Handbook on Mobility maps. Mobility Monograph, 2, University of Nottingham (1976).

James, G.A., Armstrong, J.D. & Campbell, D.W. Verbal descriptions used by mobility teachers to give navigational instructions to their clients. New Beacon, 56, 86-91 (1973).

James, G.A. & Gill, J.M. A pilot study on the discriminability of tactile areal and line symbols. American Foundation for the Blind Research Bulletin, 29, 23-33 (1975).

James, G.A. & Swain, R. Learning bus routes using a tactual map. New Outlook for the Blind, 69, 212-217 (1975).

Kidwell, A.M. & Greer, P.S. The environmental perceptions of blind persons and their haptic representation. New Outlook for the Blind, 66, 256-276 (1972).

Leonard, J.A. & Newman, R.C. Spatial orientation in the blind. Nature, 215, 1413-1414 (1967).

Leonard, J.A. & Newman, R.C. Three types of 'maps' for blind travel. Ergonomics, 13, 165-179 (1970).

Nolan, C.Y. & Morris, J.E. Improvement of tactual symbols for blind children. U.S. Department of Health, Education & Welfare Final Report, Project No. 5-0421 (1971).

Schiff, W. Using raised line drawings as tactual supplements to recorded books for the blind. Recordings for the Blind, Final Report, Project No. RD 1571-5 (1966).

Wiedel, J.W. & Groves, P.A. Tactual mapping design, reproduction, reading and interpretation. Occasional Paper, 2, University of Maryland (1972).

ACKNOWLEDGEMENTS

The project on map design and reproduction at the Blind Mobility Research Unit is part of a research programme on the mobility of the visually handicapped which is sponsored by the Medical Research Council and the Department of Health and Social Security, and directed by Professor C.I. Howarth and the author.

Much of the practical work on this mapping project was carried out by Dr G.A.James under the supervision of the late Dr J.A. Leonard and, subsequently, the author. Figures 1 - 5 have been reproduced from Dr James' unpublished doctoral thesis.

Some of the studies reported here were carried out in conjunction with Dr J.M.Gill of the University of Warwick who was responsible for the development of a computer controlled engraving system which was used in the reproduction of the point, line and areal symbols for the experiments.

HUMAN LOCOMOTION GUIDED BY A MATRIX OF TACTILE POINT STIMULI

Gunnar Jansson

Department of Psychology, Uppsala University, Box 227, S-751 04 Uppsala, Sweden

ABSTRACT

The main aim of this paper is to sketch the present status of endeavours to construct a mobility aid for the blind with the information about the environment and the pedestrian's relations to it given in the form of a matrix of tactile point stimuli. Two series of experiments, on the externalization and the localization of objects, are also summarized. Both series were made at Smith-Kettlewell Institute of Visual Sciences, San Francisco, the first with a non-wearable device with 400 vibrators, the second with a wearable one with 1024 electrodes. The result of the first series indicated that "passive" externalization of moving targets is not always accurate even after long experience with this kind of stimulation. Localization of a small target was shown, in the second series of experiments, to be possible both in a short time and with great accuracy. Accuracy was not affected by the introduction of another target, but time was affected. The mean of the responses was not changed with an increase of the "visual angle" of the device, but the variability increased. In a general discussion the information needed for locomotion is contrasted with what is available in a device of the present type and with what is possible for the user to pick up. Especially, the need for basic research on the possibilities of the tactile system to pick up flow patterns is stressed. It is stated that the development of devices of this kind has reached a plateau and that new knowledge about the perceptual systems involved is needed before substantial progress in their use as mobility aids can be made.

AIDS FOR THE GUIDANCE OF LOCOMOTION WITHOUT VISION

In former times, the totally blind person on his own was expected to be able only to totter along groping his way with his hands and maybe using a stick. During the last few decades the long cane has been introduced increasing the independent and safe walking of the blind to a considerable degree. But the long cane does not scan the space above waist level. It sometimes does not give early enough warning and it does not give very much information about the general layout of the environment. The blind pedestrian walks, to a large extent, as in a tunnel missing much information about his surroundings.

The technical mobility aids developed so far have mainly been designed for the detection of obstacles, e.g. the Russel Path Sounder and the laser canes. To some extent the finding of landmarks and openings in barriers are also considered. Such devices are sometimes called clear-path-indicators. The Sonicguide pretends also to be a so called environmental sensor, but it is not yet clear to what extent such a claim can be justified.

The main aim of this paper is to sketch the present status of attempts to give richer information to the blind pedestrian by providing him with a matrix of tactile point stimuli, in principle the same kind of device as the reading aid the Optacon. Two of my own series of experiments with aids of this kind will also be summarized.

THE ELECTROPHTHALM AND THE TACTILE VISION SUBSTITUTION SYSTEM

The name Electrophthalm, meaning "electric eye", was first given to a device constructed by Noiszewski in 1897 (Starkiewicz & Kuliszewski, 1963). It consisted of one photocell placed on the forehead with an auditory output. The user is said to be able to localize a window and a lamp. A main point during its continued development, originated by Starkiewicz in the fifties and described internationally by Starkiewicz and Kuliszewski (1963) was to increase spatial resolution. In the latest version 300 vibrators placed on the forehead are used (Kurcz, 1974; Palacz, 1975). Efforts have also been made to miniaturize the device making it easily wearable (see Fig. 1).

In the development of the Tactile Vision Substitution System two main variants have been constructed: (1) a 400 vibrators device with the display on the back of the user seated in a stationary dental chair with a camera movable by the hands (see e.g. Bach-y-Rita, 1972); (2) a 1024 electrodes system positioned on the abdomen of the user and wearable as shown in Fig. 1 (see e.g. Collins & Madey, 1974).

The Polish and the American systems are essentially the same kind of device. There is an optical input and a tactile output, and the skin is expected to forward information given in a matrix of stimulators, vibrators or electrodes. The sites of the tactile display are different, but this is probably not an important difference. Scadden (1973) showed that loci of a tactile display are easily interchangable.

This interchangability is also a reason to describe these tactile systems as devices where active touch is operating. Even if the user is not manipulating objects in the same way as with his hands, there is a muscular system steering the input (the head muscles when the camera is mounted on the head). This active manipulation of the camera has repeatedly been stressed as very important for the successful pick up of information.

Up till now the technical development has been reasonably successful. There exist prototypes which can be used in experiments, and the subjects are reported to be able to perform many tasks with the aid of the devices. Much of this information, however, is anecdotal or is based on rather crude experiments with few subjects, and the generality of the findings must sometimes be questioned. The "what" problem has been more studied than the "where" problem, which, concerning the Tactile Vision Substitution System, may be related to the fact that the first versions were stationary. One example of experimental results is the experiments demonstrating that subjects can make form discriminations with good accuracy (White et al., 1970), but one should be aware that these results are limited to a group of twenty-five objects and that the task is much more difficult in a real life situation with a large number of alternative objects.

There have been a few locomotion studies. Collins, Scadden & Alden (1977) reported positive results from a study of the electrical system on an obstacle course indoors , but as no comparison was made with walking without the aid

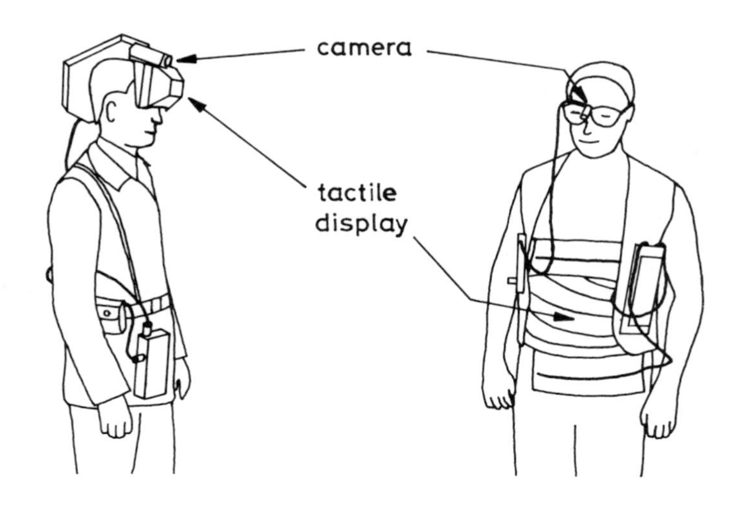

Fig. 1. The Electrophthalm (to the left) and the Tactile Vision Substitution System (to the right). After photos taken by the author in 1976.

it is uncertain how much the aid contributed. Locomotion with the Electrophthalm has been studied over the years during long training sessions (e.g. Noiszewski & Starkiewicz, 1970), but the lack of details about the studies make them sometimes difficult to interpret. That is also true of the latest report (Kurcz, 1977). It contains, however, a description of a limited use of the aid outdoors after sunset with judgments about the location of lanterns and illuminated surfaces. In spite of some success it was concluded that the device, in its present form, is not suitable for general use by the blind. The same is true also of the Tactile Vision Substitution System.

My own experimental work in this area has focused on two main problems: the externalization of the percept and the precision in the localization of objects during locomotion. The experiments, made at Smith-Kettlewell Institute of Visual Sciences, San Francisco, will be reported in detail elsewhere and are only summarized here.

EXTERNALIZATION

The light impinging on the retina does not give rise to a percept there but instead in the space outside the observer. Externalizations of the same kind have been reported also when the Tactile Vision Substitution System is used. Especially interesting is the anecdote about the accidental zooming of the camera lens to which the subject responded with an adequate avoidance movement. In spite of the fact that the stimulation was on his back he responded to an object perceived in front of him (Bach-y-Rita, 1972). It has also been reported, however, that there was no externalization in spite of weeks of training (Guarniero, 1977).

Most experiments with the dental chair version of the Tactile Vision Substitution System have been made with stationary targets but with a mobile camera. The aim of one of my experiments was to see to what extent subjects, well trained in the use of this device, correctly externalized moving targets with a stationary camera. The proximal stimuli consisted of squares moving in linear paths and changing size. Corresponding visual stimuli yield, unambiguously, percepts of objects moving in three-dimensional space (Johansson, 1964). The subjects responded by indicating the direction of the perceived motion path with a bar in front of them. If externalization was good this task could be expected to be executed accurately and rapidly. The result demonstrated that well-trained subjects could respond in this way to a majority of the stimuli, but that they sometimes did not do so. Whether the responses were based on immediate perception or on more cognitive judgments can not be decided, but the experiment shows that the externalization of stimuli of this kind is not completely accurate. Thus motion of the target is not sufficient for correct externalization.

PRECISION OF LOCALIZATION

When a person with an aid of this kind approaches an object the stimulation on his skin is a unit of stimulators corresponding to the object, the number increasing during the approach. The position of this unit within the total stimulation area depends on the direction of the camera on his head and the direction of his walking. A good strategy for approach is probably to walk in such a way that the stimulation is centered horizontally and increases in size (cf Gibson, 1974).

If the subject wants to touch or grasp the object he should move his hand, if it is "visible" to him, in such a way that the stimulation corresponding to his finger(s) or hand coincides with the stimulation corresponding to the object.

The aim of this series of experiments was to study with what precision such a task can be performed with the aid of such a matrix of tactile point stimuli.

In these experiments the wearable matrix developed at Smith-Kettlewell Institute of Visual Sciences (Collins & Madey, 1974) was used. The tactile display consists of 1024 electrodes* and it is fixed to the abdomen with elastic rollers (cf Fig. 1).

In one of the experiments the subjects started from a distance of 7 feet (214 cm) and walked towards a vertical screen where a target 3 x 3 inches (7.6 x 7.6 cm) in size was to be found in different locations within an area of 40 x 40 inches (102 x 102 cm). Two blind subjects, well trained in using the aid, both performed the task of finding the target, walking towards it and pointing

* At the time of these experiments three rows at one end of the display were out of function and of the remaining 928 (32 x 29) stimulators between 82 and 96% were simultaneously functioning. Because of all the delicate details of which such a display consists it does not seem to be possible at present to get prototypes of this type to function perfectly.

to it at a median time of 15 sec with a mean deviation from the center of the target of less than 1 inch (2.5 cm) both vertically and horizontally. This means that their mean value was within the target. The standard deviations of their responses were larger vertically (3.03 and 3.15 inches, i.e. 7.7 and 8.0 cm, respectively) than horizontally (1.47 and 1.06 inches, i.e. 3.7 and 2.7 cm, respectively). Less trained sighted blind-folded subjects had longer times but about the same accuracy.

A larger object close to the target was introduced to study the effect of masking. No effect on accuracy was obtained, but the time was about double. Apparently the extra time was used to make the necessary discriminations.

In the experiments mentioned the camera angle was about 20°. In one experiment this angle was compared with an angle of 150°. In order to be able to work with this larger angle in a reasonable way we decreased the starting distance to 5 feet (153 cm) and increased the target size to 6 x 6 inches (15.2 x 15.2 cm). Under these conditions the camera angle had no clear effect on time and mean accuracy but the variability of the accuracy was effected, the larger angle giving larger variability.

In summary, these experiments demonstrated that the task of localizing a small target can be performed rapidly and accurately with the aid of a tactile display of this kind, even by rather untrained subjects. However, when there are masking targets time increases, and when the camera angle increases the variability of the response increases. Both these effects are drawbacks when the aid is considered as a mobility aid, as the normal environment contains more than one object and a wide camera angle is wanted for locomotion.

INFORMATION NEEDED FOR LOCOMOTION

There are at least three main problems in the development of mobility aids for the blind:
(1) What information is necessary and sufficient for locomotion guided by vision?
(2) Can this information be made available to an alternative perceptual system?
(3) Can the alternative perceptual system pick up the available information?

The first and the last problems are perceptual, the second problem to a large extent technical. To reach a good solution it is necessary to consider them all. In this section the first of them will be discussed. The theoretical and experimental analysis of this problem has only started.

An analysis from the standpoint of the blind was made by Leonard (1972). He made the following list of functions needing information from the environment:
(1) orientation (both about the layout in general and about the more immediate environment)
(2) object detection (both obstacles to be avoided and landmarks to be found)
(3) detection of terrain changes
(4) posture

In the development of mobility aids special interest has been devoted to object detection. This function may be further subdivided in the following aspects:

(a) detection (that)
(b) localization (where)
(c) identification (what)

Leonard pointed out that all these functions are performed by vision at low levels of awareness and well in advance of the time when action is needed.

The analysis given above does not stress enough the dynamic character of the functions performed by vision. Walking in sighted animals is an activity continuously controlled by vision (Gibson, 1974; Lee & Lishman, 1977). The following list of locomotary functions, part of a tentative list by Gibson (1974), gives a more dynamic picture:
(1) steering (including e.g. avoiding obstacles and finding openings)
(2) approaching
(3) entering encloses
(4) keeping a safe distance
(5) dodging, warding, parrying

Common to all these activities is that the necessary information is a flow of the ambient optic array. Units in the array move in gradients of velocities which are easily picked up by the visual system. Figure 2 indicates what such a flow field looks like when the pedestrian is walking in an urban area. The picture is simplified in many ways, e.g. no account is taken of the side and up-and-down movements during walking, but it demonstrates that the flow pattern contains motions simultaneously in many different directions.

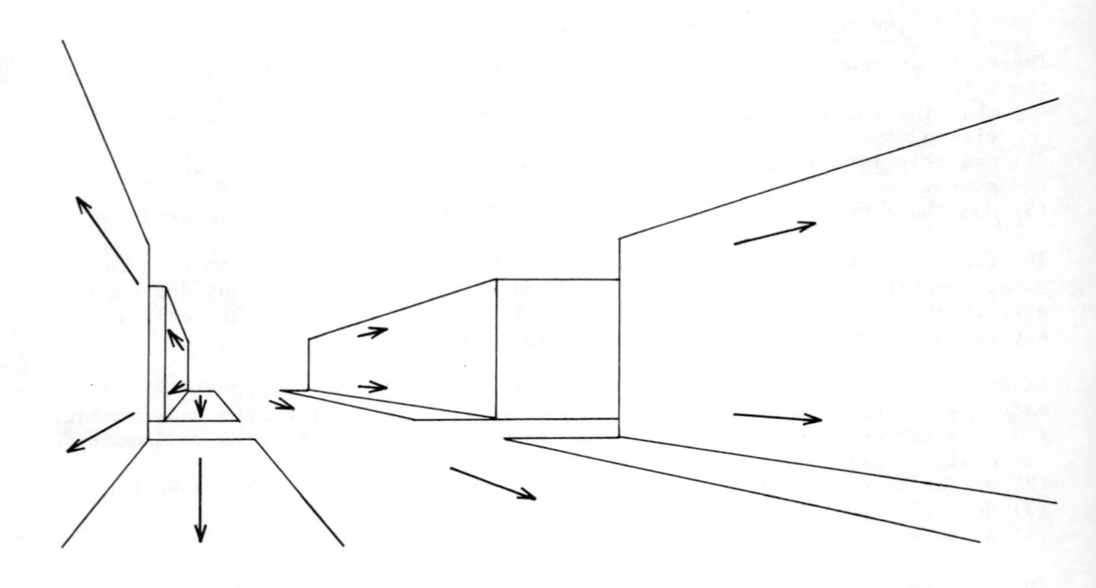

Fig. 2. A flow pattern for a pedestrian in an urban area. After similar figures in Gibson (1950).

INFORMATION AVAILABLE IN A MATRIX OF TACTILE POINT STIMULI

A well-known example of a matrix of point stimuli is a visual picture where a matrix of dots represents the original optic array. The matrix considered here differs from such a picture in its crudeness. The spatial resolution is low which means that contours are presented in a stepwise form (except in a few special cases), that textures can be represented only in a very limited way, and that texture gradients can hardly be represented at all. All these features of optical stimulation are very important for visual perception, but they can be transferred to matrices of the present kind only crudely. The technical endeavours have been much devoted to increase spatial resolution, but also the 1024 points of one of the present devices give a low resolution. A Puerto Rican device with 3600 electrodes has been reported ("Electrode matrix on skin enables blind to 'see'", 1976) but not yet demonstrated internationally. It should of course be better from this point of view, but even this resolution is much lower than the one available optically.

The resolution of light intensities is also very low, in the prototypes with vibrators restricted to two levels, on - off, except in a prototype under development where a few more are used (Holmlund, 1976). In principle it is possible to increase the number of levels, e.g. with variation in the amplitude of the vibration. With electrical stimulation there are still more possibilities (Collins & Madey, 1974; cf also Saunders, 1976), which makes this alternative theoretically attractive.

Figure 3 gives one concrete example of what the two resolution limitations may

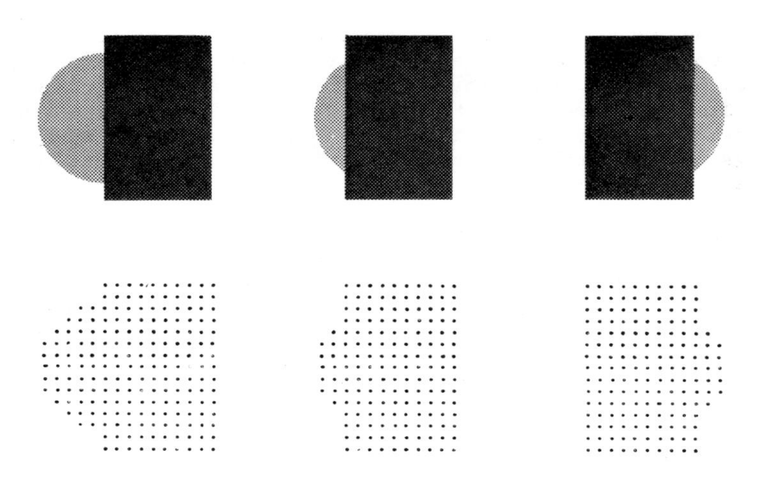

Fig. 3. An example of the effect of the limitations in spatial and "brightness" resolution in a matrix of tactile point stimuli. The upper row indicates three phases (from left to right) of the optical stimulation available when a circle passes behind a rectangle. The lower row shows what this may mean in a matrix of point stimuli.

mean. Because of the limitations in spatial resolution the circle arc in the upper row is represented stepwise in the lower row. Because of the limitations in the resolution of light intensities the two intensities in the upper row giving two easily separable surfaces are represented by only one "brightness" in the lower row with no border (occluding edge) between the two surfaces.

CAN THE AVAILABLE INFORMATION BE PICKED UP?

The possibilities of finding technical solutions to some of the limitations of the present devices seem sometimes to lead to too optimistic a view. It is of course not enough to make the information available. It must also be possible for the alternative perceptual system to pick it up. These possiblities have not yet been studied enough.

Evaluation studies concerning the present devices as mobility aids have typically been made in rather complicated environments where the subjects move around freely. This means that they are difficult to interpret theoretically. They indicate, however, that object detection (cf Leonard's list above) is possible, at least indoors. Detection (that) functions also with rather small objects and localization (where) can be made with good precision. Identification (what) can work at least with a limited number of objects.

Of the remaining functions in Leonard's list, orientation over a larger field and the detection of terrain changes have hardly been studied. Posture in the meaning of position of the hand was picked up in some of the experiments mentioned above.

The experiments up till now indicate that steering and approaching in Gibson's list can be performed under simplified conditions to some extent. But much remains to be studied. At present we know very little - if anything - about the limitations and possibilities of the tactile system to pick up the crude representations of the flow patterns that are known to be so important for visually guided locomotion. I think that much of the endeavours in the immediate future should be devoted to this problem area. That information of this kind is presented in such a way that it can easily be picked up by the tactile system seems to be a basic requirement if aids of this kind are to be as useful for the locomotion without vision as we hope.

The main alternatives used in attempts to increase the possiblities of tactile matrices have been to increase the resolution concerning space and light intensity. Saltatory phenomena (Geldard, 1975) might indicate another alternative. They demonstrate the possibility, with suitable spatio-temporal patterns of stimulation, to get percepts located in between the stimulus points. These phenomena have not yet been studied in more complex motion patterns, but the possibilities of having flow patterns represented with the aid of this kind of phenomenon ought to be studied.

CONCLUSIONS

Attempts to give the blind a mobility aid in the form of a matrix of tactile point stimuli have been successful in the sense that functioning prototypes have been built that can be used in experimental studies. Development seems now to have reached a plateau, however. More basic knowledge of what kind of information the tactile system can pick up, especially under what conditions flow patterns - as given in these matrices - can be picked up, is needed for the future development of mobility aids of this kind.

ACKNOWLEDGEMENTS

This study was supported by the Bank of Sweden Tercentenary Foundation (dnr 75/116). The experiments at Smith-Kettlewell Institute of Visual Sciences in San Francisco were possible by the generous cooperation by Dr P. Bach-y-Rita, Dr C. C. Collins, and Dr L. A. Scadden. Mr R. MacDonald assisted in performing the experiments, and Mr A. Alden, Mrs P. Apkarian-Stielau, Mr G. Holmlund, Dr D. McGuiness, and Mr P. Mizuno were helpful in different ways. The author is also indebted to Dr B. W. White for many discussions on perceptual problems with aids of this kind. A visit to Poland was paid for by Uppsala University. Interesting discussions with Dr W. Starkiewicz, Dr O. Palacz, and Mr E. Kurcz are acknowledged.

REFERENCES

Bach-y-Rita, P. Brain mechanisms in sensory substitution. New York: Academic Press, 1972.

Collins, C. C. & Madey, J. M. J. Tactile sensory replacement. Proceedings of the San Diego Biomedical Symposium, 1974, 13, 15 - 26.

Collins, C. C., Scadden, L. A. & Alden, A. B. Mobility studies with a tactile imaging device. Paper presented at a conference on Systems and Devices for the Disabled, Seattle, June 1 - 3, 1977.

Electrode matrix on skin enables blind to 'see'. Electronics, September 16, 1976, pp. 39 - 40.

Geldard, F. A. Sensory saltation. Metastability in the perceptual world. New York: Wiley, 1975.

Gibson, J. J. The perception of the visual world. Boston: Houghton Mifflin, 1950.

Gibson, J. J. An ecological approach to visual perception. Manuscript to book in preparation, 1974.

Guarniero, G. Tactile vision: a personal view. Journal of Visual Impairment & Blindness, 1977, 71, 125 - 130.

Holmlund, G. Personal communication, 1976.

Johansson, G. Perception of motion and changing form. Scandinavian Journal of Psychology, 1964, 5, 181 - 208.

Kurcz, E. Elektroftalm - a mobility aid with a tactile display. In Report on European conference on Technical Aids for the Visually Handicapped, March 1974. Stockholm: Handikappinstitutet, Report No 8, 1974.

Kurcz, E. Elektroftalm EL 300 - a photoelectric instrument for the discernment of simple images through the medium of forehead skin. Mimeographed report. The PZO Polish Optical Works, Warsaw, Poland, 1977.

Lee, D. N. & Lishman, R. Visual control of locomotion. Scandinavian Journal of Psychology, 1977, 18, 224 - 230.

Leonard, J. A. Studies in blind mobility. Applied Ergonomics, 1972, 3, 37 - 46.

Noiszewski, K. & Starkiewicz, W. Bau und Anwendung eines Mehrkanals-elektrophthalms. Medizinal-Markt/Acta Medicotechnica, 1970, 18, AM 68 - 71.

Palacz, O. Mehrkanaliger Elektrophthalm nach Starkiewicz. Augenoptik, 1975, 92, 134 - 138.

Saunders, F. A. Recommended procedures for electrocutanous displays. Paper presented at the Workshop for Functional Electrical Stimulation, 1976.

Scadden, L. A. Tactile pattern recognition and body loci. Perception, 1973, 2, 333 - 336.

Starkiewicz, W. & Kuliszewski, T. The 80-channel elektroftalm. Proceedings of the International Congress on Technology and Blindness. New York: The American Foundation for the Blind, 1963.

White, B. W., Saunders, F. A., Scadden, L. A., Bach-y-Rita, P. & Collins, C. C. Seeing with the skin. Perception and Psychophysics, 1970, 7, 23 - 27.

INDEX